Guojian Wang, Junjie Yuan
Multicomponent Polymers

Also of interest

Advanced Materials
van de Ven, Soldera (Eds.), 2020
ISBN 978-3-11-053765-9, e-ISBN 978-3-11-053773-4

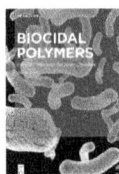

Biocidal Polymers
Pal Singh Chauhan, 2019
ISBN 978-3-11-063855-4, e-ISBN 978-3-11-063913-1

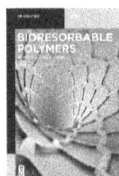

Bioresorbable Polymers.
Biomedical Applications
Devine, 2019
ISBN 978-3-11-064056-4, e-ISBN 978-3-11-064057-1

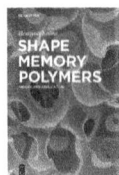

Shape Memory Polymers.
Theory and Application
Kalita, 2018
ISBN 978-3-11-056932-2, e-ISBN 978-3-11-057017-5

Guojian Wang, Junjie Yuan

Multicomponent Polymers

Principles, Structures and Properties

DE GRUYTER   同济大学 出版社
TONGJI UNIVERSITY PRESS

Authors
Prof. Guojian Wang
Department of Polymer Materials
School of Material Science and Engineering
Tongji University
Shanghai
China

Junjie Yuan
Department of Polymer Materials
School of Material Science and Engineering
Tongji University
Shanghai
China

ISBN 978-3-11-059632-8
e-ISBN (PDF) 978-3-11-059633-5
e-ISBN (EPUB) 978-3-11-059441-6

Library of Congress Control Number: 2020945806

Bibliographic information published by the Deutsche Nationalbibliothek
The Deutsche Nationalbibliothek lists this publication in the Deutsche Nationalbibliografie;
detailed bibliographic data are available on the Internet at http://dnb.dnb.de.

Preface

Since the beginning of the twentieth century, polymer materials have been developed rapidly. This is obviously due to their excellent performance and simple processing method. However, when compared with inorganic materials and metal materials, polymer materials also have many shortcomings, such as the strength is not high enough and the stability is not good enough; hence, they need to be modified.

Copolymerization is the initial modification method of polymer materials. Many useful polymer materials have been developed through copolymerization. However, the copolymerization process is relatively complex and inefficient, and the monomers that can be used for copolymerization have been studied basically, so there is not much room for further development. Therefore, people need to find a simpler modification method. In this context, the polymer blending modification method has attracted more and more attention. Multicomponent polymer is a general term for a large class of polymer materials prepared by chemical or physical methods. At present, the blending of polymer materials has become one of the main means of modification, especially since the second half of the twentieth century, many important discoveries and inventions have been made in polymer material industry. For example, the toughening modification of plastics and the development of thermoplastic elastomers are all related to the blending modification of polymer materials. Therefore, it is necessary to offer multicomponent polymer courses for undergraduate and graduate students majoring in polymer materials.

Based on years of experience in the research and teaching of the structure and properties of multicomponent polymers, the authors compiled and published the postgraduate textbook of *Structure and Properties of Multicomponent Polymers* in 2010, which has been welcomed by teachers, students and engineers.

In 2012, *Structure and Performance of Component Polymers* was listed as the excellent course construction project of graduate students by Tongji University. In order to cooperate with the construction of excellent courses, the textbook *Structure and Properties of Multicomponent Polymers* was revised to adapt to the rapid development of multicomponent polymer field, and renamed as *Multicomponent Polymers – Principle, Structure and Properties*. Today, we hope that this textbook can be used out of China for the study of students and as reference for scientists and technicians all over the world. Therefore, with the help of Tongji University Press in China and De Gruyter Press in Germany, we compiled this English version of textbook.

This textbook is compiled by the authors who have experience in teaching and scientific research of multicomponent polymers for many years. The content is divided into two parts. The first part introduces the basic theoretical knowledge of multicomponent polymers, and the second part focuses on the properties of multicomponent polymers. In the process of compilation, we have consulted a large number of relevant works and literature at home and abroad, and have received many inspirations and lessons from them. We would like to express our deep

https://doi.org/10.1515/9783110596335-202

gratitude to the authors of these works and literature. It is hoped that it can not only be used as a teaching material for master's degree students majoring in materials science and engineering, but also be used as a reference for teachers, post-graduates, senior undergraduates, researchers and technicians in related fields in teaching, scientific research and engineering technology.

Due to the wide range of contents and large amount of information involved in this textbook, and the author's lack of talent and learning, although the authors try to be correct and accurate in the process of compilation, there must be many omissions and errors in the book. We hope that readers will be free to correct them.

Authors at Tongji University, China

October 2019

Acknowledgements

The publication of this textbook was supported by Tongji University Press in China and De Gruyter Press in Germany. The master's degree students Zhenyu Huang and Kai Yang of the School of Materials Science and Engineering of Tongji University have made great contributions to the translation of Chinese characters, to whom we owe many thanks.

https://doi.org/10.1515/9783110596335-203

Contents

List of Figures

https://doi.org/10.1515/9783110596335-205

List of Tables

https://doi.org/10.1515/9783110596335-206

Chapter 1
Introduction

1.1 Abstract

Today, synthetic polymers have become indispensable materials for industry, agriculture and people's life. However, with the rapid development of modern science and technology, polymer materials have been put forward with increasingly extensive and harsh requirements. For example, polymer materials are expected to be both high-temperature resistant and easy to form, as well as have excellent toughness, high hardness and not only good performance but also low cost. Using a single homopolymer is often difficult to satisfy a variety of requirements.

In order to obtain polymer materials with excellent comprehensive properties, the traditional method of polymer modification is mainly copolymerization. Copolymerization modification has played a very important role in the development history of polymer materials. In fact, most of the polymers currently used are copolymers, and it has been difficult to find traces of homopolymers in use. But up to now, it is not only time consuming to develop new copolymers, but also the effect of modification is sometimes not obvious. Compared with copolymerization, blending modification of polymers is usually simple and efficient. In addition to the development and synthesis of new polymers, multicomponent composition of existing polymers, namely blending modification, has become an effective way to develop polymer materials and have attracted increasing interest and attention in recent years.

Practice has proved that many important properties of polymer materials (such as mechanical properties, flame retardancy, machinability and impermeability.) can be improved by this way. For example, important toughening plastics such as high impact polystyrene (HIPS) and acrylonitrile–butadiene–styrene (ABS) resin, thermoplastic elastomer, styrene–butadiene–styrene (SBS) triblock copolymer and multiblock polyether–polyester, polyether–polyurethane (PU) and polyester–PU are all very important multicomponent polymers that have been commercially produced and applied. By means of chemical grafting, block or physical blending, more products are produced by adding another component to one polymer material to improve its specific properties. Many polymers with special functions (such as conductivity, piezoelectricity and nonlinear optics) are often blended with other polymers in practical applications.

The main advantages of polymer multicomponent composites are reflected in the following aspects:

(1) By balancing the properties of each polymer component, taking advantage of each other's strengths and eliminating the weaknesses in the performance of each single polymer component, high-performance polymer materials were obtained. For example, blending polypropylene (PP) with polyethylene (PE) can

https://doi.org/10.1515/9783110596335-001

overcome the disadvantage of poor stress cracking resistance of PP and obtain blends with excellent comprehensive properties.

(2) A small amount of polymer can be used as a modifier of the other polymer, and the modification effect is remarkable. For example, the impact strength of rigid and brittle polymers such as polystyrene (PS) and polyvinyl chloride (PVC) can be increased by 2–10 times by adding 10–20% rubber polymers. For another instance, ethylene vinyl acetate copolymer can be used as a long-term plasticizer for PVC.

(3) Blending can improve the processability of some polymers. For example, refractory and insoluble polyimides can be injected into polyphenylene sulfide with good melting fluidity. In order to improve the flow properties of polycarbonate (PC), three-element blending method can be adopted. For instance, PC/polyethylene terephthalate (PET)/ethylene-vinyl acetate blends and PC/nitrile-butadiene rubber (NBR)/methyl methacrylate-butadiene-styrene (MBS) graft copolymer, in which the PC component is the matrix, NBR is the fluidity modifier and MBS is the impact modifier.

(4) Polymer blending can meet the needs of some special properties and prepare a series of new polymer materials. For example, in order to prepare flame-retardant polymer materials, the matrix polymers can be blended with halogen-containing flame-retardant polymers; in order to obtain decorative plastics with pearl luster, different polymers with different optical properties can be blended; in order to obtain good self-lubricity, silicone resin lubricity can be used to blend with many polymers; and two kinds of resins with poor miscibility and different tensile strength can be blended and foamed into multilayer porous materials, which have beautiful natural wood grain and can be used instead of wood.

Blending technology plays an essential role in the preparation of low shrinkage molding materials. Unsaturated polyester resin molding material has a large volume shrinkage when it is cross-linked and cured under the action of heating, pressure and peroxide, so the surface roughness, poor appearance and internal cracks and bubbles are easy to cause. In order to solve this problem, adding a large quantity of fillers, step-by-step polymerization, copolymerization and other methods have been used, and none of them achieved ideal results. In recent years, by blending unsaturated polyester molding materials with 7–20% thermoplastic resins, such as PS, PE and polyamide, low shrinkage or no shrinkage molding materials have been prepared.

(5) It is conducive to the variety and serialization of polymer products. The properties of polymer alloys are mainly affected by composition, structure and aggregate morphology. The change of polymer composition or blending ratio, the addition of third component (multicomponent blending) and special additives will lead to the change of the properties of polymer alloys and forming a series of different properties that can meet various requirements and applications in different occasions. For example, ABS plastics, which itself is a blend system of

various polymer molecules, can be prepared by various methods including bulk suspension, emulsion polymerization, copolymerization blending and mechanical blending. The ABS produced by different methods is different in composition proportion, morphology and rubber particle size, and the product performance is also different. Therefore, there are many brands and varieties of ABS. In addition, ABS can be blended with a variety of polymers to form multi-component blends, such as ABS/PVC, ABS/PC, ABS/polysulfone, ABS/nylon, ABS/PU and ABS/polymethyl methacrylate (PMMA). Almost every kind of plastic or elastomer alloy has formed a series abroad, and new series and varieties are constantly introduced. It is difficult for a single-component polymer to produce such many varieties.

(6) Compared with the development of a new type of homopolymer or copolymer, the development of a new type of polymer blend has much lower investment and much faster speed. Moreover, the mechanical properties of some engineering polymer blends can compete with those of aluminum alloys, and homopolymers are far from comparable in this respect. Therefore, the development of polymer blends is very fast. From 1980s to 1990s, the annual growth rate of plastics industry was 2–4%, while that of polymer blends was 9–11%, and that of engineering polymer blends was 13–17%.

1.2 Basic concepts

Multicomponent polymer refers to the macroscopically uniform and continuous solid polymer materials, which is a blend of two or more polymers by physical or chemical methods. Multicomponent polymers are also known as polymer blends and polymer alloys in many cases.

There is a difference between multicomponent polymers and regular random copolymers. In the aggregated state of multicomponent polymers, there is generally a distinct fine phase separation structure, but in the aggregated state of random copolymers, there is generally no such phase separation structure. Even in multicomponent polymers prepared by physical methods, there are still a small number of chemical bonds between different polymer macromolecules. For example, in the melt mixing process under the action of strong shearing force, the macromolecule may break and macromolecule free radicals will be produced, resulting in a small number of block or graft copolymers. In addition, recently, chemical bonds are introduced into the blends to enhance the interfacial bonding between the components of the blends.

The initial concept of polymer blends is confined to the simple physical mixing of heterogeneous polymer components. The emergence of ABS resin in 1950s formed a new concept of contact copolymerization blends. With the in-depth study of the morphology and structure of polymer blends, it is found that the existence of two-phase structure is a common and important characteristic of this system. Therefore, polymers with complex structures generally belong to the category of polymer blends. Graft copolymers, block copolymers, interpenetrating polymer network (IPN), even homopolymers containing crystalline and amorphous phases and crystalline polymers with different crystalline structures can be regarded as polymer blends. Two different combinations of polymers are shown in Fig. 1.1.

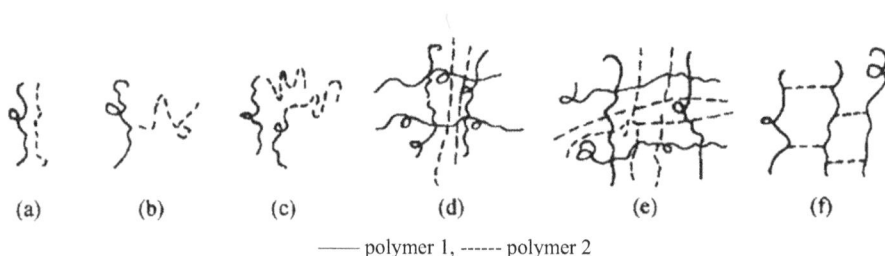

(a)   (b)   (c)   (d)   (e)   (f)

—— polymer 1, ------ polymer 2

Fig. 1.1: Schematic diagram of combinations between two polymer components. (a) Mechanical blend, (b) graft copolymer, (c) block copolymer, (d) semi-IPN, (e) IPN and (f) cross-linking blend.

In this sense, the concept of multicomponent polymers is more scientific than polymer blends.

Polymer blends are often called polymer alloys. They are equivalent, which is common in plastic engineering. From the point of view of materials, the preparation route and purpose of this material are very similar to those of metal alloys. Unlike most metal alloys, which are compatible homogeneous systems, most polymer alloys are incompatible heterogeneous systems. Therefore, polymer alloy is only a borrowed term, but it is not very strict in terms of scientific sense.

In order to consider the current generic names and for the convenience of narration, the concepts of multicomponent polymers, polymer blends and polymer alloys will be used indiscriminately in this textbook, and the names of polymer blends will be used more frequently.

The morphology and structure of polymer blends are influenced by thermodynamic compatibility between polymer components, blending methods and process conditions. In various literatures on the morphology and structure of polymer blends, different expressions such as "compatibility" and "miscibility" often appear. Although they are not unified yet, they are generally expressed as "miscibility" to represent thermodynamic dissolution (when blending, the mixed free branching $\triangle Gm < 0$), while "compatibility" is based on whether the blends with uniform and stable morphology can be obtained, regardless of whether the blends are thermodynamically soluble or not.

Therefore, even in thermodynamically incompatible blends, forced and well-dispersed mixing can be achieved depending on external conditions, and polymer blends with good mechanical properties and stability can be obtained; this polymer blends can still have good compatibility. "Compatibility" has engineering meaning, so it is also called "engineering compatibility" or "process compatibility."

There are many types of multicomponent polymers, but they generally refer to the blends of plastics with plastics, as well as the mixing of rubber in plastics and plastics in rubber. For the blend system which mixes a small amount of rubber in plastics, it is generally called "rubber toughening plastics" because of its great improvement in impact performance. The main purpose of the system mixing plastics in rubber is to enhance the mechanical properties of rubber, so it is often called "plastic reinforced rubber."

The types of polymer blends can be divided into binary and multicomponent polymer blends according to the number of polymer components contained; the matrix resins in polymer blends include polyolefin blends, PVC blends, PC blends and polyamide blends. According to the performance characteristics, polymer blends with high-temperature resistance, low-temperature resistance, flame retardancy and aging resistance were selected.

Blends of the same series of polymers (such as high-density PE and low-density PE blends) or blends of different molecular weight grades of the same polymer are often called homologous polymer blends.

In recent years, the names of engineering polymer blends and functional polymer blends appear. The former refers to polymer blends based on engineering plastics or with engineering plastic characteristics; the latter refers to polymer blends with special functions (such as antistatic, high barrier and ion exchange.) in addition to general properties.

Therefore, in the field of multicomponent polymers, the basic concepts and technical terms are still confused and need to be further unified with the development of the discipline.

1.3 Main types and representation methods of multicomponent polymers

1.3.1 Main types of multicomponent polymers

Multi component polymers can be classified by various methods. Several commonly used classification methods are introduced further.

1.3.1.1 Classification according to thermodynamic compatibility

According to the thermodynamic compatibility of each component in the blend system, the multicomponent polymers can be divided into homogeneous polymer blends and heterogeneous polymer blends.

1. Homogeneous polymer blends

Homogeneous polymer blends are multicomponent systems consisting of different polymers. Thermodynamic compatibility among components is achieved at the molecular level.

For example, PS/polyphenyl ether (PPO), PVC/polycaprolactone and blends of PVC and a series of polyacrylates are typical examples of homogeneous polymer blends. An important characteristic of this kind of polymer blends is that the blending results in the average of some important properties, that is, the properties of the blends are between those of the homopolymers.

2. Heterogeneous polymer blends

Heterogeneous polymer blends refer to the phase separation of the copolymers and the existence of two phase or multiphase structures. Most polymer blends are of this type. For example, HIPS containing PS and polybutadiene (PB) is a typical representative of such alloys. In the system, plastics and rubber constitute two phases; plastics are the main component, forming a continuous phase, also known as matrix; rubber forms a dispersed phase, which is distributed in the matrix in the form of colloidal particles, also known as microarea. These kinds of blends, because plastics are matrix, retain the characteristics of strong and hard plastics. At the same time, because of the presence of rubber particles, the blends showed good toughness. Therefore, such polymer blends can achieve the most favorable combination of the properties of each polymer component.

1.3.1.2 Classification according to the composition of polymer blends

1. Rubber-toughened plastics

The elastic rubber particles are added into brittle plastics to achieve the role of toughening plastics. In addition to the HIPS mentioned earlier, the blends of PP with a small amount of ethylene propylene rubber (EPR) and PVC with a small amount of chlorinated PE are of this type. They are two-phase systems consisting of plastic as matrix and rubber as dispersed phase. Rubber plays a toughening role relative to plastic phase.

2. Plastics reinforced rubber

It is also one of the purposes of polymer blending to strengthen rubber with lower mechanical properties by rigid plastics. Typical examples are SBS thermoplastic elastomers. The chemical composition of SBS is basically the same as that of HIPS, but their phase structure is different. SBS is based on PB and PS as dispersed phase. In this way, the system maintains the soft and elastic characteristics of rubber. The presence of PS in the plastic phase strengthens the material and acts as a physical cross-linking agent. Plastics can also be added to general rubber. For example, when a small amount of PP is added to EPR and a small amount of PE is added to BR, the 2-phase structure system is composed of plastic as disperse phase and rubber as continuous phase. At that condition, plastics play a reinforcing role in rubber.

3. Rubber blended with rubber or plastics blended with plastics

If the thermodynamic incompatibility of the blends consisting of different rubber or plastics is found, the continuous phase is usually composed of higher content components and the dispersed phase is composed of lower content components. The main purpose of blending is to improve some properties of polymer, such as excellent low-temperature flexibility, good elasticity and good wear resistance of *cis*-butadiene rubber. However, its strength is low and its antiskid property is poor. When a small amount of natural rubber (NR) or SBR is added, its shortcomings can be improved. For another example, adding a small amount of PE in PC not only improves the impact strength of PC but also improves the processing performance.

1.3.1.3 Classification according to the chemical bonds between components

According to whether there are chemical bonds between polymer blends, they can be divided into two categories. One is physical blends without chemical bonds between different polymers, and the other is chemical blends with chemical bonds between different polymer molecular chains (segments).

1. Physical blends

The blends obtained by physical mixing of different kinds of polymers in mixing equipment by means of stirring and shearing are called physical blends or mechanical blends. The result of physical blending is to maximize the dispersion between polymers and form a stable system. The main methods of mechanical blending are melt blending, solution blending and emulsion blending, in which melt blending is the most common method.

The present research shows that physical blending is not completely without chemical reaction. In fact, under intense mechanical shear, a small amount of polymers may be degraded, resulting in macromolecular free radicals, and then form graft copolymers or block copolymers. Therefore, physical blends are often accompanied by certain chemical blends.

2. Interpenetrating polymer network

IPN refer to polymer blends formed by two or more cross-linked polymers interacting and entangling with each other. In this system, there is no chemical bond between the two polymers. The two polymers form a three-dimensional continuous network with tiny aggregates and penetrate each other. They cannot be separated by simple methods. Therefore, even in the case of incompatibility between the two polymers, forced mixing can be achieved. So, this kind of blends can be regarded as physical blending through chemical methods. It is an important variety of polymer blends, but there are not many industrialized products up to now.

3. Block copolymer and graft copolymer

Block copolymers are copolymers formed by head-to-tail connections of two or more macromolecules with different chemical structures and properties. Each macromolecule chain has at least dozens of repetitive units.

Graft copolymers are copolymers formed by chemical linking of two or more macromolecules with different chemical structures and properties. Similarly, each macromolecule chain has at least dozens of repetitive units.

In these two types of copolymers, there are only a few chemical bonds between various macromolecule chains, so each macromolecule basically maintains its own chemical and physical properties, forming an independent phase region. Therefore, there are two typical multicomponent polymers, which can be regarded as polymer blends prepared by chemical methods. Block copolymers and graft copolymers have unique properties and have become the most concerned systems in multicomponent polymers.

4. Cross-linked blends

In such blends, one polymer acts as a cross-linking agent for another polymer or these polymers act as cross-linking agents for each other. That is to say, the two polymers are cross-linked by chemical bonds. This kind of cross-linking system is different from the traditional cross-linking system using small molecular curing agent because the two parts of the cross-linking system are polymers. When the degree of cross-linking is not very high, all kinds of polymers can move independently to form phase regions, so it is also a typical kind of polymer blends prepared by chemical methods. The cured products formed by the reaction of epoxy resin and polyamide belong to this type.

1.3.2 Representation of multicomponent polymers

In order to express the composition of multicomponent polymer simply and clearly, binary polymer blends are marked as A/B (X/Y), where A is the name of matrix resin, B is the name of another polymer blended into matrix resin, and X and Y correspond to the mass fraction of A and B. The expression method of polymer blends is analogous. For example, PP/PE (85/15) means adding PE to PP, whose mass ratio is 85:15, while PP/PE/EPR (85/10/5) means using PP as matrix. The other two groups are PE and EPR binary blends, and the mass ratio of the three is 85:10:5. Sometimes, when there are copolymers in the blend components and it is necessary to label them, the following representation can be used: A/B = $m{:}n$ (X/Y), where B is a copolymer and $m{:}n$ is the ratio of the number of two monomer segments in copolymer B. For example, PVC/styrene-acrylonitrile = 75:25 (95/5), which means that 25 wt% copolymers of styrene-acrylonitrile containing 95 wt% of styrene and 5 wt% acrylonitrile was added to 75 wt% PVC. It is easier to label polymer blends by abbreviating the English name of polymers, such as PP/PE for polypropylene/polyethylene blends.

1.4 Preparation of multicomponent polymers

The preparation methods of multicomponent polymers can be divided into two types: physical and chemical methods.

1.4.1 Physical blend methods

Physical blending, also known as mechanical blending, is a method of blending different types of polymers in mixing equipment. Common temperature processes generally include mixing and dispersing. In the blending operation, through the energy supplied by various mixing machines (mechanical energy, thermal energy, etc.) and under the action of stirring and shearing, the particles of the mixed materials are continuously reduced and dispersed, eventually forming a uniformly dispersed mixture. Because of the large size of polymer particles, in the process of mechanical blending, the blending mainly depends on convection and shear, and diffusion plays a secondary role.

In mechanical blending operations, usually only physical changes occur. However, under the strong mechanical shear, a small amount of polymers can be degraded to produce macromolecular free radicals, and then form graft copolymers or block copolymers. Therefore, mechanical blending process is often accompanied by a certain process of mechanochemistry.

Physical blending includes dry powder blending, melt blending, solution blending and emulsion blending.

1. Dry powder blending
Two or more different fine powdered polymers are mixed in a common plastic mixing equipment to prepare polymer blends. Commonly used mixing equipment includes ball mill, mixer, kneading machine and so on. All kinds of additives can be added to the powder mixture. The blended material can be directly molded or extruded and then molded into finished products. The effect of dry powder blending is generally not good, so it should not be used alone. However, as the initial mixing process of melt blending, it has some practical value for the blending of insoluble and refractory polymers.

2. Melt blending

Melt blending is the process of dispersing and mixing polymer components above their viscous flow temperature to prepare polymer blends. The melt blending method has the advantages of good blending effect and wide application. It is the most commonly used blending method.

3. Solution blending

Solution blending is a method of preparing polymer blends by adding various polymer components into common solvents (or dissolving separately and mixing), stirring evenly, then heating and vacuuming to remove solvents or adding precipitators, finally precipitating and separating them. This method has little practical significance in industrial production except for the direct application of blends in solution state (such as the preparation of coatings and adhesives). But there are many applications in laboratory preparation.

4. Emulsion blending

Different kinds of polymer emulsions were put together and mixed evenly, and coagulant was added to coprecipitate to form the blend. This method can be applied when the raw material polymer is emulsion or the blend is applied in emulsion form.

Mechanical blending of polymers mainly depends on various mixing, kneading and mixing equipment. The properties of blends are closely related to the mixing efficiency of blending equipment. A series of high efficient mixing and extrusion equipments have been developed in order to achieve efficient mixing and dispersing effect and prepare blends with excellent properties. These mixing and extrusion equipments strengthen the shear and convection effects and improve the blending effect from three aspects: increasing the shear rate, prolonging the mixing time and strengthening the separation and disturbance of mixtures. At present, high efficiency mixing extrusion equipment mainly includes mixing single screw extruder and mixing twin screw extruder.

1.4.2 Copolymerization blending method

Copolymerization blending method is a chemical method which can be divided into graft copolymerization blending and block copolymerization blending. Graft copolymerization blending is more important in the preparation of polymer blends.

The basic method of graft copolymerization blending is to prepare polymer 1 and then dissolve it in another monomer 2, so that monomer 2 can be polymerized and copolymerized with polymer 1. The prepared polymer blends usually consist of three components: polymer 1, polymer 2 and graft copolymer with polymer 2 on the framework of polymer 1. The ratio of the two polymers, and the length, quantity and distribution of the graft chain have a decisive influence on the properties of the blends.

The presence of graft copolymers improves the miscibility between polymers 1 and 2, which enhances the interaction between phases. Therefore, the properties of polymer blends prepared by copolymerization blending method are better than those of corresponding mechanical blends.

Copolymerization blending method has developed rapidly in recent years. Some important polymer blends, such as HIPS, ABS resin and MBS resin, are all prepared by this method.

1.4.3 Interpenetrating polymer network

IPN is a kind of multiphase polymer blend material in which two or more polymers are interwoven into the interwoven network by chemical method. IPN technology is an important method to prepare polymer blends.

The preparation method of IPN polymer is close to that of graft copolymerization blending, and the chemical structure of interphase is close to that of mechanical blending. Therefore, IPN can be regarded as a mechanical blend realized by chemical means.

The IPN composed of X polymer A and Y polymer B is simplified as IPNx/yA/B.

IPN has different types, such as step-by-step, synchronous, interpenetrating network elastomer (IEN) and latex IPN (LIPN), which are prepared by different synthetic methods.

1. Step-by-step IPN

The step-by-step IPN is prepared by synthesizing cross-linked polymer 1, swelling it with monomer 2 containing initiator and cross-linking agent, and then following in-situ polymerization and cross-linking of monomer 2. For example, the cross-linked polyvinyl acetate (PEA) was synthesized first, then the equivalent styrene monomer containing initiator and cross-linking agent was used to swell it. After the swelling was uniform, the styrene was polymerized and cross-linked to produce white leather IPN50/50PEA/PS.

Since the first common IPN is based on elastomer as polymer 1 and plastic as polymer 2, it is called inverse IPN when plastic is used as polymer 1 and elastomer is used as polymer 2.

If only one polymer in the two polymer components of IPN is cross-linked, it is called semi IPN.

The step-by-step IPN mentioned earlier refers that the swelling of monomer 2 to polymer 1 has reached equilibrium state, so the IPN prepared has a macroscopic uniform composition. If the monomer 2 is rapidly polymerized before the swelling equilibrium is reached, the macroscopic composition of the product has a certain gradient because the concentration of monomer 2 gradually decreases from the surface to the inside of the polymer 1. The product is called gradient IPN.

2. Synchronous IPN

If the two kinds of polymer network are generated at the same time without order, it is called synchronous IPN (SIN). The preparation method is to mix the 2 monomers together so that the two monomers can be polymerized and cross-linked in a noninterference mechanism or manner. This can be achieved if on monomer conducts addition polymerization reaction while another monomer conducts condensation polymerization, such as SIN epoxy/acrylate, a SIN composed of epoxy resin and acrylate.

Half SIN is also commonly referred to mesenchymal polymer. The reaction of forming half SIN is called mesenchymal polymerization.

3. Interpenetrating network elastomer

This kind of IPN is made by mixing two kinds of linear elastomer latex, coagulating and cross-linking at the same time, which is called IEN. For example, IEN PU/polyacrylic acid (PAA) was prepared by mixing, coagulating and cross-linking PU latex and PAA latex.

4. Latex IPN

When IPN, SIN and IEN are thermosetting materials, emulsion polymerization can be used to solve the problem that they are difficult to process. In LIPN, IPN is prepared by emulsion polymerization. The cross-linked polymer 1 was used as "seed" latex, and monomer 2, cross-linking agent and initiator were added to make monomer 2 polymerize and cross-link on the surface of "seed" latex particles. The IPN prepared by this method has core–shell structure. Because interpenetrating network is limited to the range of latex particles, it is also known as micro-IPN. LIPN can be molded by injection or extrusion, which is suitable for processing thermoplastic materials and can be made into thin films.

1.4.4 Block copolymer and graft copolymer

According to the principle of polymer chemistry, common copolymerization methods can only be used to prepare random copolymers and alternating copolymers but not block copolymers and graft copolymers. These two kinds of copolymers need special synthetic methods. This textbook will give a thematic discussion on the synthesis and properties of these two kinds of multicomponent polymers; the details will not be given here.

1.5 Development of blending modification of polymers

The history of polymer blends can be traced back to 1846, when Hancock mixed NR with Gutta Percha rubber to make raincoats and proposed the idea of mixing the two polymers to improve product performance.

The first commercially produced polymer blend was a blend of PVC and NBR, which was commissioned in 1942. NBR is mixed with PVC as a long-term plasticizer. In the same year (1942), Dow Chemical Company sold Styralloy-22, which is an IPN of styrene and butadiene. The term "polymer alloy" was first proposed. In 1942, a mechanical blend of NBR and styrene-acrylonitrile copolymer, that is, A-type ABS resin, was developed. ABS resin is known as the modified material of PS. This new polymer material is strong and tough, and overcomes the weakness of PS. In addition, ABS resin is one of the most important engineering plastics because of its good corrosion resistance, easy processing, and it can be used to prepare mechanical parts. Therefore, ABS resin has attracted great interest, which has opened a new polymer science field of polymer blending modification.

The blend of ABS and PVC and PC is a very essential polymer material, which was put into the market in 1969. Up to now, more than 25% of industrial polymer blends contain ABS components. Polymer blends containing ABS resins occupy a large proportion of the market. Such blends accounted for 74% in Europe, 65% in Japan and 69% in North America.

In 1960, it was found that adding PS to PPO or PPE could improve its processability. PPE/PS blend is thermodynamically compatible; its density is higher than the weighted average density of each component, and its performance shows obvious synergistic effect. The copolymer was put into operation in General Electric company in 1965 and its trade name is Noryl. Now it has become an important engineering material.

In 1962, mechanical blends of EPR and PP were produced. Shortly after, the polymer blends were also prepared by block copolymerization blending method.

In 1972, Goodrich company introduced epoxy resin toughened by carboxyl-terminated liquid NBR, a commercial product of thermosetting resins toughened by rubber. Toughening technology is gradually applied to nylon, PC, PMMA and other polymer systems.

In 1975, Du Pont developed the ultra-high toughness polyamide, named Zytel-ST. This is a blend made by adding a small amount of polyolefin or rubber to polyamide. The impact strength of the blend is much higher than that of polyamide. This discovery is very important. It is now known that in other engineering plastics, such as PC, polyester resin (PET) and polyformaldehyde, a small amount of polyolefin or rubber can also greatly improve the impact strength.

The toughening modification of polymer materials is an important part in the development of polymer blends. Since the introduction of impact PS in the 1940s, the toughening and modification of plastics has been a research hotspot. The study of toughening mechanism began in the 1950s, and the early toughening mechanism of brittle polymers toughened by elastomers was formed. In the 1960s and 1970s, the toughening mechanism of elastomers has been widely recognized, and the research on the toughening mechanism has gradually shifted to the toughening mechanism of blends based on relatively tough polymers. Japanese scientists first proposed the toughening mechanism of rigid particles in 1984 by studying PC/acrylonitrile-styrene resin (AS) blends. This new concept has promoted the research of rigid polymer particles toughening plastics in recent years.

In recent years, theoretical research and industrial practice of polymer blending modification are in full swing. There are tens of thousands of patents published every year in the world, and more papers are published. The latest developments in the field of polymer blending modification can be summarized from the literature at home and abroad:

1. Progress in the study of morphological structure. By controlling the morphology and structure of polymer blends, better or more distinctive polymer blends were designed and manufactured; for example, dispersed phase lamination gave polymer blends some new functions (barrier, antistatic, etc.); a small amount of elastomers with toughening effect improved the toughness of brittle matrix markedly in the morphology of the network; crystalline polymers refine spherulites and acquire better toughness by blending with noncrystalline polymers or other crystalline polymers. Through these studies, there has been a new development in the understanding of the morphology and structure of polymer blends.

2. Progress in toughening mechanism of polymer blends. Before 1970s, the study of rubber toughening polymer and its toughening mechanism mainly focused on brittle polymer matrix. Since the 1980s, the toughening mechanism of tough polymer matrix has been studied. It is considered that the dissipation of impact energy mainly depends on the plastic deformation caused by shear yield of matrix, rather than on the crazing as brittle matrix. As a matrix, thermoplastic polymers can be divided into brittle matrix and tough matrix, and their rubber toughening mechanism includes two dominant mechanisms, which is an important progress. In recent years, the research on toughening plastics with rigid polymer particles and its mechanism has been very active, forming a new branch. The necessary conditions for rigid polymer particles toughening include high toughness of matrix polymer, good interface bonding between rigid particles and matrix and proper concentration of rigid particles. Before the yield deformation of the rigid particles occurs, the impact can be transferred to the tough matrix by the rigid particles, which weakens the plastic deformation.

3. Progress of compatibilization technology. Since the 1970s, besides nonreactive compatibilizers expanding their varieties and applications, reactive compatibilizers have sprung up. This promotes the compatibility of many incompatible blends and lays the foundation for the development of a series of new polymer alloys.

In addition, IPN, dynamic vulcanization, reactive extrusion, molecular compositing, block copolymers, graft copolymers and hyperbranched polymers are the representatives of new polymer blending technologies.

Exercises

1. Try to draw a diagram of all the combinations between the two polymer components.

Drawing Tips:
—— polymer 1, ----- polymer 2

(a) Mechanical blend, (b) graft copolymer, (c) block copolymer, (d) semi-IPN, (e) IPN and (f) cross-linking blend

2. What are the common classification methods of multicomponent polymers?

Answer:
(1) Classification according to thermodynamic compatibility: homogeneous polymer blends and heterogeneous polymer blends.
(2) Classification according to the composition of polymer blends: rubber-toughened plastics, plastic reinforced rubber, rubber blend with rubber or plastics blend with plastics.
(3) Classification according to the chemical bonds between components: physical blends, IPN, block copolymer and graft copolymer, cross-linked blends.

Chapter 2
Physical and chemical principles of multicomponent polymers

2.1 The concept of compatibility between polymers

Corresponding to the solubility of low molecule substance, the miscibility of polymers is also known as compatibility, which is the thermodynamic solubility of the polymers. The compatibility of polymers, originating from the concept of compatibility of components in emulsion system, is an expression of the polymer blending processing performance and means the ability of polymers to disperse well with each other and obtain good performance and structure stable blends. Compatibility and miscibility are not completely consistent, for example, the viscosity ratio of two polymer melts is not directly related to their thermodynamic compatibility, but it is an important parameter for their miscibility. Therefore, the former is also known as thermodynamic compatibility, while the latter is called technologic compatibility or engineering compatibility. In general, good miscibility between polymers is the basis for good compatibility.

To what extent polymers mixture is thermodynamically compatible? The answer to this question is much more complex than the solubility of small molecule substance. It is a problem that the experimental criterion for compatibility is not clear. At present, in both scientific research and engineering applications, the method of determining glass transition temperature (T_g) is widely used to distinguish the compatibility between polymers. If the blends have two T_g, the blends can be identified as incompatible systems, but if only one T_g is detected, it is difficult to determine whether the compatibility of polymers is good or not, which involves the criteria for the compatibility between the polymers. If judged by the standard of small molecule substance solubility, the polymer should mix randomly at the chain level to achieve the true miscibility, as the chain is equal to the size of the small molecule. In fact, T_g only reflects the motion characteristics of polymer segments. Therefore, the mixing of polymers at the chain level is usually called compatibility, that is, the basic unit of mixing is chain segment. This method can be used to judge the compatibility between polymers if the accuracy of determination is able to achieve the size of chain segment, which is based on the understanding that the compatibility is determined by glass transition temperature. Therefore, people who engaged in engineering and technology believe that if the size of microregion in the blend system is less than 10–15 nm (equivalent to the size of chain segment), it is a compatible system. But for those engaged in basic theoretical research, such a compatible scale seems too large. With the development of modern testing technology, we can detect the existence of 2–4 nm dispersed microphase, and the "yardstick" of compatibility will also be

https://doi.org/10.1515/9783110596335-002

smaller. Therefore, the real solution to the compatibility criterion of polymer blends needs further development of experimental technology. Up to now, there is no definite criterion. It is not surprising that the different methods of tests for the "scale" of specific objects are different, and sometimes different methods for the compatibility of the same blends have different conclusions.

There is a concept of the degree of compatibility when two polymer blends are completely separated and precipitated from thermodynamics compatibility in macroscopic. It is not as simple and clear as the mixture of small molecule solvents, either soluble or insoluble. A lot of polymer blends with two phase structures are thermodynamically incompatible systems, but they are not completely separated system; it is a system with a kind of polymer dispersed evenly in the matrix of the other polymer in the form of microregion. There exists a transition region at the interface of two phases (in which two kinds of polymers molecules coexist). That means the two kinds of polymer molecules are compatible in the transition region, but the whole blends are phase separated, so this kind of blend is compatible in a certain degree. Most polymer blends are the blend system with some degree of compatibility. They are relatively stable in dynamics, and there are no separation phenomena in using; the compatibility of this sense is the processing compatibility. It focuses on evaluating the miscible degree, uniform dispersion and macrostability of the two polymers from the perspective of technology. Compared with the original polymers, the mechanical properties of the blends which meet the requirements of processing compatibility are improved.

In addition, the compatibility is also evaluated by the improvement degree of the mechanical properties of the blends, and those blends in which mechanical properties are obviously improved are called the mechanical compatibility system. In fact, there are not substantial differences between processing compatibility and mechanical compatibility; they are concepts in which compatibility is considered from different perspectives. It should be noted that the compatibility of the blends is usually evaluated as good compatibility or poor compatibility, most of which are aimed at mechanical compatibility or process compatibility.

It should also be pointed out that the methods of enhanced blending operation or prolonging blending time between the polymers cannot be seen as a way to change the compatibility of the polymers. These measures only meet the conditions for the complete mixing of the two polymers, and it is not possible to change an incompatible system into a compatible system. For example, the melt blending of polystyrene (PS) and polymethyl methacrylate (PMMA) is always incompatible, no matter how long the mixing time is and how much blending strength is. However, the blend of PS and polyphenylene oxide (PPO) is still a kind of homogeneous polymer blends even if the speed of the mixer is low enough and the storage time after the blending is long enough. Because compatibility is determined by the thermodynamic properties of polymers, compatibility is an invariant unless special measures are taken.

The miscibility between polymers is an important basis for choosing suitable blending methods, and it is also one of the key factors to determine the morphology and properties of blends, so it is necessary to make a more systematic explanation.

2.2 Thermodynamic analysis of miscibility of polymers

The compatibility of the two polymers is determined by their thermodynamic properties. According to the thermodynamics principle, in order to make the two polymers compatible, the mixed free energy (ΔG_m) of the blend system must meet the following condition:

$$\Delta G_m = \Delta H_m - T\Delta G_m < 0 \tag{2.1}$$

where ΔH_m and ΔS_m are molar mixing heat and mixed entropy, respectively; T is absolute temperature. For polymer blends, if there is no special interaction between two kinds of polymer molecules (such as forming hydrogen bonds), $\Delta H_m > 0$ in the mixing process which means it is a heat absorption process. According to formula (2.1), the positive mixing heat is disadvantageous to the compatibility of polymers. Although entropy is increased in the mixing process, it is known from the study of polymer physical chemistry that when polymers are mixed, the increase of entropy is very limited, and a polymer made up of x chain links has a much smaller contribution to the entropy of the system than that of x small molecules. Therefore, the entropy term is often not enough to overcome the contribution of the thermal term to the ΔG, that is to say, in most conditions, it cannot meet formula (2.1), so most polymer blends are incompatible systems.

Flory has used lattice model to discuss the thermodynamics process of polymer dissolved in small molecule solvents and derived the expression of mixed free energy of polymer solution:

$$\Delta G_m = RT(n_1\ln\Phi_1 + n_2\ln\Phi_2 + \chi_1 n_1\Phi_2) \tag{2.2}$$

where subscript 1 represents the solvent, 2 represents the polymer, n represents the molar number, Φ represents the volume fraction and χ_1 is the polymer solvent interaction parameter. The first two items in brackets represent the entropy term, and the last one represents the heat term. The entropy value is negative, because $\Phi < 1$, which is beneficial to the compatibility of the two polymers. The contribution of heat term to ΔG depends mainly on χ_1 value; the greater the value, the more unfavorable to compatibility.

If the relationship between polymer and solvent system is applied to polymer–polymer system, a simple two-polymer blend system can be described quantitatively.

If polymer A and polymer B are mixed, according to formula (2.2), the mixing free energy should be:

$$\Delta G_m = RT(n_A \ln \Phi_A + n_B \ln \Phi_B + \chi_{AB} n_A x_A \Phi_B) \tag{2.3}$$

where χ_{AB} is called the interaction parameter of polymer A and polymer B, and x_a is the number of chain links of polymer A.

If the molar volume of the polymer chain is V_R, and the total volume of the system is V, then the following formula is obtained:

$$n_A = V\Phi_A/V_R x_A$$
$$n_B = V\Phi_B/V_R x_B \tag{2.4}$$

where x_B is chain number of polymer B. Substituting formula (2.2) into eq. (2.3), the following Scott equation is deduced:

$$\Delta G_m = \frac{RTV}{V_R}\left(\frac{\Phi_A}{x_A}\ln \Phi_A + \frac{\Phi_B}{x_B}\ln \Phi_B + \chi_{AB}\Phi_A\Phi_B\right) \tag{2.5}$$

The analysis shows that the compatibility between polymer A and polymer B depends on the relative value of entropy term and heat term in formula (2.5). For the entropy term, the larger the x_A and x_B are, the smaller the absolute value is. That is to say, the higher the polymerization degree is, the more unfavorable the compatibility is and for the heat item, the larger the χ_{AB} is, the more unfavorable the compatibility is. The formula also shows that when the degree of polymerization of polymers is very high, the entropy term becomes very small, and the free energy of the system is mainly determined by mixed heat. In other words, mixing heat plays a more important role in the compatibility of polymer blends.

From the earlier discussion, it can be seen that the interaction parameter χ_{AB} is the key to decide whether the system is compatible. If the χ_{AB} of the system changes, the shape of the ΔG–Φ curve determined by formula (2.5) also changes, and the compatibility changes accordingly. Figure 2.1 is a curve drawn from the volume fraction Φ of ΔG_m of polymer B when $x_A = x_B = 100$. When the value of χ_{AB} is very small (e.g., equal to 0.01), the ΔG_m value is less than zero in the whole composition range, and curve (1) has a minimum value. At this point, the system is homogeneous in the whole composition. This system, which can form a homogeneous phase with any composition, is called a completely compatible system; when the value of χ_{AB} is larger (e.g., 0.1), the ΔG_m is greater than zero in the whole range of composition and the curve has a maximum value. The free energy of any composition blend is higher than the free energy of the pure polymer A or the polymer B. In this case, the system is unstable, and phase separation is inevitable, so it is an incompatible system; when the value of χ_{AB} is between the above two curves (e.g., 0.03), the situation is more complex. As shown in curve (3) of Fig. 2.1, although ΔG_m is less than zero in the

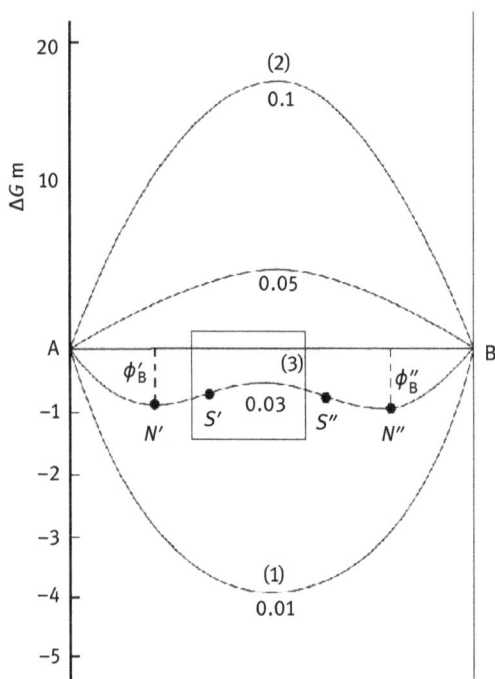

Curve (1): $X_{AB} = 0.01$; curve (2): $X_{AB} = 0.1$
Curve (3): $X_{AB} = 0.03$, $x_A = x_B = 100$

Fig. 2.1: The effect of X_{AB} value on the $\Delta G_m - \Phi$ curves.

whole range of composition, the curve has two minimum values (its corresponding composition is Φ'_B and Φ''_B). When the system is composed of $\Phi_B > \Phi''_B$, or $\Phi_B < \Phi'_B$, the system is homogeneous; when the total composition is between Φ'_B and Φ''_B, the system will be separated into two phases of Φ'_B and Φ''_B, because two stable phases with smaller free energy can be formed after the system is in phase separation situation. In other words, when there are two minimum values in the ΔG_m curve of the system, it is no longer homogeneous in the whole composition range. This system, which is separated into two phases in a certain range and remains as homogeneous phase outside this range, is called partial compatible system.

2.3 Critical conditions for phase separation

From the earlier discussion, it is known that $\Delta G_m < 0$ does not guarantee the complete compatibility between two kinds of polymers. The $\Delta G-\Phi$ curve of fully compatible system has only one minimum, and the curve is concave up. This phenomenon needs to mathematically meet the following two conditions:

$$\text{necessary condition: } \Delta G_\mathrm{m} < 0 \tag{2.6}$$

$$\text{sufficient condition: } \partial^2 \Delta G_\mathrm{m} / \partial \Phi_\mathrm{B}^2 > 0 \tag{2.7}$$

The system that satisfies only the necessary conditions is partially compatible system.

Figure 2.2 is the ΔG–Φ_B curve and phase diagram of partially compatible systems. The two minimum values on the ΔG–Φ_B curve are called binode. When the temperature is T_1, the corresponding two node points are composed of Φ_1' and Φ_0''. When the temperature changes to T_2, the composition of two node points is Φ_2' and Φ_2''. A series of double nodes N_1' and N_1'', N_2' and N_2'' can be obtained by changing the temperature. The binodal curve (V solid line) is formed when these binodes are connected, as displayed in Fig. 2.2(b).

In the ΔG–Φ_B curve of partly compatible systems, there are two inflection points S' and S'' besides binode, which are named spinodes. In the same way, changing the temperature, the composition Φ'_s and Φ''_s of the spinodes also change, and connect the spinodes at these different temperatures to form a spinodal curve (the V-shaped dotted line), as shown in Fig. 2.2(b).

It is observed from the experiment results that the system is homogeneous when the composition is located outside the binode curve. When the composition is within the binode curve, the system is in phase separated state, meanwhile, the system becomes incompatible when the composition is within the spinodal curve. Therefore, the binodal curve is also known as the metastable curve, and the spinodal curve is also called the unstable curve.

With the change of temperature, the binode and the spinode may converge to one point. This point is called the critical point (C). The temperature at the critical point is called the critical intermiscible temperature (T_c), and the composition of the system is called the critical composition. The critical condition for the phase separation of system is the condition of the conjunction of binode and spinode. In mathematics, two conditions must be satisfied at the same time:

$$\frac{\partial^2 \Delta G_\mathrm{m}}{\partial \Phi_\mathrm{B}^2} = \frac{\partial \Delta \mu_\mathrm{A}}{\partial \Delta \Phi_\mathrm{B}} = 0 \tag{2.8}$$

$$\frac{\partial^3 \Delta G_\mathrm{m}}{\partial \Phi_\mathrm{B}^3} = \frac{\partial^2 \Delta \mu_\mathrm{A}}{\partial \Delta \Phi_\mathrm{B}^2} = 0 \tag{2.9}$$

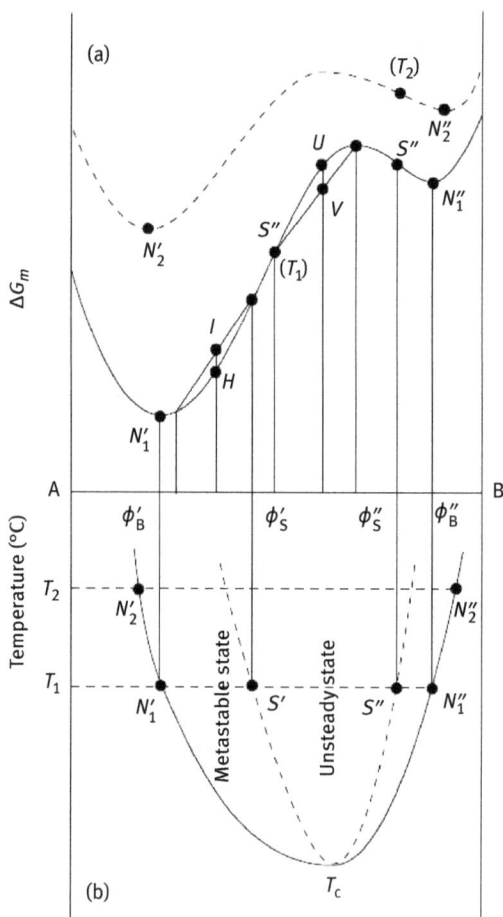

Fig. 2.2: $\Delta G_m - \Phi$ curve (a) of polymer blends and corresponding binode curve and (b) spinodal curve.

When two and three derivatives of formula (2.5) are brought into formula (2.7), the following formula is obtained:

$$\chi_c = \frac{1}{2}\left(\frac{1}{x_A^{\frac{1}{2}}} + \frac{1}{x_B^{\frac{1}{2}}}\right)^2 \tag{2.10}$$

where the subscript c indicates critical value, and χ_c is critical interaction parameter between polymers, which is commonly used as a criterion for the compatibility of polymer blends. If the χ_{AB} of polymer blend is $< \chi_c$, the system is compatible; the system is incompatible if $\chi_{AB} > \chi_c$. Formula (2.10) shows that χ_c is only determined by the degree of polymerization of the two polymers in system and has nothing to do with the properties of the polymer when there is no special interaction between the polymers involved in the blend. The smaller the degree of polymerization, the bigger the χ_c value, which is more favorable for compatibility. When $x_A = x_B$, formula (2.10) can be simplified to

$$\chi_C = \frac{2}{x_A} \tag{2.11}$$

From formula (2.10), it can be calculated that when $x_A = x_B = 100$, $\chi_c = 0.2$, the critical value is very harsh, and x_{AB} of most polymer pairs is higher than this value. Therefore, most polymers cannot form homogeneous system. Using the formula (2.11), we can obtain the critical molecular weight of the system from homogeneous system to heterogeneous system, and its critical chain number is

$$x_A = \frac{2}{\chi_c} \tag{2.12}$$

That is to say, when the degree of polymerization of A and B two polymers is greater than χ_c, the system is heterogeneous, and the system is homogeneous when the degree of polymerization of A and B two polymers is less than χ_c.

In addition, the relationship between critical composition (Φ_c) and molecular weight of polymer can be derived from the critical condition of phase separation:

$$(\Phi_A)_C = \frac{x_B^{\frac{1}{2}}}{x_A^{\frac{1}{2}} + x_B^{\frac{1}{2}}} \tag{2.13}$$

$$(\Phi_B)_C = \frac{x_B^{\frac{1}{2}}}{x_A^{\frac{1}{2}} + x_B^{\frac{1}{2}}} \tag{2.14}$$

2.4 Miscibility of polymer and phase diagram of two component blends

The polymers have high degree of polymerization, and the enthalpy of mixing is very small, so they can be truly completely miscible in thermodynamics. However, most of polymers are incompatible or partially compatible. Table 2.1 lists some common examples of completely compatible and completely incompatible polymer pairs.

It is very convenient for partly compatible systems to investigate compatibility between polymers by phase diagram. Figure 2.3 shows the basic types of phase diagrams of polymer/polymer two element blending system.

In Fig. 2.3, (a) presents an arbitrarily proportional compatibility system, (b) is the system with utmost critical intermiscible temperature (utmost critical solution temperature, UCST), (c) represents the system with lowest critical intermiscible temperature

Tab. 2.1: Completely miscible and immiscible polymer pairs at room temperature.

Miscible polymer pair		Immiscible polymer pair	
Polymer 1	**Polymer 2**	**Polymer 1**	**Polymer 2**
Nitrocellulose	Polyvinyl acetate	Polystyrene	Polyisobutylene
Nitrocellulose	Polymethyl methacrylate	Polymethyl methacrylate	Polyvinyl acetate
Nitrocellulose	Polyacrylate methyl ester	Natural rubber	Butadiene styrene rubber
Polyvinyl chloride	Methylstyrene–methacrylonitrile–ethyl acrylate copolymer (58: 40: 2)	Polystyrene	Polybutadiene
Polyvinyl acetate	Polyvinyl nitrate ester	Polymethyl methacrylate	Polystyrene
Polystyrene	Poly(2,6-dimethyl-1,4-benzyl ether)	Polymethyl methacrylate	Cellulose triacetate
Polystyrene	Poly(2,6-diethyl-1,4-phenylene ether)	Nylon 6	Polymethyl methacrylate
Polystyrene	Poly(2-methyl-6-ethyl-1,4-phenylene ether)	Nylon 66	Polyethylene glycol terephthalate
Polystyrene	Poly(2,6-dipropyl-1,4-benzyl ether)	Polystyrene	Polyethyl acrylate
Isopropyl polyacrylate	Isopropyl polymethacrylate	Polystyrene	Polyisoprene
Poly(α-methylstyrene)	Poly(2,6-diethyl-1,4-phenylene ether)	Polyurethane	Polymethyl methacrylate
Poly(2,6-dimethyl-1,4- benzyl ether)	Poly(2-methyl-6-phenyl-1,4-benzyl ether)		
Polyvinyl butyral	Styrene maleic acid copolymer		

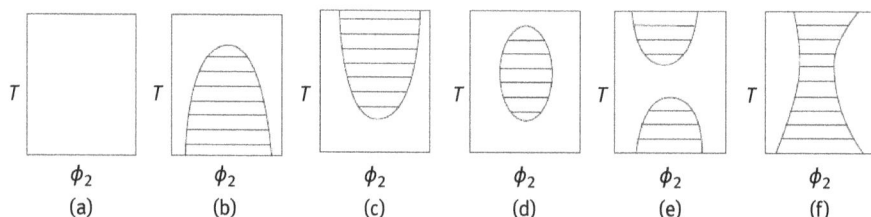

Shadow – phase separation area; ϕ_2 – volume fraction of polymer 2; T – absolute temperature

Fig. 2.3: Basic types of polymer/polymer phase diagrams.

(lowest critical solution temperature, LCST), (e) represents the system with UCST and LCST at the same time and (a) and (f) represent the system with locally incompatible regions.

Table 2.2 lists some examples of polymer blend systems with UCST and LCST.

Tab. 2.2: The phase diagram type of the blend system in Tab. 2.1.

Polymer1	Polymer2	Phase diagram type
Polystyrene	Polyisoprene	UCST
Polystyrene	Polyisobutylene	UCST
Poly(dimethylsiloxane)	Polyisobutylene	UCST
Propylene oxide	Polybutadiene	UCST
Polystyrene	Polyethylene methyl ether	LCST
Styrene acrylonitrile copolymer	Polycaprolactone	LCST
Styrene acrylonitrile copolymer	Polymethyl methacrylate	LCST
Polyvinyl nitrate ester	Polyacrylate methyl ester	LCST
Ethylene vinyl acetate copolymer	Chlorinated butyl rubber	LCST
Polycarbonate	Polycaprolactone	LCST
Polyvinylidene chloride	Polyacrylate methyl ester	LCST
Polyvinylidene chloride	Polyethyl acrylate	LCST
Polyvinylidene chloride	Polymethyl methacrylate	LCST
Polyvinylidene chloride	Poly (ethyl methacrylate)	LCST
Polyvinylidene chloride	Polyvinyl ketone	LCST

The mutual solubility and phase diagram types of polymer/polymer are closely related to their molecular weight distribution.

The phase diagram is very useful for analyzing composition and volume fraction of polymer phases. The composition and volume ratio of the blends phases can be calculated if the phase diagram and the initial composition of a polymer pair are given.

2.5 Phase separation mechanism

When phase separation in partially compatible polymer blends occurs, there are two different phase separation mechanisms.

As mentioned earlier, when the composition of the system is within the range of the spinodal curves of the phase diagram, it is unstable, and phase separation can automatically occur. As shown in Fig. 2.2(a), the curvature of the curve between the two inflection points is negative, that is, $\partial^2 \Delta G_m / \partial \Phi_B^2 < 0$. Phase separation occurs at any point U in the range of curve; the phase composition of the two separated phases must be on both sides of U, and the total free energy of the system will decrease. Therefore, phase separation spontaneously happens. This phase separation mechanism is called spinodal curve phase separation mechanism (type SD). When the system is in the range between the spinodal curve and the double node curve, it is in a metastable state and does not spontaneously occur to phase separation, because in the two ranges of $N'S'$ and $N''S''$, the curvature is positive, that is, $\partial^2 \Delta G_m / \partial \Phi_B^2 > 0$. At any point H on this curve, if the H occurs to phase separation, the total free energy after phase separation will increase. Therefore, in this case, phase separation cannot happen spontaneously to the system, which requires a slight activation, resulting in a large fluctuation in the concentration of the system, so as to form a "billet nucleus" (composed of Φ' and Φ''). On this basis, it gradually grows up and eventually forms two-phase system. This phase separation mechanism is called nucleation and growth mechanism (type NG). The total free energy of the system is still reduced because the system directly forms the dispersed billet nucleus under activated conditions.

There is no energy barrier in the phase separation process of the spinodal line mechanism. If a very small concentration fluctuation occurs, the phase separation can quickly and spontaneously happen. Figure 2.4 is a schematic diagram of the

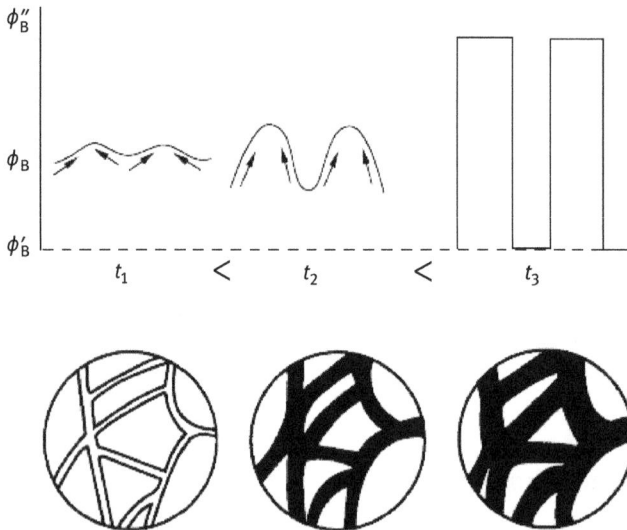

Fig. 2.4: Schematic diagram of the mechanism of phase separation of spinodal curve. (a) Change of one-dimensional concentration and (b) change of two-dimensional concentration.

spinodal curve phase separation mechanism. The Φ_B in diagram is the concentration of the given system before phase separation. Φ'_B and Φ''_B are the concentrations of two phases after phase separation. When the concentration of components in one part of the system is slightly elevated, the surrounding molecules of the same component spontaneously spread from low concentration to high concentration zone and form a reverse diffusion. The reason for reverse diffusion is spontaneous diffusion, which results in reduction of the system's free energy. As the diffusion is continuously carried out, the composition of the two phases is gradually changed. Finally approaching the two phase equilibrium concentration Φ'_B and Φ''_B required by the binodal curve. Another characteristic of this kind of phase separation mechanism is that the two-phase structure of the system that is basically fixed when the fluctuating concentration field is established. As time goes on, the composition of B in the dispersed phase gradually increases, as shown in Fig. 2.2. At the same time, due to the simultaneous phase separation in all parts of the system, the dispersed phase will overlap and form an interlaced network structure. The mechanism of spinodal curve can form three-dimensional continuous structure. This structure endows the blends with excellent mechanical and chemical stability. It is the reason that some polymer blends possess obvious synergistic effect. In many cases, when the content of a polymer is less, for example, about 10%, the mechanism of the spinodal curve can also form a droplet/matrix morphology, but the fine structure of the dispersed phase is often different from that of the NG type.

The phase separation caused by nucleation and growth mechanism is different from the earlier mechanism. In the metastable region, the slight fluctuation of concentration will not lead to phase separation. Only when the system is activated and can overcome the activation energy of forming the "billet core" barrier, the phase separation can be realized. The activation factor can be impurity, vibration and supercooling effects. Once the system is activated, the concentration abruptly changes, and the two phases with concentration of Φ'_B and Φ''_B (the concentration of two node points) is formed from the beginning, as shown in Fig. 2.5. As the time goes on, the nucleus of the billet is gradually increased. The nucleation process is as follows: the polymer concentration of the nucleus is Φ'_B, and the concentration of the continuous phase adjacent to the nucleus is Φ''_B, while the concentration of the continuous phase matrix far away from the nucleus is still Φ_B. In this way, the molecules of the main body polymer B diffuse to the low concentration (Φ'_B) zone, which is the normal concentration gradient diffusion (the diffusion coefficient is positive). Subsequently, these molecules spontaneously aggregate on the billet cores and make the billet cores big and big. This process has been carried out until the two phases meet the equilibrium state required by the lever principle. The concentration of two phases is invariable in the whole process, and the change is only the relative quantity of two phases.

Different from the spinodal curve phase separation mechanism, the formation and growth process of the nucleus is slow, and the interface of the two phases moves

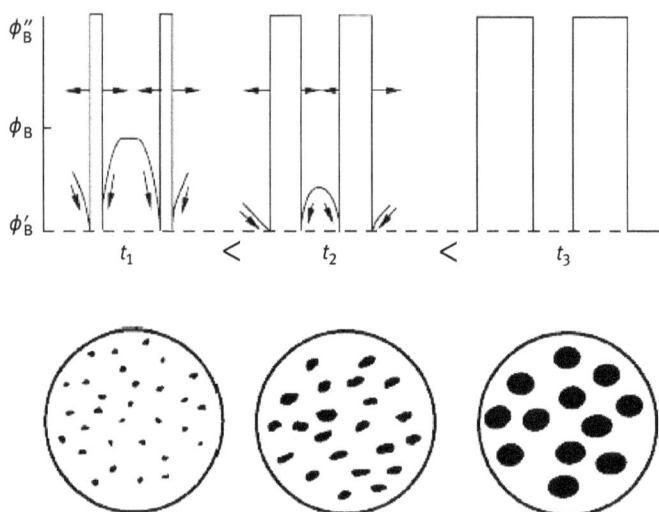

Fig. 2.5: Schematic diagram of the mechanism of nucleation and growth phase separation.
(a) Change of one-dimensional concentration and (b) change of two-dimensional concentration.

along with the time, and the size of the dispersed phase gradually increases. The microarea size and the distance between the microdomains depend on the diffusion rate and diffusion time of the polymer. The more the diffusion rate is, the faster the nuclear growth is. Generally, the diffusion rate increases with the increase of temperature, and the nucleation rate decreases with the increase of temperature. This is similar to the crystallization process of polymers, and at a suitable temperature, the growth rate of the dispersed phase microdomains is the fastest. In the early stage, the size of dispersed phase mainly depended on its own growth. Later, it could be condensed into large microregions through small microregions. At the same time, as the size of the dispersed phase grows, the distance between dispersed phase microregions gradually increases, and a clearer interface is formed between the two phases, and the interface of the spinodal curve phase separation is blurred. Considering the improvement of interface strength, it is more advantageous to form a fuzzy interface. The morphology and structure formed by the NG-type mechanism phase separation are mainly bead droplets/base bodies, that is, one phase is a continuous phase and the other is dispersed in it in the form of spherical particles.

If the blending system can finally reach equilibrium, the results of the two-phase separation mechanisms should not be different. But in fact, due to the high viscosity of polymer blends, it is difficult for the system to achieve a real equilibrium state for the dynamic factor. Therefore, the two-phase separation mechanisms often lead to the formation of different microdomain morphology (Figs. 2.2(b) and 2.5(b)).

2.6 Application of solubility parameter theory in blends

The use of solubility parameter δ_P to describe the interaction between liquids has been widely recognized. The solubility parameter is defined as the square root of cohesive energy density (CED):

$$\delta_P = (\Delta E_V/V)^{\frac{1}{2}} = (CED)^{\frac{1}{2}} \tag{2.15}$$

where ΔE_V is the internal energy and V is the volume.

The solubility parameters originate from liquid system and have been gradually extended to the polymer–solvent system, and the solubility parameters of many polymers (Tab. 2.3) have been measured. Due to the different methods of determining solubility parameters, the experimental values given in different literatures are quite different.

Tab. 2.3: Solubility parameters of some polymers.

Polymer	The numerical range of the polymer manual /$MP_a^{\frac{1}{2}}$	Coleman calculation value/$MP_a^{\frac{1}{2}}$	Small calculation value /$MP_a^{\frac{1}{2}}$	Van Krevelen calculation value /$MP_a^{\frac{1}{2}}$
Teflon	12.7	–	12.3	13.1
Poly (dimethylsiloxane)	14.9–15.6	–	12.5	–
Polyisoprene	14.7–17.6	16.6	16.5	16.5
Polyisobutylene	15.2–20.5	16.6	16.5	17.0
Polyethylene	15.8–18.0	–	16.9	17.2
Polypropylene	18.8–19.2	–	15.4	17.0
Polymethyl methacrylate	18.6–26.3	18.4	18.2	18.7
Polymethacrylic acid	20.1–21.3	19.6	19.6	19.0
Polyacrylic acid	18.4–18.5	18.6	18.8	18.5
Polystyrene	15.6–21.1	19.4	18.7	19.2
Polyvinyl chloride	19.2–22.1	20.3	19.5	19.7
Polyacrylonitrile	25.3–31.5	28.2	26.0	27.8
Polybutylene terephthalate	21.9	23.5	21.9	21.3
Polyepoxide	20.2	19.2	–	–

Tab. 2.3 (continued)

Polymer	The numerical range of the polymer manual / $MP_a^{\frac{1}{2}}$	Coleman calculation value/$MP_a^{\frac{1}{2}}$	Small calculation value /$MP_a^{\frac{1}{2}}$	Van Krevelen calculation value /$MP_a^{\frac{1}{2}}$
Polyepoxide propane	–	17.4	16.0	19.1
Polyphenylene ether	–	19.1	19.2	–

Small, Coleman and Van Krevelen have found that the solubility parameters of polymers can be calculated by group contribution method as follows:

$$\delta_\mathrm{p} = \rho \left| \frac{SF_i}{M} \right| = \frac{SF_i}{v} \tag{2.16}$$

F_i is the molar attraction constant and M is the molar mass of the repeating unit in polymer, v is mole volume and v can also be calculated by the group contribution method:

$$v = \sum^{v_r} \ \mathrm{or} \ v = \sum^{v_g} \tag{2.17}$$

where v_r contributes to the molar volume group of the rubbery state amorphous polymer, while v_g is the molar volume group contribution of the glassy state amorphous polymer. Later, Colemen proposed a molar volume value v^* which was not affected by the rubber state and the glassy state. Table 2.4 lists some groups of F_i, v_r, v_g and v^*.

Tab. 2.4: F_i, v_r, v_g, v^* of a part of groups.

Group	F_i				v_r	v_g	v^*
	Small value	Hoy value	Van Krevelen value	Coleman value			
–CH$_3$	438	303	419	446	22.8	23.9	31.8
–CH$_2$–	272	268	280	270	16.45	15.85	16.5
$-\mathrm{CH}{\diagup\atop\diagdown}$	57	176	139	47	9.85	9.45	1.9
${\diagdown\atop\diagup}\mathrm{C}{\diagup\atop\diagdown}$	–190	65	0	–109	4.75	4.6	–14.8

Tab. 2.4 (continued)

Group	F_i				v_r	v_g	v^*
	Small value	Hoy value	Van Krevelen value	Coleman value			
—CH=	227	250	223	231	13.9	–	13.7
>C=	39	172	82	37	–	–	−2.4
—C$_6$H$_5$	1504	1397	1516	1504	64.65	72.2	75.5
C$_6$H$_6$	1346	1442	1377	1333	65.5	61.4	58.8
—O—	143	235	256	194	8.5	10.0	5.1
—CO—	563	538	685	536		13.4	10.7
—COO— (ester)	634	669	512	610	24.6	23.0	19.6
—COO— (acid)	634	669	512	610	21.0	18.25	19.6
—Cl	532	419	471	540	18.4	19.9	23.9
—CN	839	726	982	872	–	19.5	23.6
—F$_3$	–	–	–	–	–	–	–
—CF$_2$—	307	235	–	–	–	–	–
—NH$_2$	–	406	204	563	–	–	18.6
>NH	–	368	286	292	–	–	8.5
—S—	460	428	–	–	15.0	17.8	–

Taking PS, PPO and polyisoprene as an example, the solubility parameters calculated by group contribution method are shown in Tab. 2.5.

It can be seen from Tab. 2.5 that the solubility parameter calculated by group contribution method is very similar to that in Tab. 2.3, which shows that the method has certain accuracy. Second, the solubility parameters of the PS and PPO are very close, indicating that the compatibility between them should be very good. However, the solubility parameters of PS and polyisoprene are quite different, indicating that the compatibility between them is not good. In fact, PS and PPO are good compatible systems, but PS and polyisoprene are indeed incompatible systems.

The group contribution method is an empirical method, and the compatibility between the two kinds of polymers is influenced by temperature, molecular weight and other external effects. Therefore, the solubility parameter is only a reference value for estimating compatibility between polymers, and it is not an accurate basis.

Tab. 2.5: Calculation values of solubility parameters of polystyrene, polyphenylene oxide and polybutadiene.

Group	Polystyrene			Polyphenylene ether			Polyisoprene		
	Unit number	F_i	v^*	Unit number	F_i	v^*	Unit number	F_i	v^*
$-CH_2-$	1	270	16.5	–	–	–	2	540	33
$-CH\big\backslash^{/}$	1	47	1.9	–	–	–	–	–	–
$-C_6H_5$	1	1,504	75.5	–	–	–	–	–	–
$-CH_3$	–	–	–	2	892	63.6	1	446	31.8
$-O-$	–	–	–	1	194	5.1	–	–	–
C_6H_6	–	–	–	1	1,333	58.8	–	–	–
$-CH=$	–	–	–	–	–	–	1	231	13.7
$\big\backslash_{/}C=$	–	–	–	–	–	–	1	37	-2.4
Total	–	1,821	93.9	–	2,419	127.5	–	1,254	76.1
		19.4			19.0			16.5	

2.7 Methodology for the compatibility between polymers

Thermodynamic compatibility means uniformly mixing at molecular level. However, compatibility is a measure of the degree of dispersion of polymer components, which is closely related with measuring methods. So, it means the uniformity under the actual measuring condition. Different methods always lead to different conclusions. For example, according to T_g result, polymer blends polyvinyl chloride (PVC)/nitrile butadiene rubber (NBR)-40 is a homogeneous system because the system only shows one glass transition temperature. However, according to the result of electrical microscopy, it is a heterogeneous system with small phase domain. Therefore, in practice, the conclusions on the compatibility between polymers via various characterization methods are not completely consistent. This situation is general in many cases. For example, oil and water are dispersed to form ultrafine emulsion. Although it is not thermodynamically stable, a certain stability is still exhibited, which is called metastable state. The system is transparent, and the liquid drop cannot be seen under optical microscopy. Homogeneous properties are shown in many ways. But the heterogeneous characters can be observed under electron microscope or through light scattering method. Therefore, even for simple mixing of liquids, it is often difficult to conclude whether the mixture system is homogeneous or heterogeneous. "homogeneous" and "heterogeneous" are not completely absolute concepts, and they are only meaningful in thermodynamic statistics. Whether a system is homogeneous or not depends on the criteria of identification – spatial scale or time scale. It is very common to draw different conclusions from different identification methods.

There are many ways to study the compatibility between polymers. Methods to judge the compatibility by solubility parameters and Huggins–Flory interaction parameter χ_{AB} based on thermodynamics have been previously described. In addition to thermodynamic methods, T_g method, infrared spectrometry, inverse gas chromatography and viscosity method can be used to judge the compatibility of polymer blends. The most commonly adopted method in engineering is the T_g method.

2.7.1 Glass transition temperature method

Measuring the compatibility of polymers by glass transition temperature method is mainly based on the following principles: the T_g of the polymer blends has a direct relationship with the mixing degree of two polymers chains. If the two polymer components are completely compatible and the blend is homogeneous, the system only exhibits one T_g. If the two components are completely incompatible and form two phase structures with obvious interface, the components exhibit two T_g, which are equal to the T_g of each polymers, respectively. The T_g of partially compatible systems falls between the earlier two limiting cases.

When there is a certain degree of chain segment mixing between the two polymers, there is a certain degree of diffusion between them, and the interface layer takes up a position that cannot be ignored. Although there are still two glass transition temperatures, they are closer to each other, and the proximity extent depends on the degree of mixing. The greater the degree of mixing of polymers chain segments, the closer the two temperatures approach each other. In some cases, the interfacial layer may also exhibit a less obvious third T_g. Therefore, according to the T_g of polymer blends, not only the compatibility between components can be deduced, but also the information about morphology and structure of polymer blends can be obtained.

It was also presented that the compatibility of polymers could be estimated by the width of glass transition temperature interval of blends. For pure polymers, $T_w = 6\,°C$; for completely compatible polymer blends, $T_w = 10\,°C$ approximately; for partially compatible blends, $T_w > 32\,°C$. In some cases, T_w is also related to the composition of the blends.

Recently, Lipatov has proposed a semiempirical formula to characterize the phase separation degree of polymer blends. The temperature corresponding to the peak in the diagram is the corresponding T_g, as shown in Fig. 2.6.

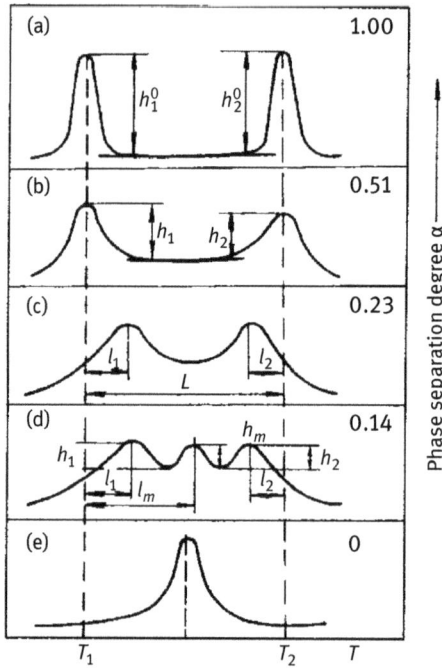

Fig. 2.6: Relationship between glass transition temperature and phase separation degree of polymer blends.

The phase separation degree α is defined as follows:

$$\alpha = [h_2 + h_1 - (l_1 h_1 + l_2 h_2 + l_m h_m)/L]/(h_1^0 + h_2^0) \tag{2.18}$$

where

l_1, l_2 represent the temperature shift of the T_g of the two pure polymer components of the blend.

h_1, h_2 represent the height of the transition peaks corresponding to the two T_g of the polymer blend.

h_1^0, h_2^0 represent the height of the two transition peaks corresponding to the two pure polymer components of the blend.

L is the temperature difference between two pure polymer components of the blend.

The subscript m represents the intermediate phase (interface phase), so h_m represents the height of the T_g transformation peak of the intermediate phase, while l_m represents the temperature difference between T_g of the intermediate phase and T_g of the pure component.

The intermediate phase does not exist in all the blends. In addition, eq. (2.18) does not apply to the case shown in Fig. 2.6(e) when there is only one transformation peak, suggesting it is completely compatible.

$l_1 h_1$ and $l_2 h_2$ are measurements of the area under the corresponding T_g transformation peak. When $l_1 + l_2 = l$, the micromultiphase morphological structure begins to appear, as shown in Fig. 2.6(b), the eq. (2.18) is simplified as

$$\alpha = (h_2 + h_1)/(h_1^0 + h_2^0) \tag{2.19}$$

Dynamic mechanical properties tests are commonly used to study the compatibility of blends. Figure 2.7 shows the tangent loss curve of the PVC/polyamide 6 (PA-6) blends compatibilized with three different compatibilizers and the curve without compatibilization. Curve A is the tangent loss curve of the PVC/PA-6 blends without compatibilization. Because of the poor compatibility between PVC and PA-6, the curve presents two tangent loss peaks, which correspond to the T_g of PVC (82 °C) and T_g of PA-6 (55 °C). Curves B, C and D are the tangent loss curves of the PVC/PA-6 blends compatibilized with polyethylene (PE)-g-maleic anhydride (MAH), acrylonitrile–butadiene–styrene (ABS)-g-MAH and ethylene vinyl acetate (EVA)-g-MAH, respectively. All the three lines have only one loss peak, indicating that all the three compatibilizers have good compatibilization effect on PVC/PA-6 system.

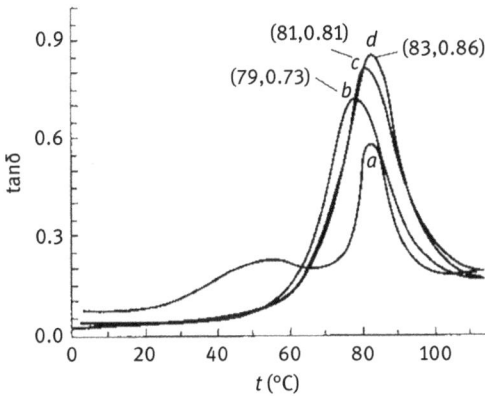

Fig. 2.7: Tangent loss curve of PVC/PA-6 blends.

Differential scanning calorimetry (DSC) measurement can be used for the study of compatibility of crystalline polymers. As shown in Fig. 2.8, curves 1 and 4 are the DSC curves of pure PA-12 and polypropylene (PP), respectively. Curve 3 is the DSC curve of PP/PA-12 blend with mass ratio of 80/20. Curve 3 shows two melting crystallization peaks, indicating that the two components are incompatible. However, the melt crystallization temperature of PP increases after blending, indicating that PP and PA-12 had heterogeneous nucleation effect on the other polymers, respectively.

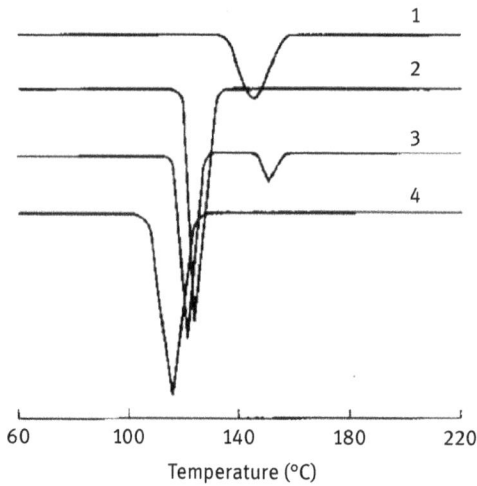

Fig. 2.8: DSC curves of PP, PA-12 and PP/PA-12 blends.

After compatibilizer PP-g-MAH was added into the system, the two melting crystalli-zation temperatures converge into a more intensive peak, indicating that PP-g-MAH could significantly improve the compatibility of PP/PA-12 blends.

2.7.2 Other methods

1. Infrared spectroscopy method

For a blend with certain compatibility, due to the interaction between the compo-nents, the infrared spectra of blends of bands causes some deviation compared with that of single component. The migration occurs mainly in the band position of some groups. For example, when the hydrogen bonds are formed between the blend com-ponents, the absorption peaks tend to shift, and the width of the peaks increases.

2. Electron microscopy method

The morphology of the blends can be directly observed by electron microscopy and can be used to study the compatibility among the blend components. In general, when there is good compatibility between the blend components and a certain thick-ness of interface transition layer is formed, the interface between the two phases can be observed in the electron microscope. In addition, when the same blending process is adopted, the size of dispersion phase with good compatibility of the blend is smaller. Electron microscopy is generally not used alone and can be used as an auxil-iary means of other characterization methods.

3. Interface thickness method

When the components of the blends are compatible, the interface layer will be formed at the boundary of the two phases. Therefore, the investigation of the thickness of the boundary layer is also an effective method to study the compatibility of the two phases. Good compatibility means large interface layer thickness, and poor compatibility means thin interface layer, or even no interface layer. The thickness of the interface layer is measured by small-angle X-ray scattering, small-angle laser scattering and ellipsometry.

4. Interfacial tension measurement method

The interfacial tension between polymer phases is closely related to their compatibility. The solubility parameters are similar, the compatibility is good and the interfacial tension is small. Therefore, the compatibility of polymers can be judged by measuring the interfacial tension between polymers. The interfacial tension of polymers is mainly determined by two methods: thermodynamics method and kinetics method. The thermodynamic methods include the suspension method and the rotating liquid drop method. The kinetic methods include the rupture silk thread method, the embedded fiber retracting method, the two-dimensional deformed droplet retraction method and the three-dimensional deformed droplet retraction method. Besides, there are other methods that include small amplitude dynamic shear oscillatory rheology and surface tension calculation method.

5. Transparency measuring method

In general, the compatible polymer system is homogeneous, and its optical properties are homogeneous with only one refractive index, so it is usually optically transparent. However, the incompatible polymer system is opaque. Nevertheless, this method is crude and cannot be used as an absolute criterion for judging the compatibility of polymers blends. Because if the refractive indexes of the two components in the incompatible blends are the same, or the size of the dispersed phase is less than the visible light wavelength, it may still be transparent.

6. Common solvent method

The two polymers are dissolved in the same solvent, and the blend solution is prepared at a certain concentration. If the solution can be stable for a long time without stratification at a certain temperature, it is a compatible system. Otherwise, it is an incompatible system. Because there are many factors affecting the solubility of polymers, this method could only characterize the compatibility of polymers with the same condition including polymer concentration, the type of solvent and the temperature.

7. Dilute solution viscosity method

When polymer blends are dissolved in a common solvent to form a blend solution, the intercomponent interaction (attraction or rejection) will affect the viscosity of solution. The dilute solution viscosity method is to characterize the compatibility between polymers through the change of viscosity. At different concentration of solution, the composition of the polymer is graphed with viscosity. If it is linear, it shows that the polymer is completely compatible on molecular level. If it is nonlinear, it is a partially compatible system. If a "S" curve is presented, the two kinds of polymers are completely incompatible.

2.8 Methods for improving the compatibility between polymer blends

2.8.1 The significance of improving the compatibility of polymer blends

From the point of comprehensive properties, the polymers with large differences in structure are often selected to prepare polymer blends, but the compatibility between the polymers with large structural differences is usually not satisfied. It is of great significance to solve this contradiction and to obtain polymer blends with good performance.

Because of the poor compatibility of most polymers, it is often difficult for the blend to reach the required dispersion. Even with the aid of external conditions, the two polymers are dispersed uniformly during the blending process, and the separation phenomenon occurs during the process of use, which leads to the instability of the blends and the degradation of the properties. The best way to solve this problem is to adopt the so-called "compatibilization" measures. The compatibilization has two meanings: one is to make the polymers easy to disperse each other to get the macroscopic and uniform blend products; the other is to improve the properties of the phase interface between the polymers and increase the adhesion between each other, so that the blend has a long last and stable performance.

Before introducing the methods to improve the compatibility, the structural characteristics of the compatible polymer blends are discussed, which will help to choose the appropriate methods to improve the compatibility of the incompatible systems.

2.8.2 Structural characteristics of compatible polymers

The relationship between compatibility of polymer blends and thermodynamic properties of polymers has been discussed previously. The Van der Waals force between polymers is not conducive to compatibility between polymers. The ΔH between nonpolar polymers is greater than zero, and it is difficult to satisfy the basic conditions of thermodynamic compatibility. Even polymers that are very similar in molecular struc-ture are incompatible in some occasions, such as PS/poly-α-methylstyrene and poly-isoprene/polybutadiene. But if there are special interactions among the polymers including hydrogen bonds, the interaction of acid with alkali, charge transfer com-plexation and so on, a negative ΔH of the mixing process is formed, that is, the mix-ing process becomes exothermic, which helps to make the blend system compatible.

PVC and polyacrylate are almost compatible, and their compatibility comes from the formation of hydrogen bonds between ester group and PVC, as shown in Fig. 2.9.

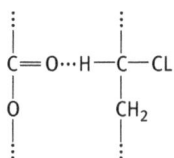

Fig. 2.9: Schematic diagram of the formation of hydrogen bond between PVC and polyacrylate.

Polycarbonate (PC) is compatible with many kinds of polyesters. It has been proved that the mixing heat of the bisphenol A-type PC and the aliphatic polyester is negative, and they are compatible. The aliphatic PC is incompatible with the aliphatic polyester, and their mixing heat is positive. It is the π–π complexation between the electrons of ester groups in polyesters and aromatic ring of PC that determine whether the mixing heat of this kind of blend is positive or negative, so the special interaction only exists between the aromatic PC and the aliphatic polyesters.

The formation of salts or complexes between polymers with opposite ionic charges can lead to the compatibility between them, which is a strong interaction. A typical example is the interaction between PS sulfonate and poly(trimethyl benz-amine) styrene (Fig. 2.10).

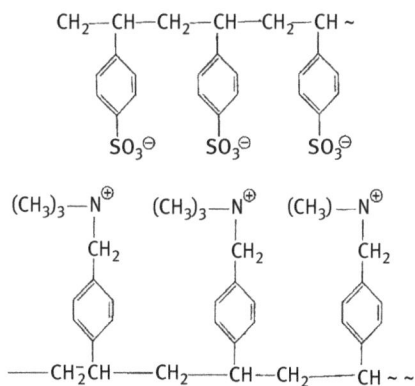

Fig. 2.10: The complexation between polystyrene sulfonate and poly(trimethyl benzamine) styrene.

As a result of this interaction, the properties of the blend are quite different from that of the homopolymer. For example, these polyelectrolytes can be dissolved in water, acid, alkali and some strong polar solvents, and the complexes formed by them are insoluble in acids and bases.

Blending polymers with electron-rich group and electron-deficient group will lead to charge transfer between molecules, thus resulting in compatibility of the system. The polymers with the electron-rich group are the electron donor, and the polymers with electron-deficient groups are electron acceptor. In addition, there will be a strong interaction between them. The mechanical properties of the blends will be obviously improved even if they are not completely compatible.

Polyvinylpyrrolidone is a water-soluble polymer which exhibits alkalinity because of its amide group. Polyacrylic acid is also a water-soluble polymer. The two polymers are well compatible when they are blended, but the resulting blends are insoluble in water. It indicates that the two polymers have formed a new complex through acid–base action.

2.8.3 Compatibilization and its methods

Due to the poor compatibility between most polymers, compatibilization must be adopted in order to improve the dispersion of the two components. There are many ways to produce compatibilization, such as adding compatibilizer, adding macromolecule cosolvent, introducing hydrogen bond or ion bond between polymer components, forming interpenetrating polymers.

1. Adding compatibilizer
Compatibilizers refer to the additives which can promote the combination of the two incompatible polymers to get stable blends with the aid of intermolecular forces. The compatibilizer molecules often have similar structural characteristics with the blend composition, which are compatible with the specific parts of the polymer components. Therefore, the interfacial tension between the two components can be reduced and the compatibility between the components can be increased. The role of the compatibilizer is equal with the role of emulsions in the colloid chemistry and the role of coupling agent in the polymer composites.

2. Compatibilization is induced by chemical reaction during mixing process
It is well known that the macromolecular chains of the rubber will undergo free radical cracking and rebinding in a high-speed shear mixer. A similar phenomenon occurs when polyolefin is strongly mixed, forming a small amount of block or graft

copolymers resulting in compatibilization. In order to improve the efficiency of this process, some free radical initiators such as peroxide are added to promote the production of free radicals.

In the process of mixing, condensation polymers can also produce obvious compatibilization due to chain exchange reactions. For example, polyamide 66 (PA-66) and polyethylene terephthalate (PET) can produce obvious compatibilization in the mixing process due to catalytic ester exchange reaction.

It is also an effective compatibilization method to promote the cross-linking interaction of the blend components in the mixing process. Crosslinking can be divided into two categories: chemical cross-linking and physical cross-linking. For example, chemical cross-linking of low-density PE/PP is achieved by radiation. In this process, a copolymer with compatibilization is first formed, and the desired structure is formed under the action of the copolymer. Then, the cross-linking structure is stabilized by continued cross-linking. Crystallization can be attributed to physical cross-linking, such as PET/PP and PET/nylon 66. Due to the crystallization of the orientational fiber, the morphology of the blends formed is stable and the compatibilization is achieved.

3. Introduction of interaction groups between polymer components

Introduction of ionic groups or ionic-dipole interaction in polymer components can enhance compatibilization. For example, 5% mol $-SO_3$ H group is introduced into PS structure, while the ethyl acrylate is copolymerized with 5% mol vinyl pyridine and then the two components are blended to produce stable blends with excellent properties.

The complexation effect of electron donor and electron acceptor can also increase compatibilization. The blends with such specific interactions often exhibit LCST behavior.

4. Cosolvent method

Two insoluble polymers often form a true solution in a cosolvent. When the solvent is removed, the interface between phases is very large, thus the weak polymer–polymer interaction is enough to stabilize the formed structure.

5. Interpenetrating polymer network (IPN) method

IPN technology is a new method to produce compatibilization. The principle is to combine two kinds of polymers into a stable interpenetrating network, thus producing obvious compulsive compatibilization.

2.8.4 Improving compatibility by changing chain structure

1. Introduction of polar groups to the molecular chains by copolymerization

The intermolecular interaction can be significantly increased by introducing polar groups to the molecular chain by copolymerization, and the compatibility of polymer blend can be improved apparently. PS is a polymer with weak polarity, and it is difficult to be compatible with other polymers apart from PPO and polyvinyl methyl ether. However, acrylonitrile, which contains strong polar groups, could react with styrene to form copolymer (SAN). SAN can form compatibilization system or mechanically compatible polymer blends with many polymers. For example, in PC/SAN blends, the blend shows good toughness when the PC content reaches 30%. However, the PC/PS system has no obvious toughening effect when PC content is below 70%. In addition, polymers such as SAN, polycaprolactone (PCL), PMMA, nitrocellulose and polysulfone can also form compatible or mechanically compatible systems.

PVC is incompatible with polybutadiene, but it can form a compatible system with NBR, a copolymer of butadiene/acrylonitrile with AN content of 30–40%. PVC is also incompatible with PE, but it can be well compatible with some polar ethylene copolymers, such as EVA (VAc content of 65%), ethylene/CO copolymer, EVA/CO terpolymer and ethylene/SO_2 copolymer. These copolymers can form dipolar–dipolar interactions or hydrogen bonds with PVC due to their polar groups, which make them become compatible homogeneous systems.

2. Chemical modification of polymer molecular chains

Chemical modification of the polymer is also an important way to change the polarity of polymer molecular chains. A typical example is chlorinated PE (CPE) made of PE after chlorination. CPE can be compatible or mechanically compatible with many polymers. For example, PE is incompatible with polymethacrylate, PVC, PCL and so on. While CPE is compatible with most polymethacrylate, and CPE/PCL and CPE/PVC can both be compatible or mechanically compatible polymer blends.

In addition, through the way of sulfonation, chlorosulfonation, hydrogenation, hydrogen chloride addition and epoxidation of polymers, the chain structure of the polymers can be modified and the compatibility with other polymers can be improved. For example, PE and NBR are incompatible, while chlorosulfonated PE (CSPE, PE after chlorosulfonation) can be mechanically compatible with NBR in the blends.

Grafting copolymerization of polymers is another effective method for chemical modification of macromolecular chains. For example, PB has no obvious toughening effect on PS in PS/PB blends. However, the impact strength of high-impact PS containing PB-g-PS is over ten times higher than that of PS, which is related to the improvement of the compatibility because of graft copolymer. Similarly, EVA-g-PVC, CPE-g-PVC and other graft polymers are blended with PVC and their mechanical compatibility is also significantly improved.

3. Introduction of specific interaction groups to molecular chains through copolymerization.

Introducing specific interaction groups to polymer molecular chains through copolymerization can make them form hydrogen bonds, ionic bonds, dipoles and acid–base interactions with other polymers.

The copolymerization of styrene and hydroxyl monomers leads to the introduction of hydroxyl units into the PS molecular chains, which greatly improves the compatibility with polyester and polyether polymers. These hydroxyl-containing monomers are given in Fig. 2.11.

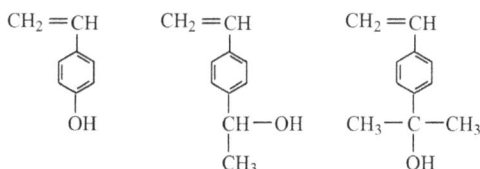

Fig. 2.11: Examples of the monomer containing hydroxyl styrene.

The higher the hydroxyl groups content introduced to PS, the higher the density of hydrogen bond formed between carbonyl group and ether group and the more benefit is for compatibility. For example, PS is completely incompatible with polymethyl ethyl acrylate (PEMA), but it is possible to form a compatible system with PEMA as long as 1% mole hydroxyl units is introduced to the PS molecular chains. It can be seen how the hydrogen bond is effective in improving the compatibility.

Ions with opposite charges are introduced to incompatible polymer blends to produce ion–ion interaction, which can be made into compatible polymer blends. For example, PS/polyethylacrylate (PEA) is an incompatible system, but the sulfonate ion is introduced into the PS to form Polystyrene sulfonate (PSSA) and the vinyl pyridine ion is introduced to PEA to form PEAVP. When the molar concentration of ionic groups is more than 5%, it can become a compatible blend system because of the interaction between the two ions. Both PSSA and copolymer of ethyl acrylate and vinyl pyridine (PEAVP) are dissolved in tetrahydrofuran alone, but the precipitates are formed after mixing their tetrahydrofuran solution, and there is a very strong ion–ion interaction between them. Similar examples include PS/polyimides (PI), PS/polyisobutylene, which are also incompatible systems. If sulfonated ions are introduced into PS, vinyl pyridine ions are introduced into PI or PB; their blends will become compatible systems.

Table 2.6 shows some examples of increased compatibility resulting from specific interactions between polymers.

Tab. 2.6: Polymer blends with specific interaction.

Polymer blends	Interaction type	Polymer blends	Interaction type
PVC/PCL	Hydrogen bond	PEO/bisphenol A polyhydroxy ether	Hydrogen bond
PAA/PEO	Hydrogen bond	PVEM/bisphenol A polyhydroxy ether	Hydrogen bond
PA/ABS	Hydrogen bond	PPO/PS	Π–hydrogen bond
Hydroxylated PS/PEMA	Hydrogen bond	PVF_2/PMMA	Dipole–dipole
PBT/PCV	Hydrogen bond	PVF_2/PVAc	Dipole–dipole
EVA/PVC	Hydrogen bond	PMMA/PC	π–π complexation
PMMA/PVC	Hydrogen bond	PC/PBT	π–π complexation
PS/PCL	Hydrogen bond	Sulfated PS/cationized PVAc	Ion–ion complexation

4. Formation of IPN or cross-linking structure

As long as a polymer is soluble in another monomer (oligomer) or two kinds of oligomers that are mutually soluble, the interpenetrating network structure of the polymers can form and the phase structure is permanently kept, which is the compelling soluble interaction. The same polymer blends composition that is mechanically incompatible or exhibits incompatibility under mechanical blending shows excellent performance after the formation of IPN, which has been proved by numerous examples.

In incompatible polymer blends, chemical bonds between two components can be formed by chemical cross-linking, which can also improve the compatibility. For example, after the vulcanization of SBR/BR blends, only one T_g was displayed, which eliminated all the characteristics before the vulcanization. The compatibility of PE/PVC, PE/natural rubber, PE/NBR and other systems has also been improved through the cross-linking process.

5. Changing the molecular weight

The compatibility of polymer pairs is closely related to their molecular weight. The blending process of rubber/rubber and rubber/plastics has proved that the using of multisection plasticizing is beneficial to the dispersion and compatibility of the two components, because the multisection plasticizing can reduce the molecular weight of the polymer more effectively.

2.8.5 The application of compatibilizer

1. Type of compatibilizers
In an incompatible blend system, the mechanical compatibility of the blend is greatly improved because of the existence of a third component. The third component is called compatibilizer. The effect of the compatibilizer in the blends is mainly to reduce the interfacial tension between two phases, to improve of the stability and dispersion of the dispersed phase and to increase the interfacial drilling force, thus making a mechanically incompatible system into a polymer blend with excellent comprehensive properties. Therefore, the effect of the compatibilizer can be summed up as follows: (1) reducing the interfacial energy between the phases, (2) promoting the phase dispersion in the process of blending, (3) preventing the condensation of the dispersed phase and (4) strengthening the cementation between the phases.

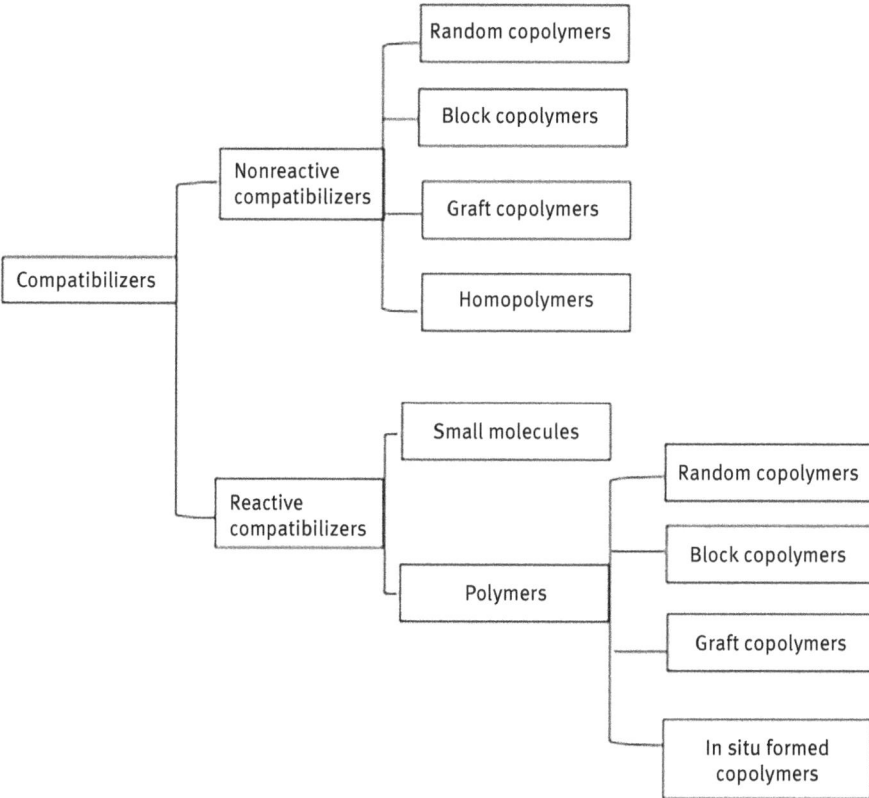

Fig. 2.12: The type of compatibilizers.

Most of the substances used as compatibilizer are copolymers. These copolymers have common structural characteristics of the two components of the blends and have specific interactions with the components of the blends. Therefore, compatibilizer tends to disperse on the interface between two phases in the blend system, and the dosage is very little. However, it has a significant effect in improving the compatibility between two components.

There are different types of compatibilizers. The common compatibilizer types are shown in Fig. 2.10. At present, the types of compatibilizers are still under development.

2. Nonreactive compatibilizers

The common types of nonreactive compatibilizer have three main categories including AB-type, AC-type and CD-type compatibilizers.

(1) AB-type compatibilizer

This compatibilizer has a common chain segment of two kinds of blends, for example, PS-g-PMMA is a compatibilizer for PS/PMMA, PB-g-PS is a compatibilizer for PB/PS and EPM is a compatibilizer for PE/PP. Block copolymer, graft copolymer and random copolymer can all be used as compatibilizer. CPE is not a copolymer, but it should contain the corresponding structure of PVC and PE, so it possesses a good compatibilization effect on PVC/PE system. The addition of 20% CPE can transform PVC/high-density PE (HDPE) (50/50) from a typical brittle system into a typical ductile system, as shown in Fig. 2.13. Because Cl atoms in CPE are randomly distributed, their actual molecular structures are similar to those of PE/PVC random copolymers.

Both PE and PS are general plastics of large varieties. It is very significant if cheap PE can be used to toughen the brittle PS, however, they do not have the mechanical compatibility. Therefore, finding the suitable copolymer as compatibilizer is the key to solve this problem. It is difficult to obtain block copolymers by direct copolymerization of ethylene and styrene monomer. Usually, other ways are used to synthesize similar copolymers. For example, the PS-b-PB copolymer is obtained by anionic polymerization and then it was completely hydrogenated, so that the PB segment could be transformed into the PE segment. In order to prevent crystallization of PE segment, a certain amount of poly-1,2-butadiene should be added to the monomer. The PB block in the hydrogenated copolymer is a copolymerized chain segment of ethylene-butene, which is an excellent compatibilizer for PE/PS, and its compatibilization effect is shown in Fig. 2.14.

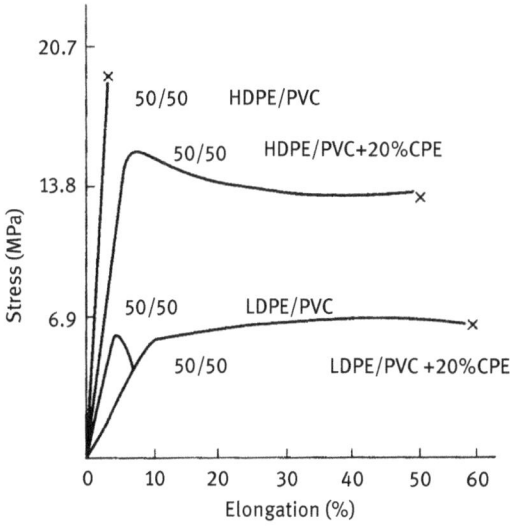

Fig. 2.13: The effect of CPE to the stress–strain of PE/PVC blends.

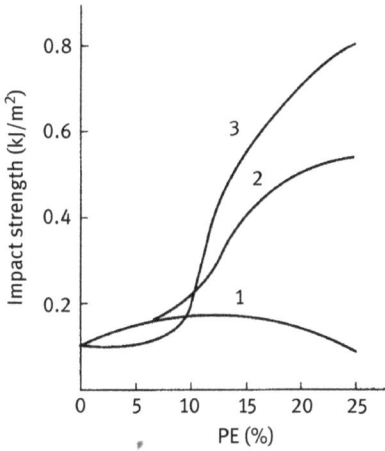

Fig. 2.14: The effect of compatibilizer to the impact strength of PS/PE blend system.
1. PS/PE blend system; 2. PS/PE blend system with 5% graft copolymer; 3. PS/PE blend system with 5% block copolymer

(2) AC- and CD-type compatibilizers

This kind of compatibilizer does not have the same chemical structure with at least one of the components in the blends, which mainly depends on the formation of specific interactions between the compatibilizer and the component polymers or the formation of chemical bonds to improve the compatibility of the two components. For example, PE and PPO are incompatible, but adding a proper amount of PE-b-PS co-polymer in PE/PPO blends can improve their compatibility, because PS and PPO have good compatibility. PP-g-MAH can be used as a compatibilizer for the PA/EPDM system, because the branched chain MAH can react with PA. BR-b-PMMA can be used as a compatibilizer for the SAN/SBR system, because the syndiotactic PMMA has a strong interaction with SAN. Another example is SEBS (partially hydrogenated SBS)-g-MAH is a CD type compatibilizer of Nylon/PPO, because PS segments are com-patible with PPO, while MAH branched chains can form chemical bonds with nylon.

PCL is one of the ideal objects for preparing this type of compatibilizer to form copolymer components, because it can form a compatible system with many poly-mers, such as PVC, SAN and chlorinated polyether. As a copolymer component, it can be used as compatibilizer for many systems. For example, PCL-b-PS can be used as compatibilizer for PVC/PS, PVC/PPO and other systems. Table 2.7 lists com-mon nonreactive copolymer compatibilization systems.

Tab. 2.7: Nonreactive copolymer compatibilization systems.

Matrix phase	Dispersion phase	Compatibilizer
PE, PS	PS, PE	S-B, E-EP, S-I-S, S-I-HBD, S-EB-S, S-B-S, PS-g-PE
PP	PS	S-EB-S
PE, PP	PP, PE	EPM, EPDM
EPDM	PMMA	EPDM-g-PMMA
PS	PA-6, EPDM, PPE	PS-PA-6, S-EB-S
PET	HDPE	S-EB-S
PF	PMMA, PS	PF-g-PMMA, PF-g-PS
PVDF	PS-PPE	PS-g-PMMA, PS-b-PMMA
PVC	PS, PE, PP	PCL-g-PS, CPE
SAN	SBR	BR-g-PMMA

3. Reactive compatibilizer

The compatibilization principle of reactive compatibilizer is completely different from that of nonreactive compatibilizer. The reactive compatibilizer is mainly a group of polymers containing functional groups that can react with the blend components. It can form new chemical bonds with polymer components of the blend to achieve com-pelling compatibilization when it is added into the system, which can be called chemi-cal compatibilization. This compatibilizer is particularly suitable for compatibilization of polymers with poorly compatible but easily reactive functional groups.

For example, copolymers catalyzed by partially hydrolyzed EVA and MAH can be used as a reactive compatibilizer for nylon/polyolefin blends. The −COOR, −OH and −COOH groups in the copolymer molecules can be "anchored" on the phase interface with a strong chemical and physical (ionic bond) interaction with nylon molecules, and the carbon chain can be intertwined with the polyolefin component, showing a stronger compatibilization.

The application examples of some reactive compatibilizers are listed in Tab. 2.8.

Tab. 2.8: Application examples of some reactive compatibilizers.

Matrix phase	Dispersion phase	Compatibilizer
ABS	PA-6/PA-66 copolymer	SAN/MA copolymer
PP, PA-6	PA-6, PP	EPM/MA copolymer
PE	PA-6, PA-66	Carboxylated PE
PP, PE	PET	PP-g-MA, Carboxylated PE
PA-6	EPM	SMA, EPM-g-MA
PPE, PS	EPDM or acid esters of EPM	PS and zinc stearate

Some components involved in the blend contain mutually reactive functional groups, and the reaction compatibilization can also occur in the reaction. This type of reactive compatibilization can be called the "in situ" reactive compatibilization. Such reactive compatibilization systems should have the following characteristics:
(1) There should be enough dispersion between the blends.
(2) The reaction rate between the functional groups of the two components should be fast enough.

According to the types of reactions, the in situ reactive compatibilization can be divided into four types:
(1) Chain scission reaction occurred in the molecular chains of the blends, resulting in the formation of macromolecular free radicals, and then rejoining to form block copolymers, graft copolymers or random copolymers.
(2) The end functional groups of a polymer react with functional groups on the main chain of another polymer to form graft copolymers.
(3) The functional groups of the two polymer chains react with each other to form graft copolymers or cross-linked copolymers.
(4) The two kinds of polymers form an ionic bond with each other.

The chemical reactions used in the "in situ" reactive compatibilization include the reaction of acid anhydride with the amino group, the reaction of epoxy group with carboxyl group, hydroxyl group, anhydride, amino group, the reaction of oxazoline with carboxyl group, and the salt-forming reaction between the molecular chains.

Exercises

1. Try to draw the basic types of phase diagrams of polymer/polymer two element blending system.

Drawing tips:
In Fig. 2.3, (a) presents an arbitrarily proportional compatibility system, (b) is the system with utmost critical intermiscible temperature (UCST), (c) represents the system with lowest critical intermiscible temperature (LCST), (e) represents the system with UCST and LCST at the same time and (a) and (f) represent the system with locally incompatible regions.

2. What are the principles of measuring the compatibility of polymers by glass transition temperature method?

Answer:
The T_g of the polymer blends has a direct relationship with the mixing degree of two polymers chains. If the two polymer components are completely compatible and the blend is homogeneous, the system only exhibit one T_g. If the two components are completely incompatible and form two phase structures with obvious interface, the components exhibit two T_g, which are equal to the T_g of each polymers, respectively. The T_g of partially compatible systems falls between the above two limiting cases.

3. What are the methods for improving the compatibility between polymer blends?

Answer:
(1) Adding compatibilizer,
(2) compatibilization induced by chemical reaction during mixing process,
(3) introduction of interaction groups between polymer components,
(4) cosolvent method and
(5) IPN method.

Chapter 3
Phase structure of multicomponent polymers

3.1 Introduction

Multicomponent polymers are often referred to as polymer blends or polymer alloys. Strictly speaking, polymer blends are not completely synonymous with polymer alloys. The alloys may be homogeneous or multiphase. Homogeneous alloys correspond to random copolymers and miscible polymer blends, while multiphase alloys correspond to immiscible or partially miscible polymer blends. This comparison is of great inspiration to the study of polymer blends.

It is well known that about 0.8% of carbon can be dissolved in iron at 1,100 °C. If the temperature is reduced to 72 °C, then the phase separation will occur. One of this phase separations is the cementite composed of Fe_3C, that is, iron carbide; the other is the pure iron containing 0.025% carbon, that is, α-iron. The two-phase structure is called bead iron, of which two phases are layered. α-Iron is soft and extendible, while cementite is very hard. Bead iron combines the comprehensive properties of these two materials. In order to illustrate the effect of the two-phase structure on the properties of the alloy, the pearlite iron can be compared with tempered martensite. Iron carbide in tempered martensite is dispersed in the continuous phase of pure iron with spherical particles, which is obviously different from the layered structure of pearlite iron. Therefore, although the two alloys have the same composition, their properties are quite different.

The above example shows that phase structures has a significant effect on the properties of the alloy. The same phenomenon also occurs in polymer blends. For example, although high-impact polystyrene (HIPS) prepared by blending rubber with polystyrene has a certain composition, different preparation methods and different processing conditions can produce quite different phase structure, so the impact strength of products will be very different.

The phase structure of polymer blends is one of the most basic factors determining their properties, like that of metal alloys. Therefore, it is necessary to systematically study the phase structure of polymer blends.

The phase structure of polymer blends is affected by a series of factors. These factors can be grouped into three types.

1. Thermodynamic factors
These factors are interaction parameters, interfacial tension and so on. Equilibrium thermodynamics can be used to anticipate the final equilibrium structure of the blend whether it is homogeneous or heterogeneous.

https://doi.org/10.1515/9783110596335-003

2. Kinetic factors

Phase separation dynamics determine whether the equilibrium structure can be acquired or not and the degree of phase separation. As mentioned earlier, two types of phase structure can be found according to the different phase separation kinetics: the dispersive structure is formed by nucleation and growth mechanism (NG), and the interlaced continuous structure is formed by mechanism of spinodal phase separation (SD). The specific phase structure is mainly determined by the quench degree of the system. The larger the quench degree, the smaller the initial size of the aggregate. This initial size can vary from 10 to 100 nm. The structural size tends to balance the expected value of thermodynamics with the extension of time. Because of the high viscosity of polymer materials, this balance structure is generally difficult to achieve. The structure formed by phase separation kinetics can be stabilized by compatibilization method, thus improving the stability of product performance.

3. The phase structure induced by the flow field in the mixed process

This is essentially due to the different nonequilibrium structures formed by different flow parameters.

Understanding the above three aspects can give a general concept of the basic mechanism of the formation of polymer blends' phase structure, and thus understand the basic ways to control the phase structure and properties of polymer blends. From the above point of view, the basic types of phase structures, phase interface structures, mutual solubility and mixing processing methods of polymer blends are discussed in this chapter.

3.2 Basic types of phase structure of polymer blends

3.2.1 Phase structures of amorphous polymer blends

Polymer blends may consist of two or more polymers. For thermodynamically miscible blends, it is possible to form homogeneous phase structures, whereas two or more phases can be formed. This multiphase structure is the most common and complex, and is the focus of the following introduction. For the sake of simplicity, we mainly discuss the case of two components, but the basic principles involved are also applicable to multicomponent systems.

Two-phase polymer blends consisting of two polymers can be divided into three basic types according to their phase continuity: single-phase continuous structure, that is, one phase is continuous and the other phase is dispersed; two-phase interlocking or interlaced structures and mutually penetrating two-phase continuous structures.

3.2.1.1 Single-phase continuous structures

Single-phase continuous structure refers to the continuity of only one phase in the two phases or several phases of polymer blends. This continuous phase can be regarded as a dispersion medium, called matrix, and other phases are dispersed in the continuous phase, called dispersion phase. The single continuous phase structure is manifested in many forms due to the shape and the size of the dispersed-phase domains (i.e., microdomain structure) as well as their combination with the continuous phase. The relationship between continuous phase and dispersed phase in polymer blends with single continuous structure is somewhat like the relationship between sea and island, so it is often called "sea-island structure" vividly. This blend structure is the most common structural form in incompatible polymer systems.

In the complex polymer system, each phase exists in a certain form of aggregation. Because of the interleaving between phases, the relatively continuous or discontinuous phases are divided into many tiny regions, which are called phase domains or microregions. The shape, size and fine structure of phase domains of polymer blends vary according to different preparation methods, different component structures and blending proportions. The typical structures of different phase domains can be summarized into the following categories:

1. Irregular shapes of dispersed phase

In polymer blends, the dispersed phase is composed of particles with irregular shape and highly dispersed size. The products obtained by mechanical blending generally have such phase structure. In general, the larger components constitute the continuous phase, and the smaller components constitute the dispersed phase. The particle size of dispersed phase is usually 1–10 μm. For example, in the electron microscopic photos of HIPS ultrathin samples, the rubber component (polybutadiene, PB) is dispersed in the polystyrene matrix (continuous phase) with irregularly shaped particles (Fig. 3.1).

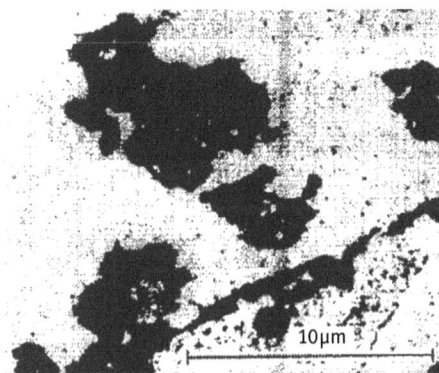

Fig. 3.1: Electron microscopic photographs of mechanical blending HIPS (black irregular particles are rubber dispersed phases).

2. Relatively regular shapes of dispersed phase

In such polymer blends, the dispersed-phase particles are relatively regular, uniform in size and shape and generally spherical. The dispersed-phase particles do not contain or contain only a very small number of continuous phases, and the interface is clearer. Bisphenol A diglycidyl ether epoxy resin toughened with CTBN [carboxyl nitrile butadiene rubber (NBR)] is an example of this phase structure (Fig. 3.2). CTBN was dissolved in low-molecular-weight epoxy resin at 50–80 °C, then cured with a curing agent. In the product, the rubber was dispersed in epoxy resin matrix with regular spherical particles, and the diameter of rubber particles was about 1 μm. Although there may be dispersed rubber in the matrix and small amount of epoxy resin dissolved in the rubber particles, the two phases are basically composed of a single component.

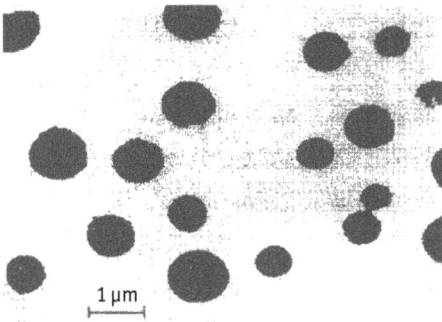

1 μm

Fig. 3.2: Phase diagram of 8.7% CTBN rubber-toughened epoxy resin (black pellets are rubber particles).

Another example of this structure is some triblock copolymers, such as styrene–butadiene–styrene (SBS) triblock copolymers. When the butadiene block is short and the content of butadiene is low (generally 20%), the butadiene block with uniform spherical particles is dispersed in the continuous phase matrix composed of polystyrene block. The diameter of spherical particles is about tens of nanometers, as shown in Fig. 3.3. When the butadiene content increases, the corresponding phase structure will change. For example, when the mass fraction of butadiene is 40%, the dispersed phase becomes a cylindrical structure. With the increase of butadiene content, butadiene will convert into the continuous phase while the styrene block becomes the dispersed phase.

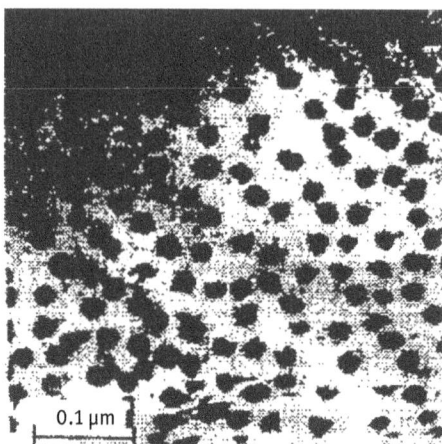

Fig. 3.3: Phase structure of the triblock copolymer (butadiene content is 20%).

3. Honeycomb dispersed phase

This phase structure is more complicated than the previous two cases. It is characterized by the fact that dispersed-phase particles contain smaller particles composed of continuous phases. Therefore, within the dispersed phase, the smaller inclusion of the continuous phase component can be regarded as the dispersed phase, while the dispersed-phase component of the particles can be regarded as the continuous phase. If the dispersed-phase particles are regarded as cells, the cell wall is composed of continuous phase components, and the cell contains smaller particles composed of continuous phase components, so it is also called "cell-like structure." If the dispersed-phase particles were cut out, their sections are similar to sausages in form or appearance, so they are also called "sausage-like structures." Most of the blends obtained by graft copolymerization have such phase structure. Figure 3.4

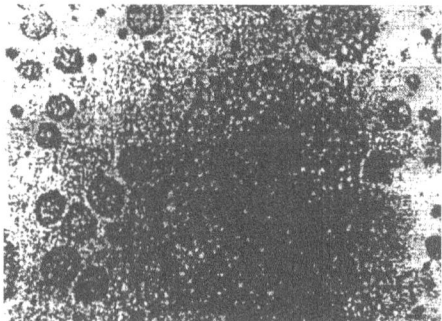

Fig. 3.4: Electron microscopic photo of G-type ABS structure (the black part is polybutadiene).

shows electron microscopic photographs of G-type (emulsion graft copolymerization) acrylonitrile (AN)–butadiene–styrene (ABS) blends. This is a photograph obtained by dyeing butadiene segments with osmium tetroxide. The black part is rubber and the white part is resin. This type of ABS is a two-phase blend composed of rubber particles and resin matrix, and the particle size of rubber particles is about 0.1–0.5 μm. The graft copolymers formed during the reaction mainly concentrate on the interface of the two phases and play the role of compatibilizer. In general, this copolymer is enveloped on the surface of dispersed particles and forms a core–shell structure. Figure 3.5 is an electron microscopic photograph of the sausage-like phase structure of grafted HIPS.

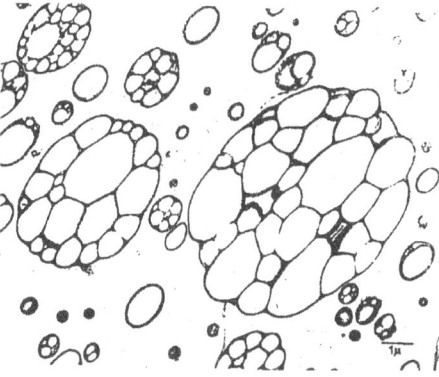

Fig. 3.5: Electron microscopic photograph of the sausage-like phase structure of HIPS.

4. Lamellar dispersed phase

This morphology means that the dispersed phase is dispersed in continuous phase matrix in a microflake shape, and when the concentration of dispersed phase is high, the flakes of dispersed phase are further formed. For example, in order to obtain polymer blends with good barrier properties, polyamides (PAs) with excellent barrier properties are dispersed uniformly in polyethylene (PE); aiming to obtain polymer blends with excellent antistatic properties, hydrophilic polymers are dispersed in microflakes and concentrated in the surface layer of the continuous phase such as PE or PA. Figure 3.6 is an electron micrograph of this phase structure. The necessary condition for the formation of this structure is that the melt viscosity of the dispersed phase is larger than that of the continuous phase polymer, and the appropriate shear rate and compatibilization technology are adopted in the blending process.

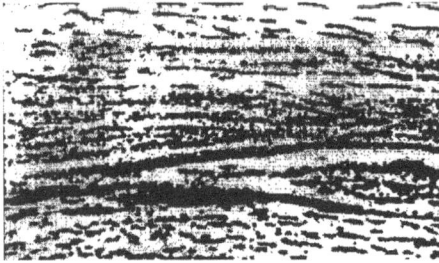

Fig. 3.6: Electron microscopic photograph of dispersed phase in lamellar polymer blends.

3.2.1.2 Two-phase interlaced layered structure

The characteristic of this kind of phase structure is that there is no continuous phase but a two-phase interlaced layered structure running through the whole sample in the system, which makes it difficult to distinguish the continuous phase from the dispersed phase. Figure 3.7 is an electron microscopic photograph of polypropylene (PP) and metallocene-catalyzed polyethylene (mEPE) blends. It can be seen from the figure that when the concentration of dispersed-phase mEPE increases to a certain extent, the system shows two-phase interlaced layered structure.

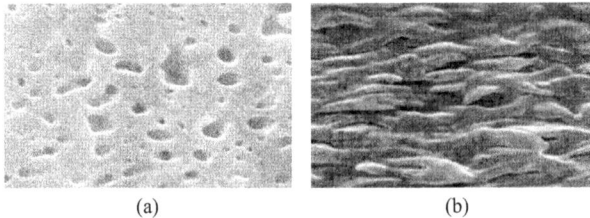

(a)   (b)

Fig. 3.7: SEM photo of PP/mEPE blends at different blending ratios: (a) PP/mEPE 60/40 and (b) PP/mEPE 50/50.

Such phase structure always forms when block copolymers produce two-phase spin separation, and the content of the two-block components is similar. For example, when the content of butadiene in SBS triblock copolymers is about 60%, a two-phase interlaced layered structure is formed, as shown in Fig. 3.8. The styrene-ethylene oxide block copolymers molded in *n*-butyl phthalate as solvents also belong to this type of lamellar structure.

Polymer blends with block copolymers as main components also easily form such morphologies. For example, the phase structure of diblock copolymers of polystyrene (PS) and polyisoprene (PS-b-PI) is interlaced in layers. When these diblock copolymers are blended with PS, because PS is compatible with PS blocks in block copolymers, the

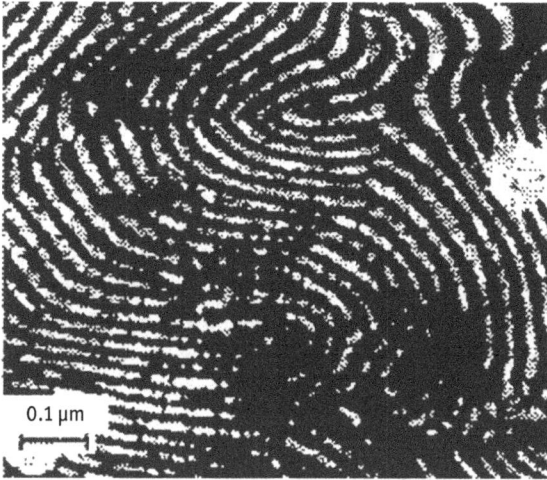

Fig. 3.8: Electron microscopic photo of phase structure of SBS (butadiene content 60%).

lamellar morphology will be expanded rather than deformed when PS dosage is small. Only when PS dosage is large, the lamellar morphology will be destroyed.

The phase structure of block copolymer A-b-B is closely related to its composition ratio. In general, the component with less content constitute dispersed phase, and larger component constitute a continuous phase. As the content of dispersed phase increases gradually, the dispersed phase changes from spherical droplets to rod-like or fibrous. When the content of the two phases is similar, a layered structure is formed. The ideal model is shown in Fig. 3.9. In principle, if the homopolymer A added into A-b-B block copolymer is compatible with phase A, its effect on phase structure is equal to increasing the proportion of A block in A-b-B (molecular weight).

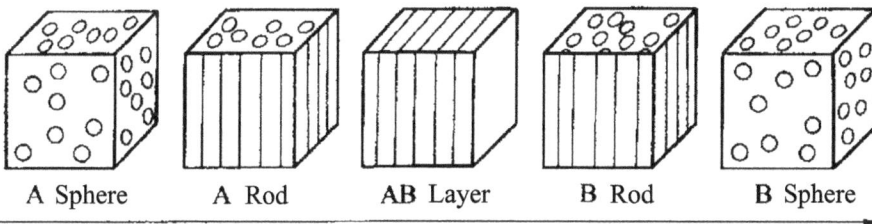

| A Sphere | A Rod | AB Layer | B Rod | B Sphere |

Fig. 3.9: Phase structure models of block copolymer (component A increased; B decreased).

Generally speaking, the compatibilized heterogeneous polymer blends and block copolymers can form bead droplets ($\varphi < 0.15$), fibrous ($0.15 < \varphi < 0.38$) or interlaced layered structures (when the content of the two components is similar) according to the variation of dispersed phase. However, the phase structure of polymer blends containing block polymers is more complex practically, unlike the model shown in Fig. 3.8. There may be various transition forms or several phase structures coexisting simultaneously. For example, spherical particles, short rods and irregular strips, blocks and fibers exist simultaneously in the same polymer blend.

From Fig. 3.8, polymer blends' structures can be reversed in a certain composition range. It turns out that the components of the dispersed phase become a continuous phase while the original component of the continuous phase becomes a dispersed phase. Within the range of phase inversion, a two-phase interlocking and cocontinuous phase structure is often formed, which improves the mechanical properties of the blends. This provides an important basis for the selection of mixing and processing conditions.

3.2.1.3 Interpenetrating two-phase continuum structure

A typical example of the interpenetrating two-phase continuum structure is interpenetrating polymer network (IPN). In IPN, two kinds of polymer networks run through each other, making the whole blend become an interwoven network. The two kinds of polymers are both phase-separated and continuous. And the size of polymer microdomain is generally small and even. When the compatibility between the two components is good, the size of the microsphere can be as small as 10 nm, and the maximum size is only tens of nanometers. The two-phase continuity of IPN has been confirmed by electron microscopy and dynamic mechanical properties. In addition, according to the Davies equation, the relation between Young's modulus and composition of two-phase continuum system is:

$$E^{1/5} = \varphi_1 E_1^{1/5} + \varphi_2 E_2^{1/5} \tag{3.1}$$

where E, E_1 and E_2 are Young's modulus of blend, component 1 and component 2, respectively, and φ_1 and φ_2 are the volume fractions of components 1 and 2, respectively. The elastic modulus of IPN basically agrees with the Davies equation, which also proves that IPN is a two-phase continuous phase structure.

The continuity of the two phases in IPN is generally different. The continuity of polymer 1 is greater than that of polymer 2, even if the content of polymer 2 is higher. The larger continuity has greater influence on the performance. For example, in a blend of *cis*-PB (polymer 1) and PS (polymer 2), PB forms a continuous phase (Fig. 3.10). It can be seen from the figure that the morphology of IPN may also have cellular structure. Polymer 1 forms cell walls and polymer 2 forms cell body with cell size of 0.05–0.1 μm. The cell walls and the internal cell body have a finer structure with a size of 10–20 nm.

The better the compatibility of the two components is, the greater the degree of cross-linking is and the smaller the phase domain of the IPN two-phase structure is.

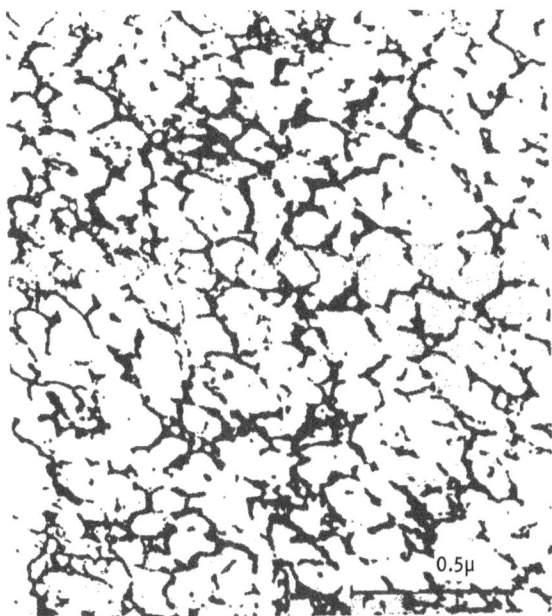

Fig. 3.10: Electron microscope photo of PB/PS IPN.

3.2.2 Morphological characteristics of blends containing crystalline polymers

All the aforementioned cases show the condition when both polymers in the blends are amorphous. The above principle also applies to the case where both polymers are crystalline, or one of them is crystalline and the other is noncrystalline. The difference is that the change of crystalline morphology and crystallinity after blending should be considered in the case of crystalline polymers.

3.2.2.1 Amorphous polymer/crystalline polymer blends

One component of polymer blends is crystalline polymer, and the other is amorphous polymer. Examples include polycaprolactone/polyvinyl chloride (PCL/PVC) blends, isotactic PS/random PS (i-PS/a-PS) blends, i-PS/polyphenyl ether blends and polyvinylidene chloride/polymethyl methacrylate (PMMA) blends. In the early stage, the phase structure of these blends was summarized as shown in Fig. 3.11: (a) the crystalline components were dispersed in the amorphous medium in the form of crystalline grains; (b) the crystalline components were dispersed in the amorphous medium in the form of spherulites; (c) the amorphous components were dispersed in the continuous phase of spherulites in the form of grains; (d) the amorphous components form larger domains distributed in the spherulites continuous phase. Figure 3.12 is an

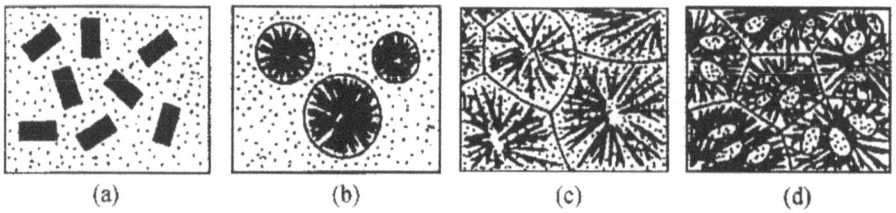

(a)   (b)   (c)   (d)

Fig. 3.11: Schematic diagram of phase structure of crystalline/amorphous blends: (a) the crystalline grains are dispersed in the amorphous medium; (b) the spherulites are dispersed in the amorphous medium; (c) the amorphous state is dispersed in the spherulite and (d) the amorphous state forms larger phase domains and distributes in the spherulite.

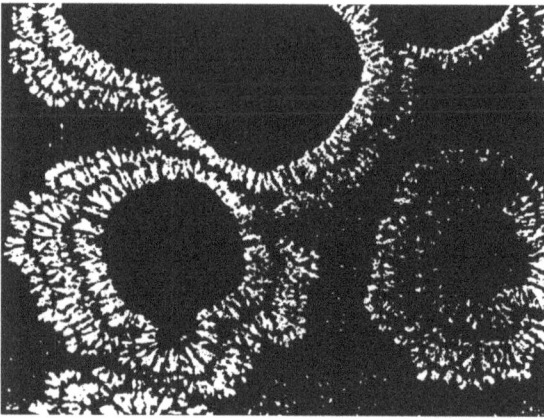

Fig. 3.12: Optical microscopic photographs of PS/PEO blends.

optical microscope photograph of polyethylene oxide (PEO) spherulites distributed in PS matrix.

According to extensive research reports in recent years, the aforementioned four kinds of crystalline structures cannot fully represent the whole morphology of crystalline/amorphous polymer blends, that is, at least four kinds should be added as follows: spherulites are almost full of the whole blend system (continuous phase), and amorphous polymers are dispersed between spherulites and spherulites; spherulites are slightly destroyed and formed dendrites dispersed among amorphous polymers; crystalline polymers fail to crystallize and form amorphous/amorphous blends (homogeneous or heterogeneous); crystallization occurs in amorphous polymers and the system is transformed into a crystalline/crystalline polymer blend system (which may also contain amorphous regions containing one or two polymers simultaneously).

3.2.2.2 Crystalline polymer/crystalline polymer blends

Examples of crystalline/crystalline polymer blends include polybutylene terephthalate (PBT)/polyethylene terephthalate (PET) blends, PE/PP blends and PA/PE blends. Because crystalline polymers still have amorphous regions, crystallinity and crystal structure are affected by many factors. The phase structures of such blends are more complex, and it is possible to present six morphologies as shown in Fig. 3.13.

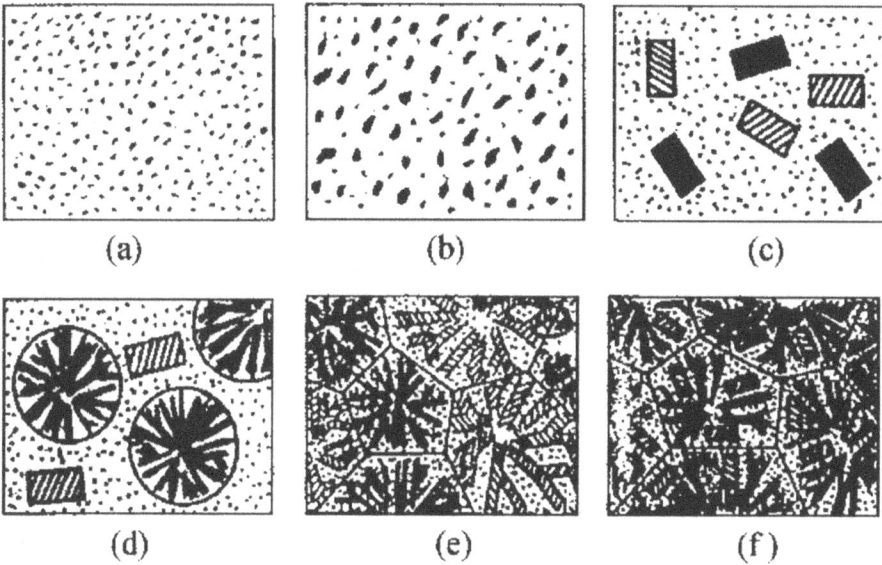

Fig. 3.13: Possible phase structures of crystalline/crystalline polymer blends.

In Fig. 3.13, the crystallinity of the original two crystalline polymers was destroyed, which is shown in (a) and (b) conditions, and amorphous blends were formed, in which (a) is formed with good compatibility and (b) is formed with poor compatibility. For example, PBT/PET melt blends, due to transesterification, formed random block copolymers, which completely lost crystallinity, and their blends showed the morphology of such amorphous blends.

Figure 3.13(c) and (d) shows that the crystalline morphologies of two crystalline polymers are relatively common. For example, PP/ultra-high-molecular-weight PE (UHMWPE) and polyphenylene sulfide/PA blends can form two-phase separated crystalline/crystalline morphologies under certain preparation conditions. Figure 3.9(c) and (d) appears to show that the crystalline state is dispersed in amorphous state, but when the blend system can fully crystallized to a high degree of crystallinity, it becomes a small amount of amorphous interlayer distributed between crystalline grains and spherulites, which is obviously different from Fig. 3.9(c), (d).

Figure 3.13(e) shows that the two crystalline polymers form spherulites separately, while the spherulites contain the components of amorphous region.

If two crystalline polymers can form eutectic, the morphology is as shown in Fig. 3.13(f). There are also amorphous and nonamorphous regions in the blended spherulites. This latter case is not shown in the figure. Many studies have confirmed the existence of eutectic in polymer blends. For example, the study of linear low-density PE (LLDPE)/UHMWPE blends obtained by solution blending shows that the melting peak of LLDPE/UHMWPE blends is single and different from the original melting peaks of two polymers, which proves the formation of eutectic.

In addition to the aforementioned various possible morphologies, according to recent extensive studies, the following morphologies have also been found in crystalline/crystalline polymer blends. If the number is continued according to the morphological category in Fig. 3.13, then: (a) one crystalline polymer forms crystals, while the other crystalline polymer is not crystallized and transforms into amorphous state. For example, for PP/UHMWPE melt blends, when crystallized at 140 °C, only PP crystallizes while PE becomes amorphous, and PE can penetrate into PP spherulites or lamellae; (b) single-component crystals coexist with two-component eutectics. In the study of high-density PE (HDPE)/LLDPE, it was confirmed that the low-regularity part of LLDPE crystallized separately, while the high-regularity part participated in the formation of eutectic with high-density linear PE. The studies of the HDPE/LDPE (low-density PE) blends lead to similar results.

In addition, epitaxial crystallization is also a special noticeable case in the morphology of crystalline/crystalline polymer blends. The so-called epitaxial crystallization is the orientational growth of a crystalline substance on another substance (matrix). The study of crystallization between crystalline polymers began in the mid-1980s. At present, most of the studies are on isotactic PP (i-PP)/HDPE and i-PP/PA blends. Taking the i-PP/PE blend as an example, when the blend film is stretched, the PP as the matrix will appear as shown in Fig. 3.14(a). In this morphology, black is the crystallization zone, marked as the light-colored zone of A, and the P molecule in the crystallization zone orientates along the stress direction, while the crystallization

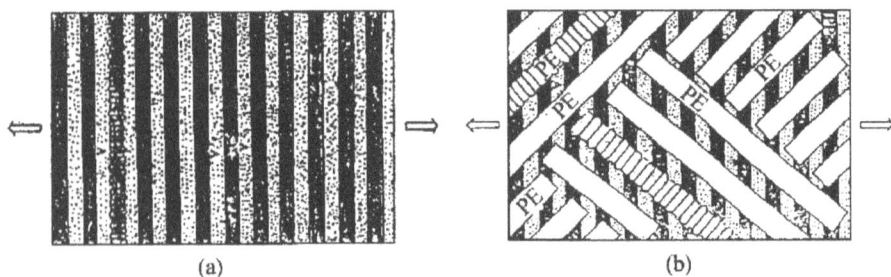

(a)　　　　(b)

Fig. 3.14: Epiphytic crystallization of i-PP/PE blend films under tension: (a) crystalline morphology of matrix PP under tension and (b) epiphytic crystallization of PE on PP crystal.

grows along the direction perpendicular to the stress, forming a "mutton string" crystallization. The growth direction of PE, another blend component, grows on the PP crystal at an angle of 452° with the growth direction of PP crystal, as shown in Fig. 3.14(b). The formation of epitaxial crystallization can significantly improve the mechanical properties of the blends. Therefore, it arouses people's interests greatly.

3.3 Factors affecting phase structure of polymer blends

3.3.1 Factors affecting continuity

Whether the polymer component is in the continuous phase or the dispersed phase has different effects on the properties of the blends. In general, continuous phase determines the basic properties of materials (such as mechanical strength, modulus and elasticity), while dispersed phase affects some special properties of materials (such as impact resistance, optical properties and permeability). Therefore, it is very important to determine the two-phase composition of polymer blends according to the requirements of material properties. It is generally believed that the main factors affecting the continuity of the phase are the following.

1. Component ratio

Typically, high content of component is easy to form a continuous phase. For example, when the proportion of ethylene propylene diene monomer (EPDM) in the EPDM/PP blend is higher than 60% and PP is lower than 40%, the continuous phase is EPDM, the dispersed phase is PP, and the macroelasticity of the blend shows rubber elasticity, which is a kind of plastic-reinforced elastomer. When EPDM content is lower than 40% and PP content is higher than 60%, the blend shows the characteristics of plastics, PP is the continuous phase, EPDM is the dispersed phase and the blend is a toughened PP plastic.

Because there are many factors affecting the morphology of blends, the composition ratio is not a sufficient condition to determine the continuous phase and the dispersed phase.

Assuming that the dispersed-phase particles in the blend system are arranged in the form of "hexagonal compact filling" of spheres with the same diameter (Fig. 3.15), the critical volume fraction of the dispersed phase can be calculated theoretically to be 74%. In other words, when the volume content of one component is more than 74%, it is impossible for the component to become a dispersed phase. Similarly, when the volume content of a component is less than 26%, the component can no longer be a continuous phase. When the component content is between 26% and 74%, whether a component becomes a continuous phase or dispersed phase depends not only on the component ratio, but also on other factors.

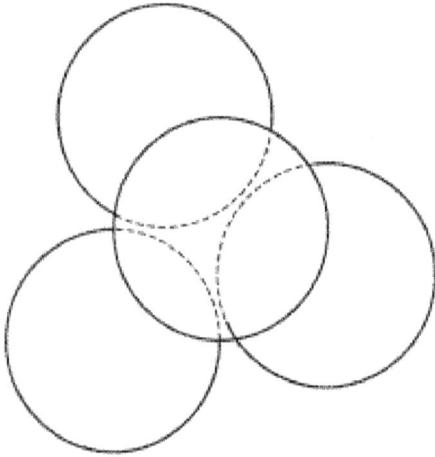

Fig. 3.15: Hexagonal compact filling sketch.

2. Viscosity difference

Under the blending conditions, the viscosity difference between the two components has a great influence on the phase continuity. Generally speaking, the components with low viscosity have relatively good fluidity and are easy to form continuous phase, while the components with high viscosity are easy to form dispersed phase, so there is often the theory of "soft-wrapped hard." However, the effect of viscosity

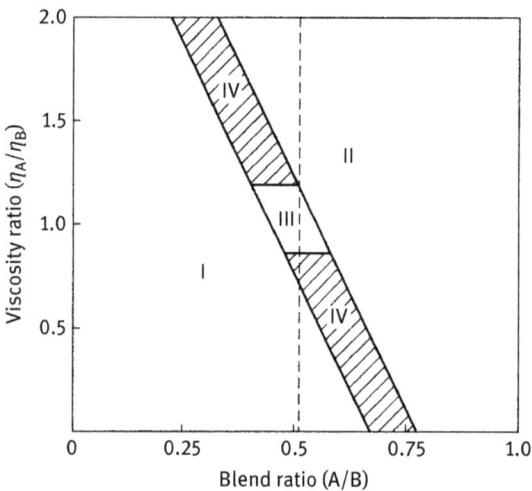

Fig. 3.16: Effect of blend ratio and viscosity ratio on continuity of blends. Phase I: component A are continuous phases; phase II: component B are continuous phases; phase III: component A and B are both continuous phases; phase IV: phase transition zone.

ratio can only play a role in a certain range of component ratios, and the relationship between them is shown in Fig. 3.16.

In addition, due to the different sensitivity of the viscosity of different components to temperature in the blending system, the phase inversion sometimes occurs during the blending process with the change of temperature. For example, in rubber–plastic blends, the viscosity of the temperature-sensitive plastic phase is much greater than that of rubber as shown in Fig. 3.17. Therefore, when $T > T^*$ (T^* is called constant viscosity temperature), the viscosity of plastic phase decreases sharply and the viscosity changes from higher than that of rubber phase to lower than that of rubber phase, which leads to phase inversion.

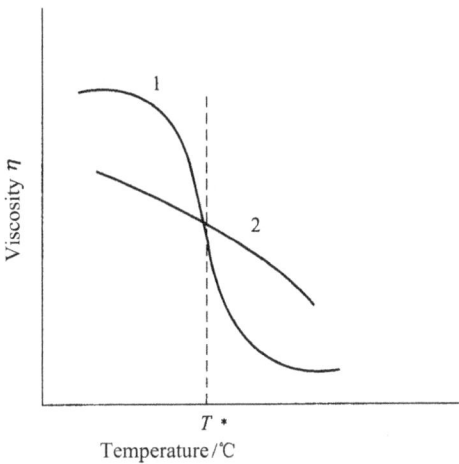

Fig. 3.17: Effect of temperature on the viscosity of polymers: (1) plastic phase and (2) rubber phase.

3. Cohesive energy density

Cohesion energy density (CED) is a measure of the intermolecular force of polymers. Polymers with high intermolecular force are difficult to extend into continuous phase in blends. The dispersed phase formed under higher shear stress has better stability. For example, in the blend system of chloroprene rubber and natural rubber, because the CED of chloroprene rubber is much larger than that of natural rubber, even if the content of chloroprene rubber is as high as 70%, it will still be the dispersed phase.

4. Solvent type

When polymer blends are prepared by solution method, the continuous phase composition will change with the variety of solvents. For example, when PS-PB-PS tri-block copolymer is used as casting film and benzene/heptane (90/10) is used as solvent, the prepared film is composed of PB block as continuous phase and polystyrene block as dispersed phase. Because benzene is the solvent of both PS and PB, heptane can only dissolve PB. The volatility of benzene is better than that of heptane, so when benzene is almost exhausted, PB is still in the solution state with heptane as solvent, and PB block will become a continuous phase. When using tetrahydrofuran/methyl ethyl ketone (90/10) as solvent, tetrahydrofuran is the common solvent of the two components, and methyl ethyl ketone is only the solvent of PS block. Therefore, tetrahydrofuran volatilizes first, methyl ethyl ketone volatilizes slowly and the continuous phase is PS.

5. Preparation process

For systems copolymerized by bulk grafting or solution grafting, the first synthesized polymer tends to form a continuous phase, while the later grafted polymer tends to form a dispersed phase. For example, in the presence of a small amount of PB, styrene was grafted onto PB. In the case of stirring-free static polymerization, PS homopolymer is most produced in the final product, and a certain amount of graft copolymers are also produced, that is, copolymers with PB as the main chain and polystyrene as the branch chain.

In this system, although the amount of PS far exceeds that of PB, the PB that first became a polymer is still a continuous phase, and PS forms a dispersed phase. Its morphology is shown in Fig. 3.18.

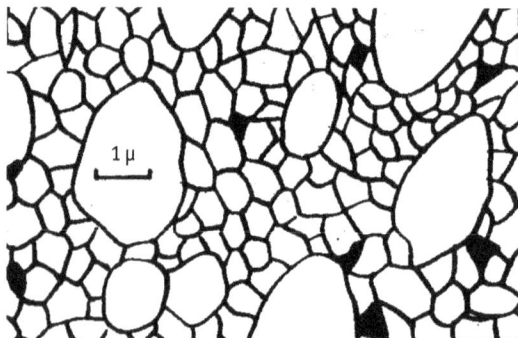

Fig. 3.18: Phase structure diagram of PS/PB graft copolymer system (black line is PB continuous phase).

However, if the reaction is carried out under stirring, the situation will be quite different. When the conversion rate of styrene monomer reached a certain degree, PS phase changed from dispersion to continuity, and further formed honeycomb phase structure.

3.3.2 Factors affecting the morphology and size of microregions

1. Effect of preparation methods

The phase structure of polymer blends is closely related to the preparation methods and technological conditions. For the same polymer blend, different preparation methods will lead to different phase structures. Due to different process conditions, the phase structure of the same preparation method will be different. For example, mechanical blending usually produces irregular dispersed phases; block copolymers often form regular microregions; blends prepared by graft copolymerization-blending method are easy to form honeycomb structure and so on. However, there are often some exceptions, such as in the mechanical blends of ethylene propylene rubber and PP, the particles of ethylene propylene rubber are regular spheres rather than irregular coarse particles of general mechanical blends. This may be due to the crystallization of PP, the low viscosity after melting and the dominant effect of interfacial tension.

When HIPS and ABS were prepared by bulk method and bulk-suspension method, the NBR particles contained 80–90% resin (PS). The formation of resin inclusions is mainly due to the influence of phase transition process. The epoxy resin toughened by rubber is prepared by the same method without phase transition, so the rubber particles do not contain epoxy resin. The ABS rubber obtained by emulsion polymerization contains about 50% volume resin, and the diameter of rubber particles is also smaller. The phase structures of ABS prepared by different methods are shown in Fig. 3.19.

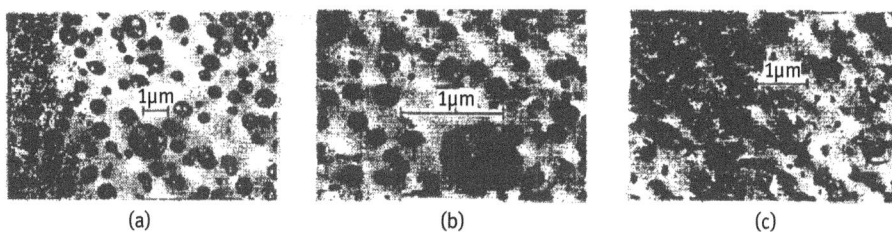

(a) (b) (c)

Fig. 3.19: Phase structure photographs of ABS prepared by three different methods: (a) bulk suspension method, (b) emulsion method and (c) mechanical blending method (dyed with osmium tetroxide, the black part is rubber phase).

2. Effect of compatibility between components

The better the process compatibility between the polymers in the blend system is, the easier their molecular chains diffuse to each other and achieve uniform mixing. The transition zone between the two phases becomes wider, the interface becomes blurred, and the size of the dispersed-phase microregion becomes smaller. In a fully compatible system, there is no phase interface, and the microregion disappears and becomes a homogeneous system. In the completely incompatible systems between the two polymers, the tendency of diffusion between the polymers is very small, the interface is obvious, the interfacial bonding force is very poor and even the macroscopic delamination phenomenon occurs. Most of the polymer alloy systems are between the two, which have a certain degree of compatibility and appropriate microarea size. Generally speaking, the smaller the size of the microregions is, the wider the transition zone between the two phases is and the better the properties of the blends are. But for different polymer alloy systems, the optimum size of microzone is different, and not all of them apply to the rule of "the smaller the microregion is, the better the performance is."

3. Effect of molecular weight
From the thermodynamic principle of compatibility discussed earlier, it can be seen that reducing the molecular weight of polymer is beneficial for improving the compatibility of polymer, and thus reducing the molecular weight of polymer is beneficial for reducing the size of microregion. For example, for the system of rubber-toughened plastics, the molecular weight of rubber has a greater impact on the size of rubber particles. The viscosity of rubber increases with the increase of its molecular weight. However, when the rubber viscosity increases to a certain extent, it is difficult to break the rubber phase into tiny particles.

4. Effect of viscosity difference on components of blends
When the two polymers are blended, the bigger the difference in viscosity is, the more difficult for the dispersion of the dispersed phase, and the larger the particle size of the dispersed phase is. Generally speaking, the closer the viscosity of the two polymers is, the better the dispersion effect is and the smaller the particle size of the dispersed phase is.

5. Effect of melt elasticity
Polymer melt has not only viscous but also elastic characteristics, so elasticity is also an important factor affecting the shape and size of the microregion of the blend. Generally speaking, the components with high elasticity need high shear force to disperse, and after dispersion, regular spherical particles tend to be formed due to the shrinkage of molecular chains.

6. Effect of technological conditions
For the mechanical blending system, because of the high viscosity of the polymer, the strengthening mixing conditions have a great influence on the dispersion uniformity and microregion size of the polymer. First, it is necessary to control the appropriate

mixing temperature so that both polymers are in a melting state. Generally speaking, higher temperature is beneficial for the uniform distribution of dispersed phase, but higher temperature will lead to polymer degradation; second, appropriate mixing time should be adopted. The longer mixing time is beneficial for dispersion of dispersed phase, and the size of microzone decreases in a certain range, but it has no effect after a certain time. Therefore, blindly prolonging the mixing time is not beneficial for the dispersion effect.

When mechanical blending method is used, two-stage blending (also known as master batch method) is beneficial for reducing the size of dispersed phase, and the product performance is better than direct blending method, that is, the mixture with close plastics/rubber blending ratio is first prepared for mixing, and then diluted to a predetermined ratio. For example, the tensile strength and elongation at break and stress at constant elongation of the two-stage blended NR/PE (90/10) are increased by about 50% compared with that of the direct blended NR/PE (90/10).

In the preparation of HIPS and ABS by graft copolymerization-blending method, as mentioned above, the stirring rate of the polymerization process has an important influence on their morphology and structure. When the stirring rate is too low or the shear force is less than the critical value, the polymer formed first is the dispersed phase, and the reverse phase does not occur even when the content of the polymer formed later is much larger than that of the polymer formed first. On the contrary, if the shear force is too large, the dispersed-phase particles will be too small, the encapsulated PS is less and the toughening effect is not good. In addition, the degree of grafting also affects the size of rubber particles. With the increase of the degree of grafting, the size of rubber particles gradually decreases, which has been confirmed by a large number of experimental results. This is because the graft copolymer plays the role of compatibilizer. The high degree of grafting improves the compatibility and is conducive to the formation of smaller colloidal particles.

7. Effect of mixing equipment

When polymer blends are prepared by melt blending, the dispersion degree of the blends is closely related to the mixing equipment. In order to achieve the allowable dispersion of the two polymers in full mixing and compatibility, the conditions for intensified mixing must be provided by effective mixing equipment. Different mixing equipments have different dispersion effects on the blends, and the properties of the blends are also different. At present, there are mainly two-roll plasticizer, internal mixer, extruder and so on, which are used for melting and blending of polymers in industrial production. The mixing effect of two-roll mixer is the worst. The mixing and dispersing effect of high-efficiency mixer is higher than that of common mixer. The codirectional parallel twin-screw extruder is better than that of common single-screw extruder.

3.3.3 Influencing factors on phase structure of blends containing crystalline polymers

3.3.3.1 Crystalline polymer/amorphous polymer blends

The phase structure of crystalline polymer/amorphous polymer blends is not only related to the component content, but also affected by the crystallinity. When the crystallinity is small, the spherulites or grains are dispersed in amorphous matrix; when the crystallinity is high (and the content of crystalline components is high), the spherulites will fill the whole body.

For example, in PCL/PVC system, PCL is a crystalline polymer. When its volume fraction exceeds 60%, PCL spherulites will fill the whole system. With the increase of PVC content, the crystallinity and quantity of PCL decrease, but the grain size remains unchanged. This indicates that PVC is uniformly dispersed among the lamellae in PCL spherulites.

The blending of i-PS and amorphous PS is different. The results of small-angle X-ray scattering (SAXS) and differential scanning calorimetry (DSC) show that when the content of amorphous PS is less than 30%, i-PS crystallizes in the blend and the distance between crystalline regions does not change with the content of amorphous PS. Therefore, it can be concluded that amorphous PS does not enter into the lamellae of crystalline PS, but the amorphous PS aggregates into larger microregions and distributes between spherulites.

3.3.3.2 Crystalline polymer/crystalline polymer blends

The compatibility, cell structure and crystalline morphology of the two polymers have different effects on the blend system composed of two kinds of crystalline polymers. For example, the distance between two carbonyl groups in two copolymer monomers of adipic acid and terephthalic acid is equal, in the resulted polymer, the two chains in the copolymer can be substituted, formed cocrystals.

For example, LDPE and PCL can be cocrystallized when they are blended. Because PE and PCL belong to orthorhombic crystalline system, the lattice structure is similar and cocrystals can be formed.

Although the compatibility of PET and PBT crystalline polymers is very good, the SAXS and DSC studies have proved that there are still two crystals in the blend system. There are two T_m in the blend system, but only one T_g, which changes between the two T_g of PET and PBT with the blend ratio. Therefore, it can be determined that this is a three-phase blend system consisting of two spherulites (or grains) distributed in the same amorphous matrix. The amorphous parts of PET and PBT become one phase because of compatibility, while the crystalline parts form phase regions respectively.

3.3.3.3 Graft copolymers and block copolymers containing crystalline chains

If the main chain of the graft copolymer is crystalline, the existence of branched chain has a certain effect on the crystalline morphology and size of the copolymer. For example, in the grafting system with PE as the main chain and PS as the branch chain, the crystallinity of PE decreases with the increase of the degree of grafting.

For example, PP can form perfect spherulites from 180 to 80 °C after 4 h of slow cooling. The grain size of PP can reach 300–500 μm. However, when 18.1% PVAC was grafted onto PP chain, the spherulites of PP were obviously destroyed. When the degree of grafting reached 41.5%, the large spherulites could not be formed, and there were cracks between the spherulites. Similarly, in the system of i-PS grafted amorphous PS, the existence of branched chains greatly reduces the crystallization perfection of the main chain.

If the branched chain in the graft copolymer is crystalline, the main chain also affects the crystallization behavior of the branched chain. For example, the crystalline poly-ω-hydroxy undecanoate was grafted onto the main chain of polymethacrylic acid. When the degree of grafting was low, it was difficult to get close to each other because of the small number of branched chains, and the crystals could not be arranged regularly. When the degree of grafting increased to a certain extent, small grains and imperfect spherulites gradually formed. This indicates that the main chain has a certain binding and restrictive effect on the crystallization of the branched chain.

Conversely, there are also examples of the main chain promoting branched chain crystallization. For example, aliphatic polyesters that are not easy to crystallize are grafted onto PMMA molecular chains, which make branched chains easy to crystallize. In this case, the branched chain is usually a flexible chain, which is grafted onto the rigid main chain and is "supported" by the main chain, making the branched chain easy to form regular chain bundles, thus speeding up the crystallization process.

In diblock copolymers, the crystallinity of easily crystallized segments is not generally affected by the presence of another segment. Because at this time, the two polymer chains are connected by only one chemical bond, which has little influence on each other. However, in the case of multiblock copolymers, the crystallinity of easily crystalline segments is often bound, the crystallinity is reduced and the size of crystal region is also reduced. For example, the crystallinity of PP/PE diblock copolymers (PP degree of polymerization: 23,000, PEO degree of polymerization: 12,500) is evident by X-ray diffraction, while the crystallinity of corresponding quadruplex block copolymers is greatly reduced. The crystallinity of i-PS/PEO quadruplex block copolymers with similar degree of polymerization could not be observed.

3.4 Interfacial layer of polymer blends

3.4.1 Concept of interface layer

In addition to the independent phase regions of the two polymers, there is actually a third region in the blends, that is, the interfacial layer between the two phases. The interfacial layer is also known as the transition zone, in which the bonding of two phases and the mutual diffusion of two polymer chains occur. The structure and composition of the interfacial layer directly affect the bonding strength between the two polymers, and have a decisive influence on the properties, especially the mechanical properties of the polymer blends.

If the interface between two phases in polymer blends is clearly separated and the molecules or segments in the two phases do not penetrate each other, the interaction force between the two phases must be very weak. Such systems usually do not have good strength. Therefore, it is very important to improve the interface condition and form a certain thickness of the interface layer.

Two-phase molecular chains coexist in the interfacial layer of process compatible blends, where the two polymer molecular chains diffuse and permeate each other to form a transition zone. It plays an important role in stabilizing the morphology of the two-phase structure, improving the interfacial cohesion and improving the mechanical properties of the blends.

The degree of interfacial diffusion of two macromolecular chains in mechanical blends mainly depends on the solubility parameters, interfacial tension and molecular weight of the two polymers.

Generally speaking, if the solubility parameters of the two polymers are similar, the two molecules will diffuse easily and the interfacial layer is thicker; the completely incompatible blend system will not form the interfacial layer. The surface tension of the two polymers is close and the interfacial tension is small, which is beneficial to the wetting and diffusion of the two-phase polymer molecules, and the interfacial layer is easy to form.

The schematic diagram of the interfacial layer structure of mechanical blends can be shown in Fig. 3.20. The two molecular chains fully contact and penetrate into each other in the interfacial layer, and interact with each other by the secondary force to form a strong interfacial bonding force.

The structure and composition of the interfacial layer are different from those of the independent polymer phase region. The results show that: (1) the distribution of the two molecular chains is not uniform, and a concentration gradient is formed from the phase region to the interface; (2) the molecular chains are loosely arranged than the respective phase regions, so the density is slightly lower than the average density of the two-phase polymers; (3) more impurities such as surfactants and other additives tend to accumulate in the interfacial layer, and the polymer molecules with lower molecular weight tend to migrate to the interfacial layer. The more

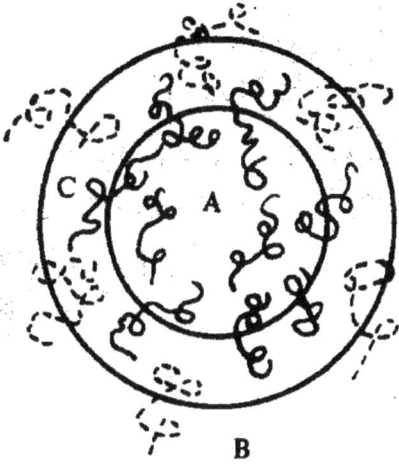

Fig. 3.20: Interdiffusion of two polymers in blends: (a) dispersed phase; (b) continuous phase and (c) interface layer.

low-molecular weight substances such as surfactants exist, the more stable the interfacial layer is, but it is not conducive to the interfacial bonding strength.

There is also an interfacial layer between the two-phase separation structures produced by block copolymers. Unlike general mechanical blends, the two blocks are chemically bonded in the interfacial layer of block copolymers. In other words, the junctions of block copolymers exist in the interfacial layer. When the content of two blocks in the block copolymer is determined, the thickness of the interfacial layer mainly depends on the compatibility of the two blocks, which is inversely proportional to the difference of the solubility parameters of the two blocks $(\delta_A - \delta_B)$ and the degree of polymerization of the two blocks.

3.4.2 Formation of interface layer

The formation of interfacial layer of polymer blends can be divided into two steps. The first step is the contact between the two phases, and the second step is the diffusion between the macromolecular chains of the two polymers.

Increasing the contact area between the two phases is undoubtedly beneficial to the diffusion of macromolecular chains and the adhesion between the two phases. Therefore, it is very important to ensure a high degree of dispersion between the two phases and to reduce the domain size appropriately in the blending process. In order to increase the contact area between the two phases and increase the dispersion degree, efficient blending machines such as twin-screw extruder and static mixer can be used; another way is to use IPN technology; the third way is the most feasible method at present, which is to use compatibilizer.

When two polymers are in contact with each other, the diffusions between chains occur. If the two polymer macromolecules have similar activity, the chain segments of the two macromolecules diffuse at the same speed, and the composition of the interfacial layer is relatively uniform. If the activity of the two polymer macromolecules differs greatly, unidirectional diffusion occurs, and obvious concentration gradient will appear. The driving force of this diffusion is the mixing entropy, that is, the thermal motion of the chain segment. If the heat is absorbed in the mixing process, the increase of entropy is eventually offset by the heat of mixing. The degree of final diffusion depends mainly on the thermodynamic solubility of the two polymers. Due to the poor compatibility of most polymers and the significant difference in the activity of different macromolecules, the results of diffusion often lead to obvious concentration gradients on both sides of the phase interface (Fig. 3.21). The interfacial layer (also called interfacial zone) between two phases is formed by the regions with obvious concentration gradients on both sides of the phase interface and the phase interface.

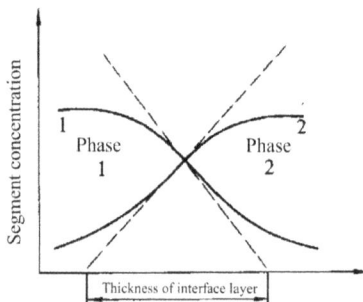

Fig. 3.21: Concentration gradients of two polymer segments in the interfacial layer. (1) Segment concentration of polymer 1 And (2) Segment concentration of polymer 2.

3.4.3 Interface layer thickness

The thickness of the interfacial layer is mainly determined by the mutual solubility of the two polymers. In addition, it is also related to the size, composition and phase separation conditions of macromolecules. There is only a slight diffusion between the segments of the polymer, which is basically insoluble. Therefore, there is a very clear and definite phase interface between the two phases, and the interface layer is thin. With the increase of mutual solubility between the two polymers, the diffusion degree increases, the interface becomes blurred, the thickness of the interface layer Δl increases and the adhesion between the two phases increases. The two completely miscible polymers eventually form homogeneous phase and the phase interface disappears.

Generally, the thickness of interface layer Δl is several tens of nanometers. For example, the thickness of PS/PMMA blend Δl was about 5 nm observed by TEM. When the phase domain is very small (i.e., highly dispersed), the volume of the interface layer can account for a considerable proportion. For example, when the diameter of dispersed-phase particles is about 100 nm, the interfacial layer can reach about 20% of the total volume. Therefore, the interface layer can be regarded as the third phase with independent characteristics.

From the thermodynamic point of view, the thickness of the interface layer depends on two factors: entropy and enthalpy. The enthalpy factor refers to the interaction energy between polymers 1 and 2. It is proportional to the square of the difference of solubility parameters δ_1 and δ_2, and the square of the difference is proportional to the interaction parameter χ_{12}. Table 3.1 lists the χ_{12} of some blend pairs and the thickness of their interfacial layers.

Tab. 3.1: χ_{12} and interfacial layer thickness of polymer blends.

Polymer pairs	χ_{12}	Interfacial layer thickness/nm
PS/PB	0.03	3
PS/PMMA	0.01	5

Of course, there is no simple relationship between χ_{12} and the difference of solubility parameters when the polarity between the two polymer components is very large. When there is a special interaction between the two polymers (such as strong polarity and hydrogen bond), the χ_{12} is even negative and the interfacial layer thickness can still reach a large value.

3.4.4 Properties of interface layer

1. Adhesion between two phases

As far as the adhesion between two phases is concerned, there are two basic types of interface layer. The first is the case where two phases are bonded by chemical bonds, such as grafting and block copolymers. The second type is the combination of two phases only by secondary force, such as general mechanical blends.

The wetting-contact theory and the diffusion theory are generally accepted for the combination of secondary forces between two polymers. According to the wetting-contact theory, the bonding strength is mainly determined by the interfacial tension. The smaller the interfacial tension is, the greater the bonding strength is. According to diffusion theory, the bonding strength is mainly determined by the mutual solubility between the two polymers. The greater the mutual solubility, the higher the bonding strength. Of course, in order for the two polymer macromolecular chains to diffuse with each other, the temperature must be above T_g.

In fact, these two theories are inherently unified, but they deal with problems differently. The interfacial tension is proportional to the square of the difference of solubility parameters. So, when the mutual solubility is good, the interfacial tension must be small.

2. Morphology of interfacial macromolecular chains

As shown in Fig. 3.22, the concentration at the tail end of the macromolecule in the interface layer is higher than that in the bulk, that is, the chain ends concentrate towards the interface. The end of the chain tends to be perpendicular to the orientation of the interface, while the whole macromolecular chain is roughly parallel to the orientation of the interface.

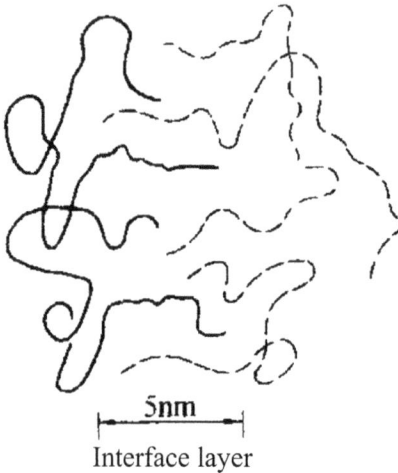

Interface layer

Fig. 3.22: Orientation of macromolecular chains and chain ends in the interface layer of polymer blends.

3. Molecular weight grading effect of interface layer

The results show that if the molecular weight distribution of the polymer is wide, the low-molecular weight part will concentrate on the interface region, resulting in molecular weight grading effect. This is due to the high mutual solubility of polymers at low molecular weight.

4. Induced crystallization effect of interface layer

In some blends based on crystalline polymers, dispersed phases can induce crystallization of continuous phases through interfacial effect, resulting in a large number of fine crystals and avoiding the formation of large spherulites, which plays an important role in improving the impact resistance of materials.

5. Density and diffusion coefficient

The interfacial polymer density may increase or decrease, depending on the interaction force between the two phases. When there are chemical bonds and strong mutual attraction, the density of interfacial layer is larger than that of bulk, but if there is no such effect, the density of interfacial layer is smaller than that of bulk. There is only a secondary force between the two phases, and the density of the interface layer is generally smaller than that of the bulk. At this time, the free volume fraction of the interface layer increases. Although the increase of free volume fraction is not very large, it increases the diffusion coefficient by three orders of magnitude.

6. Enrichment of additives
If there are other additives in the blend system, the distribution of these additives in the bulk phase and interfacial layer of the two polymers is generally different. Surface-active additives, compatibilizers and surface-active impurities will converge to the interface.

As mentioned earlier, the mechanical relaxation performance of the interface layer is different from that of the bulk layer. The interfacial layer and its volume fraction have significant effect on the properties of the blends. This is also the reason why the phase domain size has a significant effect on the properties of the blends.

The results show that the glass transition temperature of the interfacial layer is between that of the pure components of the two polymers. As the domain size decreases, the volume fraction of the interface increases, and the glass transition as the third phase becomes more obvious.

In a word, the interfacial layer can be regarded as the third phase between the two polymer components in terms of composition, structure and properties.

3.4.5 Factors affecting interface layer structure of polymer blends

3.4.5.1 Effect of intersolubility on interface layer structure of polymer blends
In many cases, thermodynamic mutual solubility is the main driving force for the uniform mixing of polymers. The better the mutual solubility of the two polymers, the easier it is to diffuse and achieve uniform mixing, and the wider the transition zone. The more blurred the phase interface, the smaller the phase domain and the greater the bonding force between the two phases. There are two extreme cases. One is that the two polymers are completely immiscible. The tendency of diffusion between the two polymer chains is very small, and the interface between the two polymers is obvious. As a result, the mixing is poor, the bonding force between the two polymers is weak and the properties of the blends are poor. In order to improve the properties of blends, appropriate technological measures should be taken, such as copolymerization-blending method or adding appropriate compatibilizers. The second extreme case is that the two polymers are completely soluble or highly soluble, when the two polymers can completely dissolve into homogeneous systems or differential dispersion systems with minimal phase domains. These two extreme conditions are

not conducive to the purpose of blending modification (especially mechanical properties modification).

Generally speaking, what is needed is that the two polymers have moderate mutual solubility, so that the blends with suitable domain size and strong binding force between phases can be prepared.

In order to illustrate the effect of miscibility on the phase structure of the blends, the following is discussed with the example of PVC/NBR blends.

The solubility parameter δ of AN–butadiene copolymer (NBR) is related to the content of AN, as shown in Tab. 3.2. The δ of PVC is 9.7.

Tab. 3.2: Relation between NBR's δ and AN Content.

AN mass fraction/%	51	41	33	29	21	0
δ	10.2	9.6	9.4	9.1	8.6	8.2

According to the measurement of dynamic viscoelastic properties and electron microscopic analysis, PVC and PB are not soluble, with large phase domains, obvious phase interfaces, weak bonding force between the two phases and low impact strength. For PVC/NBR blends, when the content of AN is about 20%, they are partially miscible systems with moderate phase domains, high bonding strength and high impact strength. When the content of AN in NBR is more than 40%, the δ of PVC and NBR is close to each other, and they are basically completely miscible. The blends are nearly homogeneous, with very small phase domains and low impact strength. Figure 3.23 fully illustrates the relationship between domain size and mutual solubility.

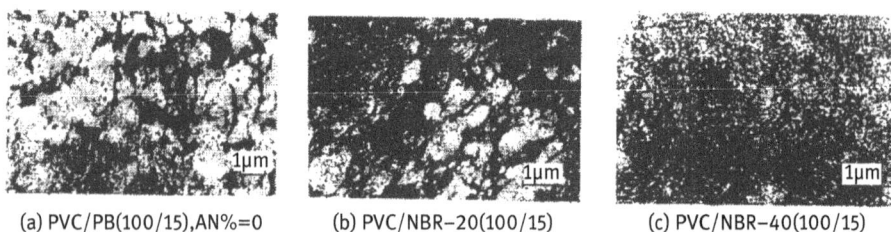

(a) PVC/PB(100/15),AN%=0   (b) PVC/NBR–20(100/15)   (c) PVC/NBR–40(100/15)

Fig. 3.23: Ultraslice electron microscopic photographs of PVC/NBR blends.

3.4.5.2 Effect of molecular weight distribution on interface layer structure of polymer blends

The molecular weight distribution of polymer also affects the interfacial layer and the binding force between the two phases of polymer blends. It has been previously mentioned that the solubility between polymers is related to molecular weight, which increases when molecular weight decreases. When the molecular weight distribution of polymer is wide, the low-molecular fraction tends to diffuse to the interface layer, which plays a role of compatibilizer to a certain extent, so it is conducive to the thickening of the interface layer, thereby improving the adhesion between the two phases. However, if there are too many low molecular weight polymers in the system and too many low molecular weight polymers are accumulated in the interfacial layer, the strength of the interfacial layer itself will be reduced and the overall strength of the material will also be reduced. Therefore, there is a correct understanding and effective control of the influence of molecular weight distribution of polymers on the interfacial layer of polymer blends.

3.4.5.3 Effect of processing on interface layer structure of polymer blends
In the flow field of blending process, the mutual solubility of polymers may change. There are two different situations: (1) stress causes irreversible changes such as precipitation, crystallization and denaturation. At this time, the mutual solubility between components decreases and the interfacial layer becomes thinner. (2) Stress causes reversible changes and increases mutual solubility. The phenomenon that stress increases mutual solubility is often called stress homogenization. Because of the deformability of dispersed-phase particles, the viscoelastic effect appears in the flow field and the stored elastic energy is the main reason for the homogenization of dispersed-phase particles. At this time, the proportion of interfacial layer in the whole sample will increase.

Exercises

(1) What are the types of phase domains with single continuous structure?
(2) What are the factors affecting the phase structure of polymer blends?
(3) What is the definition of interface layer?

Chapter 4
Toughening mechanism of multicomponent polymers

4.1 Introduction

In the development of multicomponent polymers, toughening with elastomer as dispersed relative brittle plastics is the most important part, and elastomer-toughened plastics is the most important variety in polymer blends. Among them, the most important ones are high-impact polystyrene (HIPS), acrylonitrile–butadiene–styrene (ABS) plastics, methacrylate–butadiene–styrene (MBS) plastics, polyvinyl chloride (PVC) toughened by ABS, MBS, acrylic resin (ACR), toughened polycarbonate, epoxy resin toughened by rubber and others.

Toughening refers to the improvement of impact resistance of previously brittle materials by some means. Therefore, elastomer-toughened plastics are characterized by high impact strength, which is usually 5–10 times or even tens of times higher than that of matrix resins. In addition, the impact strength of elastomer-toughened plastics is closely related to the preparation methods, because different preparation methods often make the interfacial bonding strength and morphological structure change greatly. For example, the impact strength of polystyrene (PS) toughened with polybutadiene varies greatly with different preparation methods, as shown in Fig. 4.1.

Toughening modification of polymers has undergone a process to change from elastomer to inelastic. Traditional toughening modification uses rubber elastomer material as toughening agent and disperses it into plastic matrix in a proper way to achieve the purpose of toughening, such as HIPS, ethylene–propylene copolymer, elastomer-toughened polypropylene and so on. However, there are some problems that are difficult to overcome, such as stiffness, strength and thermal deformation temperature are greatly reduced.

In 1984, Kurauchi and Ohta of Japan first put forward a new idea of toughening with inelastic materials (rigid particles). This method can improve the strength, rigidity and heat resistance of the material while improving the toughness of the matrix, without the shortcomings of the decline in processability, to achieve the purpose of both toughening and strengthening, and overcome the problems arising from the toughening of the elastomer. Therefore, it has received extensive attention. Based on the initial toughening of organic rigid particles, the toughening of inorganic rigid particles and nanoparticles were developed.

Toughening theory is the most active part in the study of multicomponent polymers. Since Merz put forward the first elastomer toughening mechanism in the mid-1950s, many scholars have been exploring new toughening systems and mechanisms in the process of studying polymer blends and toughening systems. From the initial simple qualitative explanation, they have gradually developed toward classification,

https://doi.org/10.1515/9783110596335-004

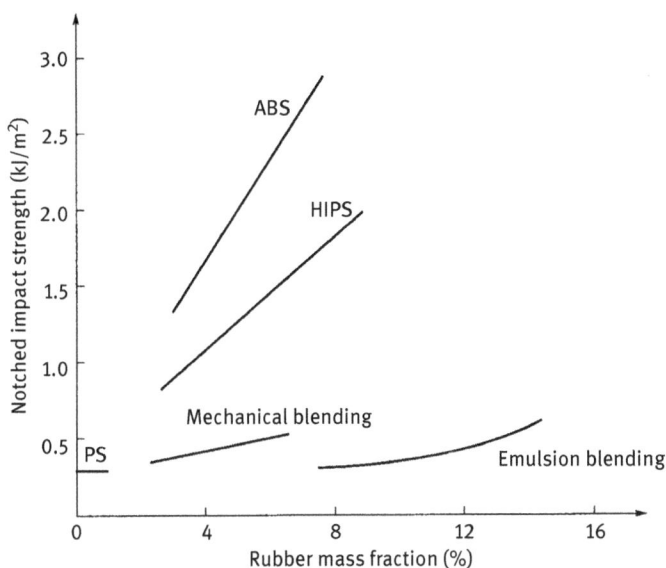

Fig. 4.1: Impact strength of toughened polystyrene prepared by different methods.

modelling and quantification. Toughening systems include simple rubber elastomers, thermoplastic elastomers, organic rigid particles and nanomaterials. The development of toughening theory can be roughly divided into the following stages: The first stage is the early toughening theory from 1950s to 1970s, which produced many theories, mainly explaining the reasons of elastomer toughening qualitatively. The second stage is the 1980s to 1990s. In this period, the research on toughening of polymer blends began to move from qualitative description to quantitative estimation. At the same time, the discovery and application of organic rigid materials and inorganic nanomaterials developed the theory of rigid particle toughening.

At present, the technology and theory of elastomer toughening and rigid particle toughening are still being developed.

4.2 Toughening mechanism of elastomer-toughened plastics

As for the mechanism of elastomer toughening plastics, many different theories have been put forward since 1950s, such as Mertz's direct energy absorption theory, Nielsen's secondary transition temperature theory, Newman's yield expansion theory, Schmitt's crack core theory, Bucknall's and Smith's multiple silver texture theory and Bragaw's craze branching theory. But these theories often only pay attention to one side of the problem. At present, it is generally accepted that the theory of craze-shear band-hole has been developed in recent years.

In order to facilitate the understanding of the theory of craze-shear zone-void, the development of toughening theory is briefly introduced.

4.2.1 Early toughening theory

1. Direct energy absorption by elastomers theory

In 1956, Mertz et al. observed the phenomena of volume expansion and stress whitening in the study of HIPS tension, and considered that microcracks were the main causes of the above phenomena. Therefore, an explanation of the mechanism of elastomer toughening plastics is put forward: when the specimen is impacted, microcracks will occur, and rubber particles will cross both sides of the cracks. To develop cracks, rubber must be stretched. Rubber absorbs a lot of energy during deformation, which improves the impact strength of plastics.

However, many experimental results and research and analysis have proved that the energy absorbed by rubber deformation during tension process is very small, accounting for only about one-tenth of the energy absorbed when the material is destroyed. Therefore, this mechanism is far from the main reason for improving the toughness of plastics. Moreover, the microcracks produced when the specimen is impacted will greatly reduce the tensile strength of the material. This mechanism is therefore questioned.

2. Secondary transition temperature theory

This theory was put forward by Niesen. He believes that the toughness of polymers is often related to their secondary transition temperatures. In elastomer-toughened plastics, the glass transition of rubber is a very strong secondary transformation, so the increase of material toughness is related to the T_g of rubber. In order to achieve effective toughening, rubber as a toughening agent must have lower T_g.

3. Yield theory

Newman and Strella noticed that the high-impact strength of elastomer-toughened plastics was mainly due to the large yield deformation of matrix resin. The reason for the large yield deformation of matrix resin was that the thermal expansion coefficient and Poisson's ratio of rubber were larger than that of plastics. The thermal shrinkage and transverse shrinkage in the cooling stage of the forming process produce static tension stress on the surrounding matrix, which increases the free volume of the matrix resin, reduces its glass transition temperature, and is prone to plastic deformation and improves its toughness.

However, further research shows that the glass transition temperature of the material caused by the above stress changes decreases only about 10 °C, far less than the degree of transforming the material into tough material at room temperature.

Moreover, some brittle plastics with high glass transition temperature can hardly even yield at room temperature. Therefore, yield theory cannot explain the toughening of many brittle materials by adding a small amount of elastomer.

4. Crack core theory

In 1960, Schmitt et al. put forward the crack core theory: rubber particles filled in toughened plastics as stress concentration point produced many small cracks rather than a small number of large cracks, which needed more energy than a small number of large cracks. At the same time, the stress field of many small cracks interfered with each other, weakening the front stress of crack development, thus slowing down the development of cracks. It will terminate the crack. The phenomenon of stress whitening is due to the formation of many small cracks.

Later, it was found that the theory had three defects: first, it failed to distinguish cracks from crazes. The small crack mentioned here is actually craze, but it does not clarify the structure and characteristics of the small crack; second, the theory only emphasizes the role of rubber particles in inducing small cracks, but fails to consider the role of rubber particles in terminating small cracks; third, it ignores the influence of matrix resin characteristics.

Nevertheless, the idea of stress concentration and induced small cracks in this theory has great inspiration and impetus to the development of toughening theory.

5. Multiple silver texture theory

In 1965, Bucknall and Smith, based on Schmitt's idea of rubber particles as stress concentrates, put forward the argument that because of the large number of rubber particles in toughened plastics, many stress concentrates cause many crazes, which can dissipate a large amount of energy. Furthermore, the idea that rubber particles or craze terminators and too small particles cannot terminate the craze was put forward.

This theory holds that in rubber-toughened plastics, the stress field is no longer uniform due to the presence of rubber particles, and rubber particles play a role of stress concentration. Under the action of tensile stress, many crazes will occur around rubber particles, especially near the equatorial plane of particles. The generation and development of a large number of crazes will consume a lot of energy, such as the plastic work done by the formation of crazes, the viscoelastic work of the growth of crazes in the direction of stress, the new surface energy produced by the formation of crazes and the fracture energy consumed by splitting chemical bonds in the middle of craze formation. In addition, due to the interaction of stress fields between many crazes, if the stress concentration at the front peak of the craze is lower than the critical value or if the craze meets another rubber particle, the craze will terminate. Rubber particles can not only cause craze but also inhibit it. The impact energy is absorbed by the large number of crazes produced when the material is impacted and rubber particles can stop the craze in time and not develop into cracks too early so that the material breaks, which protect the material from damage.

The multiple silver texture theory is based on many experimental facts. The viewpoint that rubber particles cause many crazes has been confirmed by the experimental results of electron microscopy. In HIPS, ABS and other systems, many crazes around rubber particles were observed, and the crazes often started from one particle and ended in another.

The multiple silver texture theory not only points out the dual functions of initiating and terminating the craze of rubber particles, but also points out the relationship between the matrix of plastic phase and the toughening effect. First, the interface between rubber particles and plastic phase must have a good bonding force, in order to give full play to the role of rubber particles triggering craze. Moreover, the properties of the matrix of plastic phase directly affect the effect of crazing caused by particles. When the toughness of the matrix is too good, it is not conducive to the initiation of crazing.

In addition, the multiple silver texture theory has also explained some phenomena in the process of craze deformation: stress whitening is due to the fact that the density and refractive index of the craze zone formed under tension stress are lower than those of the polymer body, and the total reflection occurs at the interface between the craze and the body; the decrease in the density of the craze is due to the existence of many holes in the craze; the transverse nonshrinkage is due to crazing in the process of cavitation and volume increase.

Therefore, the multiple silver texture theory is of great significance for the establishment of modern toughening theory.

6. Craze branching theory

In 1971, Bragaw pointed out that the occurrence of many crazes was the result of craze branching. In elastomer-toughened plastics, the craze first spreads rapidly in the matrix resin, and before reaching the limit speed (i.e., half of the sound velocity in the resin, about 620 m/s), it enters the rubber dispersed phase. Because of the lower limit speed in the rubber (e.g., at 23 °C, half of the sound velocity in *cis*-butadiene rubber is 29 m/s), the energy is released rapidly from the craze tip to the rubber dispersed phase. Strong branching occurs immediately in the resin matrix.

The theoretical calculation shows that the acceleration distance between 2 and 5 microns in the resin continuous phase is only needed to achieve such a speed. Therefore, the branching probability of the craze on the rubber particles is very high.

As a result of branching, on the one hand, the number of crazes is greatly increased, thereby increasing the energy absorption; on the other hand, the front stress of each craze is reduced, which leads to the termination of the craze.

Bragaw connected the impact strength with the sound velocity of the matrix and rubber, and concluded that the impact strength should be exponentially related to the number of rubber particles. In addition, in order to effectively branch the craze, the diameter of the rubber particles should not be less than the width of the craze, otherwise they will be embedded in the craze without any effect. For example, theoretical calculation of the width of craze in HIPS is about 0.9–2.8 μm and that in ABS is about 0.5 μm. Experiments show that the optimum particle size of rubber in HIPS is 1–10 μm, while that of rubber in ABS is 0.1–1.0 μm.

4.2.2 Craze-shear band theory

4.2.2.1 Basic ideas of craze-shear band theory

In practice, it is found that the mechanism of multiple crazes in some blends cannot be explained. For example, the HIPS/poly(phenylene oxide) (PPO) blend specimens after tension were observed by scanning electron microscopy. In addition to multiple crazes, shear band at 45° of stress were also observed. It is also found that the shear zone passes through rubber particles, indicating that rubber particles can also initiate the shear zone. At the same time, it is found that few crazes do not terminate directly in the rubber particles but in the vicinity of adjacent rubber particles, and the length of the crazes is relatively short. Therefore, it is considered that the crazes may be terminated by shear zones. Because the molecular chains in the shear zone are oriented along the direction of tension, that is, perpendicular to the craze plane, termination of the craze is more effective than rubber particles. Many experiments have confirmed that the shear zone plays a controlling role in the size of the craze, and the stress concentration effect at the tip of the craze can also lead to new shear zones. The initiation and growth of the shear zone can also consume a lot of energy. Therefore, there is interaction and synergism between craze and shear zone in polymer blends with both craze and shear zone. Both contribute to the improvement of toughness of blends. For example, there are both whitening and thin neck formation in the tensile process of ABS, which indicates that there are two deformation mechanisms of craze and shear zone, and their toughening mechanism must be a synergistic effect. For systems with better matrix toughness, such as toughened PVC, the toughening effect is mainly attributed to shear yield deformation, and the contribution of crazing is very small. These experimental facts prompted the establishment of the theory of craze-shear zone.

Bucknall et al. proposed in 1972 that rubber particles play two important roles in toughening plastics: (1) as a stress concentration center, many crazes and shear zones are induced and (2) the development of crazes is controlled and the crazes are terminated in time without developing into destructive cracks. The stress field at the end of the craze can continue to induce the shear zone to terminate the craze, and the further development of the craze can be prevented by encountering the existing shear zone. The generation and development of many crazes and/or shear zones consume a large amount of energy, which can significantly improve the toughness of toughened plastics.

According to the theory, the toughening mechanism of rubber particles mainly includes three aspects: (a) rubber particles initiate and branch a large number of crazes and bridge both sides of the crack, (b) cause matrix shear deformation to form a shear band and (c) create holes in the rubber particles and on the surface, accompanied by the stretching and shearing of polymer chains between holes and lead to plastic deformation of the matrix. These three mechanisms under impact energy are shown in Fig. 4.2.

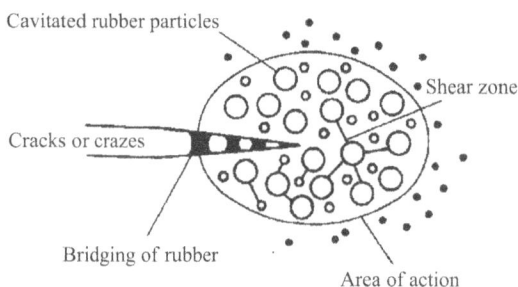

Fig. 4.2: Schematic diagram of toughening mechanism of elastomer-toughened plastics.

4.2.2.2 Initiation and branching of craze
The first important role of rubber particles is to act as a stress concentration center (if the rubber phase has good bonding with the matrix), which induces many crazes, as shown in Fig. 4.3.

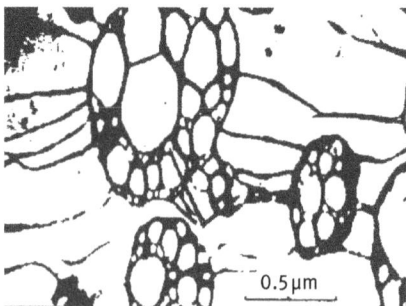

Fig. 4.3: Electron microscopic photographs of crayon induced by rubber particles in HIPS under impact.

The equatorial surface of rubber particles can cause a lot of crazes. When the concentration of rubber particles is high, many crazes can also be induced on the nonequatorial plane due to the interference and overlap of stress fields. It consumes a lot of impact energy to initiate many crazes, so the impact strength of materials can be improved.

Rubber particles can not only cause craze, but also branch craze. According to Yoff's and Griffith's crack dynamics theory, the limit velocity of crack or craze propagation in medium is about half of the sound velocity in medium. After reaching the ultimate speed, continued development leads to rupture or rapid branching and steering. According to the elastic modulus of plastics and rubber, the ultimate growth rate of craze in plastics is about 620 and 29 m/s in rubber.

Two-phase elastomer-toughened plastics, such as ABS, develop rapidly in the matrix craze. Rubber particles meet before reaching the ultimate speed. The diffusion rate drops sharply and branching occurs immediately, resulting in more new small crazes and consuming more energy, thus further improving the impact strength. Each newly formed small craze expands on the plastic matrix. According to Bragaw's calculation, these new crazes need only an acceleration distance of about 5 μm in the plastic matrix to reach the ultimate growth rate again (about 620 m/s). Then rubber particles are encountered and branched, as shown in Fig. 4.4. This estimation provides an important basis for determining the optimum distance between rubber particles and the optimum amount of rubber.

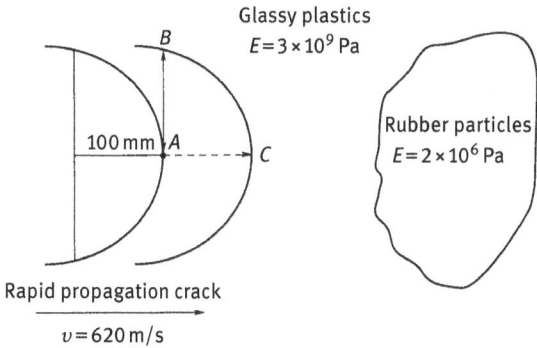

Fig. 4.4: Movements of crazes in plastics.
When the craze moves from point A to point C, a matter element near the tip of the craze moves from point A to point B.

The result of this repeated branching is to increase energy absorption and reduce the front stress of each craze so that the craze can be easily terminated.

Since the speed of craze approaching rubber particles is approximately 620 m/s, a crack or craze with a radius of 100 nm is equivalent to the effect of 109 Hz operating frequency. According to the time–temperature equivalence principle, it is estimated that the T_g of rubber phase increases by 6–7 °C for every 10-fold increase in frequency. At this time, the T_g of rubber phase increases by about 60 °C. Therefore, the T_g of rubber phase can be significantly toughened only when it is 40 to 60 °C below room temperature. T_g of general rubber is better below −40 °C. When selecting rubber, this is a problem that must be fully considered.

In addition, as shown in Fig. 4.2, rubber macromolecular chains are bridged across cracks or crazes, thus enhancing their strength and delaying their development, which is also a factor in improving their impact strength.

4.2.2.3 Shear band

Another important role of rubber particles is to initiate the formation of shear bands, as shown in Fig. 4.5. The shear band can make the matrix shear yield and absorb a large amount of deformation work. The thickness of the shear band is 1 μm and its width is about 5–50 μm. The shear band consists of a large number of irregular line clusters, each of which is about 0.1 μm thick.

Fig. 4.5: Structure of shear band.

The shear band is generally located on the plane of the maximum shear stress, which is about 45 degrees with the applied tension or pressure. In the shear band, the orientation of the molecular chains is in the direction of shear force and tensile force.

The shear band is not only an important energy-consuming factor, but also stops the craze from developing into a destructive crack. In addition, the shear band can also turn or terminate the existing small cracks.

There are three possible modes of interaction between craze and shear band, as shown in Fig. 4.6. (1) When the craze meets the existing shear band, it can heal and terminate. At this time, the high orientation of macromolecular chains in the shear band limits the development of crazes. (2) A new shear band is initiated at the tip of the craze with high stress concentration, which in turn terminates the development of the craze. (3) The shear band decreases the initiation and growth rate of the craze and changes the dynamic model of the craze.

The overall result is to promote the termination of the craze and greatly improve the strength and toughness of the material.

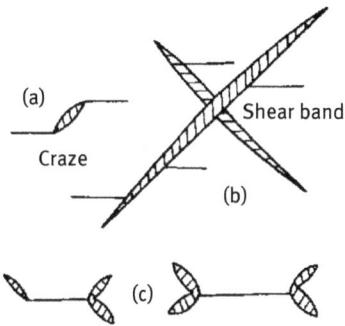

Fig. 4.6: Interaction between craze and shear band in polymethyl methacrylate and polycarbonate. (a) The shear band grows between the apex of the craze; (b) the craze is terminated by the shear band; and (c) the craze is terminated by the shear band itself.

The proportion of crazing and shear yielding is mainly determined by the following factors.

1. The properties of matrix. The brittle matrix is conducive to the formation of craze, and the greater the toughness of the matrix, the larger the proportion of shear components.

2. The properties of stress field. Generally speaking, tension increases the proportion of craze and pressure increases the proportion of shear band. Figure 4.7 shows the damage envelope of polymethyl methacrylate (PMMA) under biaxial stress. The first quadrant in the figure shows that the biaxial stress is tension, and the deformation is mainly crazing. The third quadrant is biaxial pressure, at which point only shear deformation occurs. For the second and fourth quadrants, the failure envelopes of shear and craze intersect each other, so that both craze and shear mechanisms exist simultaneously.

3. Deformation rate. The larger the deformation rate is, the higher the proportion of craze is.

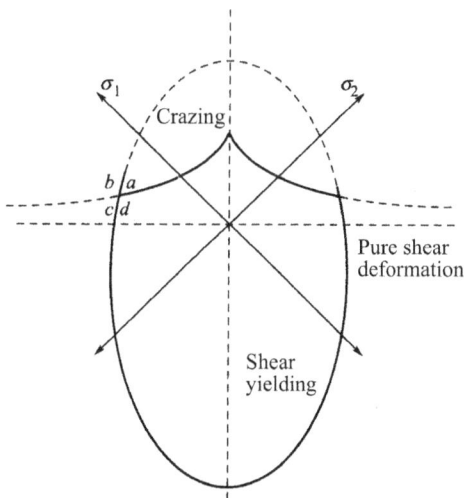

Fig. 4.7: Fracture envelope of polymethyl methacrylate under biaxial stress at room temperature.

4.2.2.4 Cavitation

Under the impact stress, the rubber particles will be cavitated in the elastomer-toughened plastic system. In other words, a large number of microholes are generated in or on the surface of rubber particles, and the diameter of the microholes is in nanometer. It is found that the cavitation can occur not only at the interface between rubber particles and plastic aggregates but also in the interior of rubber particles.

When the material is impacted, the stress concentration effect on the equatorial plane of rubber particles is the greatest, so the interface between continuous phase and dispersed particles is most prone to fall off and form voids. Interfacial voids can prevent cracks from occurring in the continuous phase, and reduce the constraints of continuous phase deformation, which makes it easy to produce forced high elastic deformation, thus absorbing a large amount of energy and improving the impact resistance.

When the morphology of epoxy resin particles toughened by carboxyl-terminated nitrile-butadiene rubber was observed by scanning electron microscopy, it was found that there were obvious cavitations in the rubber particles, as shown in Fig. 4.8. With the increase of temperature, the diameter of rubber particles increases with the diffusion of voids.

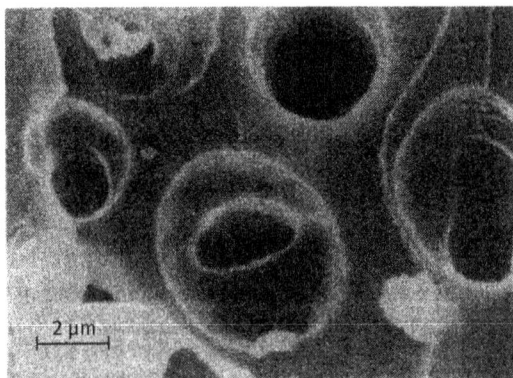

Fig. 4.8: Electron microscopic photographs of epoxy resin toughened by carboxyl-terminated nitrile-butadiene rubber.

This cavitation in rubber particles transforms triaxial stress in the matrix of crack or craze tip region into plane shear stress, which leads to shear band and is beneficial to the improvement of impact strength as shown in Fig. 4.9.

These microvoids increase the volume of rubber particles and cause the shear yield of the matrix around the rubber particles, release the static pressure and thermal stress around the particles due to molecular orientation and shear and transform

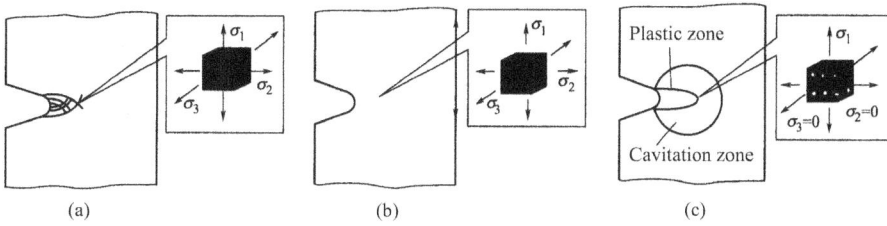

Fig. 4.9: Deformation mechanism of notched epoxy resin strip toughened by carboxyl-terminated nitrile-butadiene (CTNB) rubber. (a) Toughened epoxy resin produces triaxial tension stress at the notch front. (b) Epoxy resin toughened by CTNB rubber and rubber particles have not yet been cavitated. (c) Epoxy resin toughened by CTNB rubber. After rubber cavitation, triaxial stress changes to plane stress, and yield deformation of matrix resin occurs.

the triaxial stress in the matrix into plane stress, which results in the shear yield of the matrix. The formation of voids is not the main part of energy absorption but the plastic yield due to cavitation.

It should be pointed out that the voids generated in the stress whitening zone at the crack or craze tip are not random, but structured, that is, there are arrays with a thickness of about one to four hollow rubber particles and a length of about 8–35 particles, as shown in Fig. 4.10.

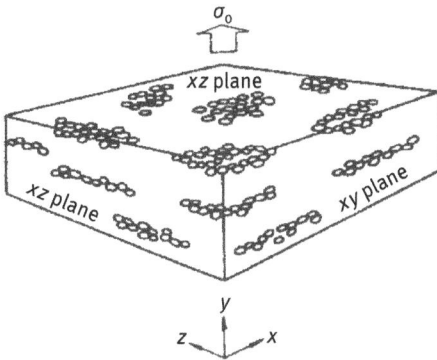

Fig. 4.10: Cavitated rubber particle array model.

Cavitation arrays are formed by the generation and development of rubber particle chains. In this particle chain, the interval between particles is about 0.05 μm. Cavitation changes the local stress state, enlarges the particle volume and forms a small plastic zone in the matrix. This process proceeds repeatedly, resulting in large plastic deformation (yield deformation).

Cavitation arrays are generated from rubber particle chains, which means that excessive uniformity of mixing is not necessarily good during blending, but a certain aggregation structure of rubber particles should be maintained. For example, when toughening PVC with ABS, the homogeneity of blending depends on temperature, and the higher the blending temperature, the more homogeneous it is. The blending homogeneity of the blends increased with the mixing temperatures of 140, 160 and 185 °C, while the impact strength of Charpy with notches decreased with the mixing temperatures of 42, 26 and 8 kJ/m², respectively. The results show that some inhomogeneity of rubber particle dispersion can improve the impact strength of the material.

4.2.3 Wu's percolation theory

In the mid-1980s, Dr. Wu of DuPont Company, USA, and others made a thorough study on the modified ethylene–propylene–diene monomer (EPDM)-toughened polyamide 66 (PA66) in the 1980s. The concept of a universal criterion of critical grain spacing for brittle–ductile transition of toughened plastics was put forward, and the thermoplastic polymer matrix was scientifically classified. The percolation model of brittle–ductile transition of plastics toughening was established, and the traditional toughening theory was pushed from qualitative to quantitative, which played an important role in the study of toughening mechanism.

The main contents of Wu's percolation theory include establishing the relationship between the toughness of blends and the structure of matrix chain, establishing a universal criterion of critical particle spacing for brittle–ductile transition and on this basis, putting forward a percolation model for brittle–ductile transition.

4.2.3.1 Relationship between toughness of blends and structure of matrix chain

The theory holds that the toughness of polymer blends is related to the chain structure of the matrix. First, the quantitative relationship between the structure parameters of the matrix chain – the entanglement density v_e and the characteristic ratio C_∞ of the chain is given:

$$v_\varepsilon = \rho_a / (3 M_v C_\infty^2) \tag{4.1}$$

where M_v is the average molecular weight of statistical unit and ρ_a is the density of amorphous region.

Flory defines v_e and C_∞ as follows:

$$v_e = \rho / M_e \tag{4.2}$$

$$C_\infty = \lim_{h \to \infty} \frac{R_0^2}{nl^2} \tag{4.3}$$

where M_e is the molecular weight of the entanglement point, R_0^2 is the mean square end distance of the undisturbed chain, n is the number of statistical units, l^2 is the mean square length of statistical units.

Because nl^2 is the mean square end distance of the free link chain, C_∞ can represent the flexibility of the real chains without disturbance. The smaller the C_∞, the more flexible the chain.

Kramer gives the relationship between the craze stress σ_z and v_e.

$$\sigma_y \propto v_e^{1/2} \tag{4.4}$$

Kambour gives the expression of normalized yield stress:

$$|\sigma_y| = \sigma_y / [\delta^2 (T_g - T)] \tag{4.5}$$

where σ_y is the yield stress, δ^2 is the cohesive energy density, T_g is the glass transition temperature, and T is the test temperature.

Wu Shouheng further gives–

$$\frac{\sigma_y}{|\sigma_y|} \propto \frac{v_e^{1/2}}{C_\infty} \tag{4.6}$$

The results show that the fracture behavior of matrix is the competition between craze and shear yield, and the magnitude of stress ratio on the left side of the upper formula quantitatively reflects the degree of competition: v_e is smaller and C_∞ is larger, matrix is easy to fracture in craze mode, and toughness is lower; v_e is larger and C_∞ is smaller, matrix is easy to fracture in shear yield mode, and toughness is higher.

According to the above research results, Wu Shouheng divided the polymer matrix into two categories: brittle matrix (craze fracture is dominant) and quasi-ductile matrix (shear yield is dominant). The boundaries of the division are roughly as follows:

When $v_e < 0.15$ mmol/cm^3 and $C_\infty > 7.5$, the polymer matrix is brittle and the impact fracture behavior is mainly craze. The brittle matrix has low crack initiation energy and low crack growth energy, so their notched impact strength is low. After rubber toughening, the energy consumption of external impact is mainly crazing. Typical examples include PS and PMMA.

When $v_e > 0.15$ mmol/cm^3 and $C_\infty < 7.5$, the polymer matrix is quasi-tough, and the impact fracture behavior is dominated by shear yield. These materials have high crack initiation energy and low crack growth energy, so they have high notch-free impact strength and low notch impact strength. After toughening by rubber, the dissipation of impact energy mainly depends on shear yield. Typical examples include polycarbonate (PC), PA, polyethylene terephthalate (PET), and polybutylene terephthalate (PBT).

When $v_e \approx 0.15$ mmol/cm^3 and $C_\infty \approx 7.5$, the impact fracture behavior is a mixed behavior of multiple craze–shear yield.

Table 4.1 shows the v_e and C_∞ values of some polymer matrices.

Tab. 4.1: Chain parameters of some polymer matrices.

Polymer	C_∞	v_e	Polymer	C_∞	v_e
PS	23.8	0.0093	POM	7.5	0.49000
SAN	10.6	0.00931	PA66	6.1	0.53700
PMMA	8.2	0.12700	PE	6.8	0.61300
PVC	7.6	0.25200	PC	2.4	0.67200
PPO	3.2	0.29500	PET	4.2	0.81500
PA6	6.2	0.43500			

It can be seen that the toughening of thermoplastic plastics is essential to improve the crack growth energy of materials.

After a detailed study of rubber toughening PA66, Wu Shouheng conserved that the impact energy was mainly consumed by the surface energy of the fracture surface, the growth energy of craze and the yield deformation energy of the matrix.

Calculations show that the impact energy consumed by craze accounted for about 25% of the total impact energy and 75% of the impact energy consumed by matrix yield when rubber-toughened PA66 was subjected to notch impact. In other words, for plastics which have certain toughness and are toughened by elastomers, the absorption of impact energy is mainly caused by shear yield deformation.

From the earlier discussion, it can be seen that the mechanism of elastomer toughening plastics is not that the elastomer absorbs energy during deformation, but that the elastomer particles act as stress concentrator in the plastics matrix to produce stress concentration effect, which leads to shear yield and crazing of the matrix and absorbing a lot of energy, thus realizing toughening. However, for elastomer-toughened brittle plastics, not all cases can have good toughening effect, brittle–ductile transition is necessary to show excellent toughening effect.

4.2.3.2 Establishing an universal criterion of critical particle spacing for brittle–ductile transition

When Wu Shouheng studied the toughening of PA66 by rubber, he assumed that the particle size of dispersed phase was a single distribution, and that there was a strong bonding force between dispersed phase and PA66 matrix. The spatial distribution of dispersed phase particles was a simple cube. The critical particle spacing

T_c (critical matrix thickness) for brittle–ductile transition of toughened plastics was deduced.

In his opinion, toughened plastics are brittle when the spacing between dispersed phases $T > T_c$, brittle-tough transition occurs when $T = T_c$, and impact strength of plastics increases sharply when $T < T_c$, but modulus decreases. T_c is a single parameter criterion for brittle–ductile transition of toughened plastics. The reason is that when rubber particles are far apart, the stress field around one particle has little effect on other particles. The stress field of matrix is a simple addition of the stress field of these isolated particles. The ability of plastic deformation of matrix is very small. The resulting craze will accelerate the growth and cannot be effectively terminated. The craze develops into crack, which leads to brittle failure of toughened plastics.

When the particle spacing is small enough, a large number of craze stress fields rapidly generated in the blends will interact, which reduces the stress at the craze tip. On the other hand, the tension perpendicular to the stress direction develops at the craze tip, and an active layer is formed between the interference craze tips. The shear yield deformation increases greatly, which makes the toughened plastics tough. Figure 4.11 shows the development of craze during the fracture of polymer blends.

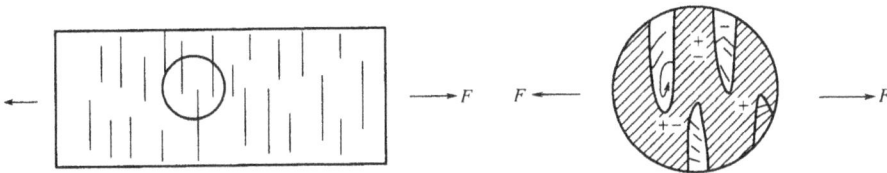

Fig. 4.11: Development of craze in the process of fracture of blends.
The right one is an enlarged picture of the interference of three craze stress fields of the left one. The place where tension develops is marked with " +" and the place where pressure develops is marked with "–."

For example, for PA66/EPDM, polypropylene (PP)/EPDM and PVC/EPDM blends, the critical particle spacing T_c PA66 > T_c PP > T_c PVC shows that PA66 is the most easily toughened, while PVC is the most difficult to toughen.

Wu Shouheng conserved that T_c is the intrinsic parameter of the system, which is determined by the properties of the polymer. On the basis of a large number of experiments, the expression of T_c is obtained as follows:

$$lgT_c = 0.74 - 0.22C_\infty \tag{4.7}$$

For rubber toughening systems based on quasi-tough polymers, when the volume fraction φ_r of rubber and the affinity between the matrix and rubber remain constant, the brittle–ductile transition occurs in a critical particle size d_c, and the d_c increases with the increase of φ_r. The quantitative relationship between the critical particle spacing T_c and the critical particle size is as follows:

$$T_c = d_c[(\pi/6\varphi_r)^{1/3} - 1]^{-1} \tag{4.8}$$

The critical particle spacing T_c is correlated with the properties of the matrix and the particle size of rubber particles by eqs. (4.7) and (4.8). Therefore, the quantification of rubber toughening theory has been greatly promoted.

4.2.3.3 Percolation model of brittle–ductile transition

In order to further explain the essence of brittle–ductile transition of elastomer-toughened plastics, Wu Shouheng et al. proposed a percolation model. He assumed that rubber particles were randomly distributed in the plastic continuous phase by spheres of the same diameter. Because the elastic modulus, Poisson's ratio and expansion coefficient of the dispersed phase rubber particles are not equal to those of the continuous phase plastic matrix, a plane stress volume sphere is formed when the rubber particles are impacted together with the surrounding $T_c/2$ thick matrix (Fig. 4.12 left). The diameter of the plane stress volume sphere is S:

$$S = d + T_c \tag{4.9}$$

where d is the diameter of rubber particles and T_c is the critical particle spacing.

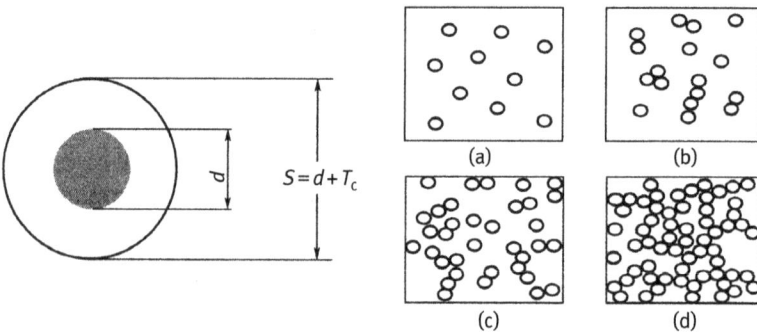

Fig. 4.12: Plane stress volume sphere of rubber particles and its percolation model. Left: schematic diagram of plane stress volume sphere around rubber particles (the middle shadows are rubber particles); right: percolation model of rubber particle Toughening.

The distance L between the spherical centers of two adjacent plane stress volume spheres is

$$L = d + T \tag{4.10}$$

T is the distance between two rubber particles.

When $L \leq S$, that is, $T \leq T_c$, the gravitational field of two adjacent plane stress volume spheres is correlated.

If the volume fraction of rubber particles in the system is φ_r, the volume fraction of plane stress volume sphere is

$$\varphi_s = (S/d)^3 \varphi_r \qquad (4.11)$$

Obviously, with the increase of φ_r, the number of associated plane stress volume spheres increases, and gradually interconnect with each other, forming percolation groups of different sizes. When φ_r increases to the percolation threshold φ_{rc}, φ_s reaches φ_{sr}, and all plane stress volume spheres are correlated to form a percolation channel (Fig. 4.12 right). At this time, brittle-ductile transition occurs.

When percolation channels are formed,

$$S = S_c = d_c + T_c \qquad (4.12)$$

S_c is the critical plane stress volume sphere diameter and d_c is the critical rubber particle size. At this point,

$$\varphi_s = \varphi_{sc} = (S_c/d_c)^3 \varphi_{rc} \qquad (4.13)$$

where φ_{sc} is the critical volume fraction of plane stress volume sphere and φ_{rc} is the critical volume fraction of rubber particles.

Combine eqs. (4.12) and (4.13) to get the following formula:

$$\varphi_{sc} = [(d_c + T_c)/d_c]^3 \varphi_{rc} \qquad (4.14)$$

The equivalent values of φ_{sc} and T_c can be calculated by substituting the d_c and φ_{rc} measured in the experiment for the brittle–ductile transition of the blend system in eq. (4.14).

4.2.4 Factors affecting impact strength of elastomer-toughened plastics

In elastomer-toughened plastics, the factors affecting the impact strength can be considered from three aspects: matrix properties, size and number of dispersed phases and interfacial adhesion.

1. Effect of resin matrix

The impact strength can be improved by increasing the molecular weight and toughness of the matrix resin. For toughened plastics such as PVC toughened by

ABS, there is an optimum content of rubber component due to the interaction between craze and shear band, as shown in Fig. 4.13.

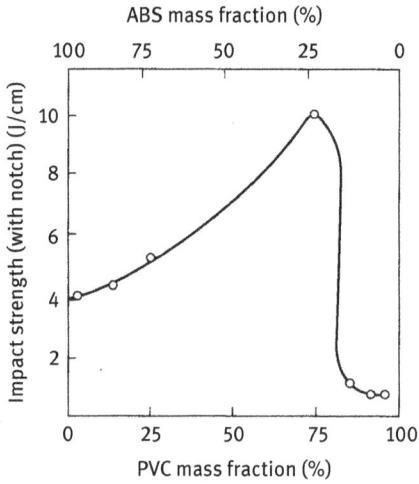

ABS mass fraction (%)

Impact strength (with notch) (J/cm)

PVC mass fraction (%)

Fig. 4.13: Relationship between impact strength and matrix composition of PVC/ABS Blends.

There should be moderate mutual solubility between matrix resin and rubber phase.
Too little mutual solubility will make the adhesion between phases insufficient; too
much mutual solubility will make rubber particles too small, even form homogeneous system, which is not conducive to the improvement of impact strength, as
shown in Fig. 4.14.

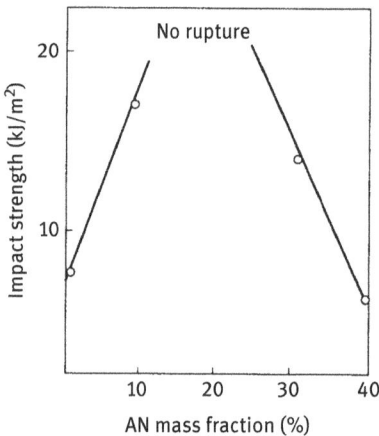

No rupture

Impact strength (kJ/m²)

AN mass fraction (%)

Fig. 4.14: Effect of AN content on impact strength of PVC/NBR (PVC/NBR = 100/15).

The molecular weight of matrix has a strong influence on the dispersion of elastomers. Increasing the molecular weight of the matrix can reduce the brittle–ductile
transition temperature and increase the impact strength. When the molecular weight

is increased, the processability of the blends will also be reduced, which may damage the comprehensive properties of the blends. Therefore, the selection of molecular weight of matrix should be appropriate.

2. Effect of rubber content

When the size of rubber particles is fixed, the rubber content increases and the number of rubber particles increases in a certain range. The rate of initiation and termination of craze increases correspondingly, which is conducive to the improvement of impact strength of materials. At the same time, the interaction of the stress field around the rubber particles is more conducive to the initiation of craze and the improvement of impact toughness. But the greater the rubber content, the better. As mentioned in the previous section, there is often a maximum rubber content for the matrix with greater toughness. More rubber content will not only not improve the toughness of the material but also affect other mechanical properties, such as strength and modulus of the material decrease greatly. Therefore, the rubber content in the blends usually has an optimum value.

3. Effect of rubber particle size

The effect of rubber particle diameter on the toughness of materials is also optimal. The optimum values for different varieties are also different, which mainly depends on the properties of matrix resins. When rubber particles are too small, they cannot stop the craze, and when they are too large, the number of crazes is too small, which is not conducive to the improvement of impact strength. The thickness of craze in PS is about 0.9–2.8 μm, so the optimum diameter of rubber particles in HIPS is 1–10 μm. When the particle size is too small, the craze will "engulf" and cannot be stopped. The size of craze varies with the matrix, so the optimum particle size of rubber varies. For a certain toughening system, there is a critical size of rubber particles, which is smaller than this size, and rubber cannot play a toughening role. For example, the critical size of HIPS is 0.8 μm, ABS is 0.4 μm and toughened PVC is 0.2 μm. From these data, it can be seen that the tougher the plastic matrix is, the smaller the critical size is.

The particle size distribution of rubber particles also has an effect. Because the large particle size is good for initiating craze and the small particle size is good for initiating shear band, the particle size distribution is in the best state.

4. Effect of glass transition temperature of rubber phase

Generally speaking, the lower the glass transition temperature of rubber phase, the better the toughening effect. Usually, the T_g of rubber phase should be below −40 °C. For elastomer-toughened plastics system, under high-speed impact load, rubber particles should be able to fully relax and deform in order to effectively trigger craze or shear band and absorb impact energy. However, under the action of high-speed load, T_g of rubber phase will increase significantly.

It has been analyzed in Section 4.2.2.3 that if the T_g of rubber is not too low in static state, the rubber particles may be in glass state under impact force, so the rubber elastic particles cannot absorb impact energy. Therefore, in order to make the rubber phase play an effective toughening role at room temperature, its T_g should be at least 40–60 °C lower than that of room temperature. In other words, T_g of general rubber should be below –40 °C.

5. Effect of rubber particle structure

In the third chapter, it was introduced that in HIPS system, the rubber dispersed phase particles also contain a large number of PS inclusions. This special structure of rubber particles has a stronger toughening effect than ordinary rubber particles. In addition to increasing the volume fraction of rubber phase, honeycomb rubber particles are different from ordinary particles in the process of mechanical deformation. Rubber particles with a large number of inclusions PS are elongated under external force, and the PS inside them is not easily deformed due to its high modulus, so the rubber particles are locally "microfibrillated." Rubber molecules in the dispersed phase are separated by PS microregions, and no large voids are produced after rubber microfibration. For the rubber particles without inclusion, the whole particle is elongated under the action of external force, while the transverse shrinkage occurs obviously. PS on the surface of rubber particles cannot be deformed immediately, resulting in voids between the outer layer of rubber particles and PS, greatly affecting the toughening effect of rubber. Therefore, the inclusion-free particles should be minimized in the HIPS system.

For ABS, the effect of particle morphology (with or without inclusion) on toughening is less obvious than that of HIPS. This is related to the high strength of microfibers formed by crazing of ABS matrix (styrene acrylonitrile, SAN). Although voids may also occur during the deformation of ABS, the craze will not break.

In addition, the degree of crosslinking of rubber also has a great impact on toughening and also has the most suitable range. Excessive crosslinking degree and high modulus of rubber phase make it difficult to play a toughening role. Excessive crosslinking degree makes it easy for rubber particles to be deformed and fragmented during processing, and is not conducive to play a toughening role. The optimum crosslinking degree of rubber phase is usually determined by experiments.

6. Effect of adhesion between rubber phase and matrix resin

Only when there is good adhesion between the two phases, can the rubber phase play an effective role. Graft copolymerization blending or block copolymerization blending can be used to increase the adhesion between the two phases. The copolymers can act as compatibilizers and greatly improve the impact strength. Adding compatibilizer directly can also effectively improve the impact strength of the system.

4.3 Mechanism of toughening plastics with nonelastomers

4.3.1 Toughening by rigid organic particles

4.3.1.1 Toughening phenomenon of rigid organic particles

Traditional toughening materials have always been based on organic elastomers, such as chlorinated polyethylene (CPE), ethylene vinyl acetate, MBS, styrene–butadiene–styrene, ACR and nitrile-butadiene rubber (NBR). The impact modification effect of elastomeric toughening materials is very good, but at the same time, it often sacrifices the precious strength, rigidity, dimensional stability, heat resistance and processability of materials. So scientists have been looking for new toughening methods to solve this problem. In 1984, when Kurauchi and Ohta of Japan studied the mechanical properties of PC/ABS and PC/acrylonitrile styrene (AS) blends, they found that the area enclosed by the stress–strain curve of PC/ABS was smaller than that of PC/AS (Fig. 4.15), and the impact strength of PC/ABS was higher than that of PC/AS (Fig. 4.15). As we know, AS is a rigid polymer, which is much more brittle than ABS (Fig. 4.16). ABS contain rubber phase, which has toughening effect on PC, which is not difficult to understand. But why is AS toughening? Moreover, the toughening effect of AS is greater than that of ABS, which is beyond the explanation of the toughening mechanism of rubber particles. Through research, Kurauchi and Ohta first proposed the toughening method of quasi-toughening plastics matrix with rigid organic filler particles. A new concept of toughening plastics with rigid organic particles was put forward, and the mechanism of this toughening method was explained by the concept of "cold drawing."

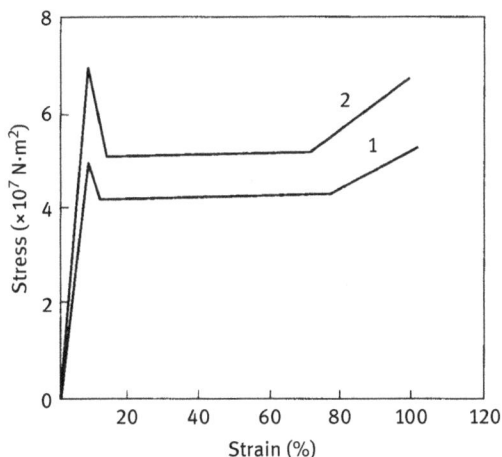

Fig. 4.15: Stress–strain curves of PC/ABS and PC/AS blends. 1, PC/ABS(70/30); 2, PC/AS(70/30).

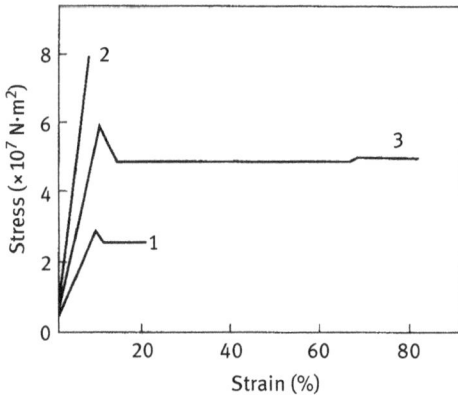

Fig. 4.16: Stress–strain curves of ABS, AS and PC. 1, ABS; 2, AS; 3, PC.

In recent years, inelastic toughening methods have been widely used in the preparation of polymer blends. For example, rigid plastics such as PS, PMMA and AS can be added to quasi-tough matrix such as PC and PA to prepare polymer blends toughened by inelastic materials. Polymers such as PVC and PMMA, which are brittle in nature, can be modified by CPE, MBS or ACR, and toughness can be further improved by adding rigid polymers such as PS. As with general polymer blends, it is important to have good interfacial bonding. For example, the interface bonding between nylon-6 and AS is not good. Styrene–maleic anhydride copolymer (SMA) should be added to improve the interface bonding. Blending with AS can significantly improve the impact strength of nylon-6.

The mechanical properties of the blends prepared by adding PS into the blends of PVC/CPE and PVC/MBS are shown in Tab. 4.2. It can be seen that PS not only improves the impact strength of PVC/CPE and PVC/MBS, but also improves or keeps the tensile strength and modulus unchanged. The optimum dosage of PS is about 3%. If the dosage exceeds the optimum dosage, the impact strength decreases sharply.

Tab. 4.2: Mechanical properties of PVC blends.[1]

Parameter	PVC	PVC/MBS	PVC/CPE	PVC/MBS/PS	PVC/CPE/PS
Impact strength (kJ/m^2)	2.5	8.4	16.2	20.6	69.5
Tensile strength (MPa)	58.4	47.5	41.0	47.1	43.7
Young's modulus (GPa)	14.1	9.8	11.0	9.8	12.1

[1]Among the 100 mass copies of polymer blends, CPE and MBS were 10 copies and PS were three copies.

4.3.1.2 Toughening mechanism of rigid organic particles

The toughening mechanism of quasi-toughening matrix with rigid organic particles is different from that of elastomer toughening of plastics. In the study of PC/ABS and PC/AS blends by Kurauchi and Ohta, it was found that although the mechanical properties of ABS and AS bulk are quite different, ABS is soft and tough, while AS is hard and brittle, the mechanical behavior of both blends is high. Electron microscopic observation showed that before stretching, ABS and AS were dispersed in PC matrix as spherical particles with a particle size of about 2 µm. After stretching, there was no craze in the blend, but the spherical particles of dispersed phase had elongation deformation, the deformation range was more than 100%, and the matrix PC also had the same size deformation. During stretching, rigid AS also seems to become tough. It is obvious that the toughening mechanism of rubber cannot be explained. Therefore, they put forward the toughening mechanism of rigid organic particles, believing that when the static pressure is strong enough to a certain value, the rigid particles in the dispersed phase of the blend will change from brittleness to toughness, which will lead to a cold-drawing phenomenon similar to that of glassy polymers. For single-component polymers, the brittleness/toughness transition has long been recognized, which is mainly related to temperature, strain rate, static pressure and other factors. Figure 4.17 is the stress–strain curve of PMMA at different static pressures. When the temperature and tension rate are constant, the shape of the curve varies with the static pressure. The yield stress and elongation at break increase with the increase of static pressure. When the static pressure is strong enough to a certain value (as shown in Fig. 4.17, the medium pressure is 102 kPa),

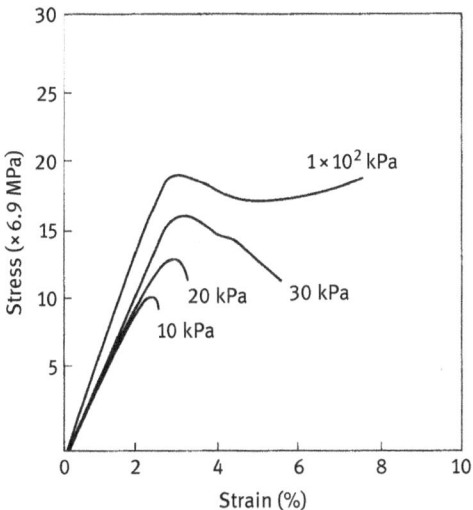

Fig. 4.17: Stress–strain curves of PMMA under different static pressures.

the cold-drawing phenomenon occurs, that is, the stress is basically unchanged, while the strain develops greatly. Similar brittle/tough transition and cold-drawing phenomena occur in PS and SAN under high static pressure.

When ABS is subjected to a sufficiently large static pressure, it also exhibits such large deformation and cold-drawing behavior, as shown in Fig. 4.18. With the increase of static pressure, the initial elongation of ABS is basically unchanged. However, when the static pressure increases to 9×10^6 Pa, the elongation at break changes abruptly and then tends to be stable, indicating that the specimen has been cold drawn.

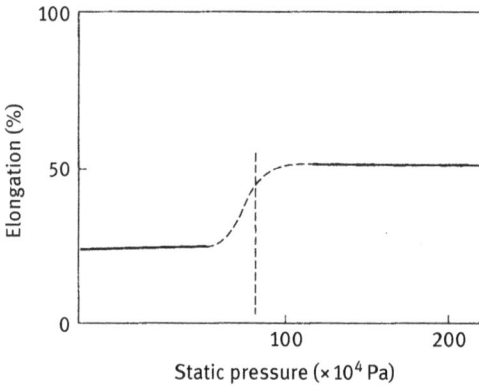

Fig. 4.18: Effect of static pressure on elongation of ABS during tension.

By summarizing the examples of toughening by rigid particles, it can be found that all blends have the characteristic that the toughness of matrix is better than that of dispersed phase, that is, the deformation ability of matrix is higher than that of dispersed phase. Therefore, the reason why rigid ABS and AS particles can toughen quasi-tough PC is due to the cold-drawing effect of ABS and AS particles. Figure 4.19 is a schematic diagram of the toughening mechanism of quasi-toughening matrix by rigid polymer particles. It can be seen from the figure that when the tough matrix is subjected to external tensile stress, the matrix is easily deformed due to the difference between the elastic modulus and Poisson's ratio of the dispersed particles and the matrix, while the rigid particles with high modulus and small Poisson's ratio cannot deform as the matrix does. In this way, on the equatorial plane of rigid particles, the compressive stress produced by the deformation of the matrix will occur. When the static pressure on the equatorial plane of rigid polymer particles is greater than the static pressure (σ_c) required for the deformation of rigid particles, rigid polymer particles will yield and produce cold tension, resulting in large plastic deformation, which consumes a large amount of energy, thus improving the impact strength. Therefore, it is considered that σ_c can be used as a criterion for the yield of rigid polymer particles.

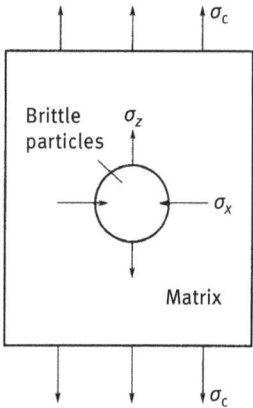

Fig. 4.19: Schematic diagram of toughening mechanism of rigid organic particles on tough matrix.

However, further studies have found that many blends consisting of quasi-tough matrix and rigid polymer particles do not show obvious cold-drawing phenomenon, and therefore do not show the toughening effect of rigid organic particles. In PC/PMMA, PC/AS, PC/polyphenylene sulfide, PBT/SAN, PA/PMMA, PA/PS and PVC/PS systems, only PC/PMMA and PC/AS show obvious toughening effect. Among them, the toughening effect of PS is obvious if PVC is blended with CPE before blending with PS, which indicates that the yield of cold drawing of rigid polymer particles cannot be judged only by elastic modulus and Poisson's ratio. In other words, it is not accurate to judge the toughening effect only according to the critical static pressure (σ_c). Inoue et al. believed that the yield of rigid polymer particles in the blend system can be achieved by means of Mises criterion of metal materials:

$$(\sigma_x - \sigma_y)^2 + (\sigma_y - \sigma_z)^2 + (\sigma_x - \sigma_z)^2 \geq 2\sigma_c^2 \tag{4.15}$$

where σ_x, σ_y and σ_z are the stress values of three axes in the rectangular coordinate system, respectively, and σ_c is the critical static pressure when yielding cold drawing occurs. When the specimen is tensioned, the compressive stress of the rigid particle on the x- and y-axes is the same, that is, $\sigma_x = \sigma_y$, then the formula can be simplified as follows:

$$(\sigma_x - \sigma_z)^2 \geq \sigma_c^2 \tag{4.16}$$

Therefore, the yield of cold drawing of rigid polymer particles in dispersed phase can be judged by their values.

The above stress analysis is carried out on a single particle without considering the interaction of particles. In fact, when the content of dispersed phase reaches a certain concentration, there must be interaction between two adjacent rigid organic particles, as shown in Fig. 4.20.

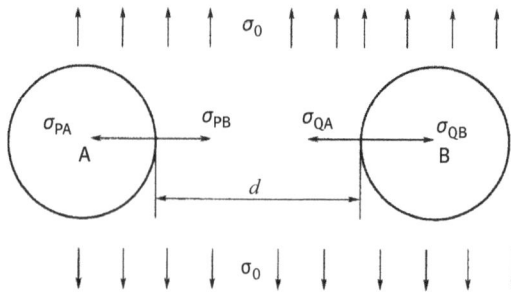

Fig. 4.20: Schematic diagram of stress interaction between adjacent rigid organic particles.

As can be seen from the graph, when particles A and B are close to each other, under the action of uniform external stress field σ_0, the compressive stress of σ_{PA} is generated at the equatorial plane of particles A, and weakens to σ_{QA} at the interface of particles B. Therefore, the actual pressure on B particles is $\sigma_{QB}-\sigma_{QA}$. Similarly, the actual pressure on particle A is $\sigma_{PA}-\sigma_{PB}$. In this way, the stress intensity of the particles will be significantly reduced with the approximation of the composition ratio of the blends and the reduction of the distance between the particles. As a result, the toughening effect will be worse. Therefore, the content of rigid polymer particles used for toughening cannot be higher than a certain value. In other words, the amount of rigid polymer particles has an optimal range. In this range, good toughening effect can be obtained. Outside this range, the impact resistance will decline sharply.

From the earlier discussion, it can be seen that the toughening object of inelastic materials is quasi-tough polymers with certain toughness, such as PC and PA. Using brittle matrix, it is not easy to yield rigid polymer particles. Therefore, it is often ineffective to toughen rigid polymer particles directly. It is necessary to use elastomer to toughen properly, and then use inelastic body to toughen further.

The greatest advantage of toughening plastics with inelastic materials is the improvement of impact strength without reducing the rigidity of materials. In addition, rigid polymers generally have good processing fluidity. Therefore, inelastic toughening system can also improve processing fluidity. These advantages make the toughening method of inelastic materials become a widely concerned field of blending modification.

4.3.1.3 Factors affecting toughening effect of rigid organic particles
Since the toughening effect of rigid polymer on the blend system is achieved by absorbing energy through its own yield deformation (cold drawing) process, it is very important whether the rigid particles can yield cold drawing when subjected to external forces. This is obviously related to the properties of matrix resin and rigid

polymer. Therefore, the factors affecting the toughening effect of rigid organic particles can be summarized as follows.

1. Toughness of matrix
Quasi-tough matrix with better toughness is beneficial to yield cold drawing of rigid particles. The blends composed of quasi-tough matrix have large deformation capacity and can produce large transverse compressive stress under external force. For example, PVC/PS system has no toughening effect, while the blend of PVC/CPE and PS has toughening effect, apparently due to the improvement of matrix toughness.

2. Rigidity of rigid polymers
Rigidity of rigid polymers as dispersed phases is not the greater the better. Practice shows that excessive rigidity is not conducive to toughening. Because the rigidity of dispersed phase polymer is too large, the stress of deformation is also required to be larger; otherwise, it is not easy to meet the conditions, so it is not easy to yield cold tension. For example, the toughening effect of SAN and many systems are composed of tough matrix, but PS has no toughening effect on most systems, which can only be attributed to the fact that PS is more rigid than SAN.

3. The content of rigid polymer particles in dispersed phase
Generally speaking, the content of rigid polymer particles should not be too large. As mentioned earlier, when the blend is subjected to external force, not only the matrix produces compressive stress on the rigid particles, but also the interaction between the rigid particles (Fig. 4.20). This kind of interaction between particles will reduce the compressive stress of particles. With the increase of the content of rigid particles and the decrease of particle spacing, the attenuation of substress becomes more significant. So the toughening effect will be worse if the content of rigid particles is large to a certain extent.

4. Binding force of two-phase interface
In order to achieve better toughening effect, the two-phase interface should have sufficient bonding strength. This requirement is applicable to any polymer blend system. For example, for PA/SAN blends, Mises criterion should have toughening effect, but the toughening effect is not ideal. The reason is that the interfacial bonding force is too poor. In fact, when the interfacial bonding force is less than σ_z, the separation will occur at the two poles of the dispersed phase particles first, which destroys the original three-dimensional stress field distribution and makes it impossible for SAN to yield and cold draw. The results show that the impact strength of PA6/SAN blends can be significantly improved by adding PS SMA as compatibilizer.

4.3.1.4 Comparison of toughening effect of rubber particles and rigid polymer particles on plastics

Comparing the toughening effects of rubber particles and rigid polymer particles on plastics, the following differences can be found:

1. There are different kinds of toughing agents. The former is rubber or thermoplastic elastomer material with low modulus, large deformation and poor fluidity, while the latter is rigid plastic with high modulus, small deformation and good fluidity.
2. The toughening objects are different. The former has toughening effect on both brittle and tough matrix, while the latter requires a certain toughness of matrix to achieve toughening effect.
3. The effect of toughening agent content on toughening effect is different. The former mostly increases the impact strength of the blend with the increase of the content of toughening agents, while the latter shows the toughening effect only in a certain range.
4. The effect of modification is different. The former improves the toughness greatly, but the modulus, strength and thermal deformation temperature of the material generally decrease to some extent; while the latter improves the toughness, the modulus, strength and thermal deformation temperature of the material increase to a certain extent.
5. The toughening mechanism is different. The former is due to the stress concentration effect of rubber particles, which results in shear band or craze deformation of the matrix, thus absorbing impact energy. The latter is due to the energy absorption of the rigid particles under the action of compressive stress of the matrix.

The two toughening mechanisms require the same bond strength between the two phases. Without high interfacial bonding strength, no composite material will have high mechanical properties. It can be seen from the analysis and comparison of the above two different toughening systems that they have their own characteristics and can complement the modification of polymer blends. Therefore, the application of two kinds of toughening systems in the study of polymer modification is not only of theoretical value, but also of great practical significance. It means that it is possible to use inexpensive large variety of polymers as raw materials, through the combination of two kinds of toughening agents toughening, to obtain low-cost high-performance multicomponent polymers.

4.3.2 Toughening of inorganic rigid particles

Inorganic rigid particle toughening polymer is a new technology developed in 1990s. In 1991, when Li Dongming and others studied $PP/CaCO_3$ system, they found that the toughness of PP could be improved obviously by adding $CaCO_3$. For example, the notched impact strength of PVC can be increased three times by adding 10% nano-$CaCO_3$. Micron $CaCO_3$ also has toughening effect on PVC, but its effect is lower than nano-$CaCO_3$.

As for the toughening mechanism of inorganic rigid particles, it is believed that the stress concentration of the matrix changes when inorganic rigid particles are added to the matrix. Tensile stress exists between the matrix and particles at the two poles, and compressive stress exists between the matrix and particles at the equatorial position. Due to the interaction of forces, the tension stress acts on the equatorial surface of particles. When the interface viscosity is weak, the splicing will occur at the two poles first, which makes the surrounding particles equivalent to forming a hole, and the stress on the equatorial surface of the hole is three times of the bulk stress. Therefore, when the bulk stress has not reached the matrix yield stress, the local point has begun to yield, that is to say, it also promotes the matrix yield, and the comprehensive effect improves the toughness of the polymer.

Another view is that inorganic rigid particles are uniformly dispersed in the matrix, resulting in stress concentration effect. When the matrix is impacted, small rigid particles can easily cause a large number of crazes around the matrix. At the same time, the matrix between particles also produces plastic deformation and absorbs impact energy, which hinders and passivates the growth of craze, and finally prevents the craze from developing into destructive cracks, thus achieving toughening effect. With the refinement of the particles, the specific surface area of the particles increases, and the interface with the plastic matrix also increases. The matrix will produce more crazes and plastic deformation when it is impacted, thus absorbing more energy, and the toughening effect is better.

Similar to the toughening of rigid organic particles, the toughening effect of inorganic rigid particles is closely related to the toughness of plastic matrix. Firstly, the toughness of matrix is improved, and then toughened by inorganic particles, better toughening effect can be obtained. For example, when PVC was blended with ACR and nano-$CaCO_3$ was added, the notched impact strength could reach 24 kJ/m^2, while the impact strength of PVC without blending was only 2.5 kJ/m^2.

Exercises

1. Give a brief account of the theory of craze-shear band.

Answer:
Rubber particles play two important roles in toughening plastics:
(1) as a stress concentration center, many crazes and shear zones are induced and
(2) the development of crazes is controlled and the crazes are terminated in time without developing into destructive cracks. The stress field at the end of the craze

can continue to induce the shear zone to terminate the craze, and further development of the craze can be prevented by encountering the existing shear zone. The generation and development of many crazes and/or shear zones consume a large amount of energy, which can significantly improve the toughness of toughened plastics.

2. What are the factors affecting the impact strength of elastomer-toughened plastics?

Answer:
(1) Resin matrix: The impact strength can be improved by increasing the molecular weight and toughness of the matrix resin.
(2) Rubber content: When the size of rubber particles is fixed, the rubber content increases and the number of rubber particles increases in a certain range.
(3) Rubber particle size: The effect of rubber particle diameter on the toughness of materials is also optimal. The optimum values for different varieties are also different, which mainly depends on the properties of matrix resins.
(4) Glass transition temperature of rubber phase: The lower the glass transition temperature of rubber phase, the better the toughening effect.
(5) Rubber particle structure: In HIPS system, the rubber dispersed phase particles also contain many PS inclusions. This special structure of rubber particles has a stronger toughening effect than ordinary rubber particles.
(6) Adhesion between rubber phase and matrix resin: Graft copolymerization blending or block copolymerization blending can be used to increase the adhesion between the two phases. The copolymers can act as compatibilizers and greatly improve the impact strength.

3. Please briefly describe the toughening mechanism of rigid organic particles on ductile matrix.

Answer:
When the tough matrix is subjected to external tensile stress, the matrix is easily deformed due to the difference between the elastic modulus and Poisson's ratio of the dispersed particles and the matrix, while the rigid particles with high modulus and small Poisson's ratio cannot deform as the matrix does. In this way, on the equatorial plane of rigid particles, the compressive stress produced by the deformation of the matrix will occur. When the static pressure on the equatorial plane of rigid polymer particles is greater than the static pressure (σ_c) required for the deformation of rigid particles, rigid polymer particles will yield and produce cold tension, resulting in large plastic deformation, which consumes a large amount of energy, thus improving the impact strength.

Chapter 5
Mechanical properties of multicomponent polymers

5.1 Introduction

The mechanical properties of polymer materials depend on temperature and time and exhibit viscoelasticity, that is, both viscous liquids and pure elastic solids. This dual mechanical behavior is due to the special molecular motion characteristics of polymers. Unlike small molecule motions, the molecular motions of polymers include not only slip between long chain molecules but also chain segment motions so the molecular motions have obvious relaxation characteristics. The differences of mechanical properties among different polymers are not only related to the molecular chemical structural factors but also to the relative molecular mass and its distribution, branching and cross-linking, crystallization and noncrystallization, crystallinity, copolymerization mode, molecular orientation, plasticization and filler molecular morphologies and aggregation state.

For multicomponent polymers, these factors undoubtedly still exist. In addition, more complex morphological and structural factors were added to the blends. In fact, the morphology and structure of the blends, such as the size of the bonding force between the two phases, the structure of the interfacial layer, the thickness of the interfacial layer, the continuity of the two phases, the domain size of the dispersed phase and the shape of the dispersed phase particles play a decisive role in the relationship between the properties of the blends and their component properties.

5.2 Relationship between properties of polymer blends and their components

5.2.1 The "mixing rule" for the properties of bicomponent polymers

The relationship between the properties of two-component blends and the corresponding properties of their components can be expressed simply by the "mixing rule." The two most commonly used relationships are

$$P = P_1\beta_1 + P_2\beta_2 \tag{5.1}$$

$$\frac{1}{P} = \frac{\beta_1}{P_1} + \frac{\beta_2}{P_2} \tag{5.2}$$

where P is one of the properties of two-component blended modified plastics; P_1 and P_2 are the corresponding properties of components 1 and 2; β_1 and β_2 represent the volume fractions of components 1 and 2, respectively.

https://doi.org/10.1515/9783110596335-005

In most cases, eq. (5.1) gives the upper limit of the properties of polymer alloys, while eq. (5.2) gives the lower limit. Since eqs. (5.1) and (5.2) do not take into account the morphology and structure of polymer alloys, sometimes the results given differ greatly from the measured values. In fact, eqs. (5.1) and (5.2) are more suitable for copolymers.

In order to establish a close relationship between the properties of polymer alloys and the corresponding properties of their pure components, a great deal of research work has been done on polymer alloys with different morphological structures and corresponding formulas have been established.

5.2.2 Relationship between properties and component properties of homogeneous blends

Homogeneous blends consist of completely compatible blends of two-component polymers, random copolymers and oligomers as plasticizers.

Due to the interaction between the two components of the blend system, the relationship between the properties of the blend and the corresponding properties of its pure components usually deviates from the simple mixing rule. Therefore, the commonly used eq. (5.3) is revised:

$$P = P_1\beta_1 + P_2\beta_2 + I\beta_1\beta_2 \tag{5.3}$$

For example, for some completely compatible blends, the glass transition temperature (T_g) can be expressed as follows:

$$T_g = \overline{W}_1 T_{g1} + \overline{W}_2 T_{g2} + K\overline{W}_1\overline{W}_2 \tag{5.4}$$

The random copolymer of vinyl chloride and vinyl acetate can be regarded as a typical example, and its T_g can be approximately expressed as

$$T_g = 30\overline{W}_1 + 80\overline{W}_2 - 28\overline{W}_1\overline{W}_2 \tag{5.5}$$

The T_g of plasticizer system can be expressed by eq. (5.6):

$$T_g = \frac{\alpha_{fp}\varphi_p T_{gp} + \alpha_{fd}(1 - \varphi_p)T_{gd}}{\alpha_{fp}\varphi_p + \alpha_{fd}(1 - \varphi_p)} \tag{5.6}$$

The middle subscripts p and d represent the free volume expansion coefficients of polymers and plasticizers near T_g, respectively, and α_{fp} and α_{fd} represent the free volume expansion coefficients of polymers and plasticizers near T_g. If $\alpha_{fp} = \alpha_{fd}$, then eq. (5.6) can be simplified to the form as follows:

$$T_g = \varphi_p T_{gp} + \varphi_d T_{gd} \tag{5.7}$$

5.2.3 Relationship between properties and component properties of single continuous structure blends

The mechanical properties of the blends with single continuous structure are related not only to the properties of the polymers involved in the blending, but also to the morphology, size, stacking mode of the dispersed phase and the interfacial bonding between the dispersed phase and the continuous phase.

The relationship between the properties of monocontinuous structure blends and component polymers is commonly expressed as follows:

$$\frac{P}{P_1} = \frac{1 + AB\varphi_2}{1 - B\psi\varphi_2} \tag{5.8}$$

where P is the property of blends, P_1 is the property of continuous phase and φ_2 is the volume fraction of dispersed phase; A, B and ψ are the coefficients, which are defined by eqs. (5.9), (5.10) and (5.11), respectively:

$$A = K_E - 1 \tag{5.9}$$

$$B = \frac{P_2/P_1 - 1}{P_2/P_1 + A} \tag{5.10}$$

$$\psi = 1 + \left(\frac{1 - f_p}{f_p^2}\right)\varphi_2 \tag{5.11}$$

where P_2 is the performance of dispersed phase, K_E is Einstein coefficient and φ_p is the maximum stacking coefficient of dispersed phase particles.

K_E is related to the shape of dispersed particles, aggregation state and interfacial bonding. For the different properties of the blends, there are different K_E. For example, K_E of mechanical properties, K_E of electrical properties and K_E of thermal properties, but the values are generally not different, so K_E of mechanical properties is often used instead. The K_E values of mechanical properties of some systems are shown in Tab. 5.1.

The definition of φ_p is

$$\varphi_p = \frac{\text{True volume of dispersed phase particles}}{\text{Masonry volume of dispersed phase particles}} \tag{5.12}$$

Equation (5.12) is a hypothesis that dispersed phase particles are "stacked" in some form. The form of "stacking" depends on the specific situation of dispersed phase particles in the blend and is related to the shape, particle size distribution and arrangement of dispersed phase particles. In fact, φ_p reflects the spatial characteristics of a specific existing state of dispersed phase particles. The values of φ_p of dispersed phase particles with different existing conditions are shown in Tab. 5.2.

Tab. 5.1: Einstein coefficient of some systems.

Types of dispersed phase particles	Orientation	Interface bonding	Stress type	K_E
Spherical	Arbitrarily	No slide	–	2.50
Spherical	Arbitrarily	Slide	–	1.0
Spherical aggregation	Arbitrarily	–	–	$2.50/\varphi_m$[1]
Cube	Random	–	–	3.1
Staple fiber	Uniaxial orientation	–	–	1.5
Staple fiber	Uniaxial orientation	–	Tensile, parallel to fiber orientation	$2\,L/D^2$
Long fiber	Random	–	Shear	$L/2D^2$

[1]The maximum stacking fraction of spherical particles in aggregates is φ_m.
[2]L is the length of fibers and D is the diameter of fibers.

Tab. 5.2: Maximum stacking coefficient φ_p of some dispersed phase particles.

Particle shape of dispersed phase	Stacking type	φ_p	Particle shape of dispersed phase	Stacking type	φ_p
Spherical	Hexagonal close-packed masonry	0.7405	Rod shape $L/D = 2$	Three-dimensional random stacking	0.671
Spherical	Face-centered cubic dense stacking	0.7405	Rod shape $L/D = 4$	Three-dimensional random stacking	0.625
Spherical	Body-centered cubic dense stacking	0.600	Rod shape $L/D = 8$	Three-dimensional random stacking	0.480
Spherical	Simple cube stacking	0.524	Rod shape L/D = 30	Three-dimensional random stacking	0.173
Spherical	Random and dense stacking	0.637	Rod shape L/D = 60	Three-dimensional random stacking	0.081
Cube	Random stacking	0.700	Fiber	Uniaxial hexagonal dense stacking	0.907

Equations (5.8)–(5.12) gives the relationship between the properties of blends and the corresponding properties of polymer components when the dispersed phase is "hard component" and the continuous phase is "soft component." Such blend systems include the blends of resin reinforced rubber, the blends of polymer with high elastic modulus in dispersed phase and polymers with low elastic modulus in continuous phase.

For blends with dispersed phase as "soft component" and continuous phase as "hard component," such as elastomer toughened plastics and low modulus and high toughness resin toughened plastics, the relationship between the properties of blends and the corresponding properties of their component polymers can be expressed as follows:

$$\frac{P_1}{P} = \frac{1 + A_i B_i \varphi_2}{1 - B_i \psi \varphi_2} \tag{5.13}$$

In this formula

$$A_i = \frac{1}{A}; \quad B_i = \frac{\frac{P_1}{P_2} - 1}{\frac{P_1}{P_2} + A_i} \tag{5.14}$$

The meanings of the symbols are the same as those defined earlier.

5.2.4 Relationship between properties of two-phase continuous structure blends and their components

Interpenetrating polymer network (IPN), block copolymers with similar content of two components and crystalline polymers all have two continuous morphologies. The relationship between the properties of such blends and the corresponding properties of their constituent polymers is commonly expressed as follows:

$$P^n = P_1^n \varphi_1 + P_2^n \varphi_2 \tag{5.15}$$

where the volume fractions of φ_1 and φ_2 are fractions of components 1 and 2, respectively; the properties of component polymers are P_1 and P_2, respectively, and n is a constant related to the system.

Another commonly used relationship is

$$\lg P = \varphi_1 \lg P_1 + \varphi_2 \lg P_2 \tag{5.16}$$

For example, for crystalline blends, they are generally composed of crystalline and amorphous phases. If the two phases are continuous, the shear modulus G_0 follows the relationship of eq. (5.15) and the n value is 1/2:

$$G_0^{1/2} = \varphi_1 G_1^{1/2} + \varphi_2 G_2^{1/2} \tag{5.17}$$

where G_0, G_1 and G_2 are the shear modulus of crystalline blends, crystalline phases and amorphous phases, respectively, and the volume fractions of φ_1 and φ_2 are the volume fractions of crystalline and amorphous phases, respectively.

The relationship between the dielectric constant of the blend and the dielectric constant of the component polymer can also be expressed by eq. (5.15), but n is generally taken as 1/3.

For the blends and block copolymers of the two polymers, when the composition of the blends and block copolymers is in the range of phase inversion, they have two interlocking phase structures, and the relationship between their properties and the corresponding properties of their components is appropriately expressed by eq. (5.16).

5.3 Glass transition of multicomponent polymers

5.3.1 Glass transition of polymers

The mechanical properties of polymer materials depend significantly on temperature. For amorphous polymers, the glass transition temperature (T_g) is an important index to measure their operating conditions. Many physical properties (such as thermal expansion coefficient, heat capacity, refractive index and electrical properties) have changed dramatically during the glass transition of polymers. For example, before and after the transition temperature range with only a few degrees difference, the modulus will change by three to four orders of magnitude. The material can change from a hard solid to a soft elastomer, which completely changes the performance of the material. As a polymer used in plastics, when the temperature rises to the glass transition temperature, it loses the properties of plastics and becomes rubber. As a material used in rubber, when the temperature drops to the glass transition temperature, it will lose the high elasticity of rubber and becomes hard and brittle. Therefore, the glass transition plays an important role in the mechanical properties of polymer materials.

There have been many debates about the nature of glass transition of polymers for a long time. Some people think that it is a thermodynamic phase transition and others think that it is not a true phase transition but a relaxation process. Therefore, many glass transition theories have emerged, such as free volume theory, thermodynamics theory, dynamics theory, modal coupling theory and solid model theory. Each theory can only solve some experimental phenomena in glass transition. At present, the more widely accepted theory is free volume theory.

The theory of free volume was first proposed by Fox and Flory, and the main work was done by Turnbull and Cohen. They believe that the volume of a liquid or solid substance consists of two parts: one is the volume occupied by molecules, called the occupied volume and the other is the volume not occupied, which is dispersed in the whole substance in the form of "holes," called "free volume." The existence of free volume provides the possibility for the conformation of polymer molecular chains to be adjusted by rotation and displacement. When the polymer is cooled, the free volume decreases gradually. At a certain temperature, the free volume reaches the lowest value and remains unchanged. At this point, the polymer enters the glass state and the chain motion freezes. Therefore, the glass state of polymer can be regarded as equal free state, and the glass transition temperature is the temperature at which the free volume reaches a certain critical value.

However, either theory recognizes that the glass transition temperature is time dependent, that is, the T_g value is related to the experimental time scale. If the experiment is carried out quickly and the time scale is short, the T_g will be high. Time and temperature are equivalent; the relationship between them can be quantitatively expressed by the activation energy of transformation. According to the activation energy values of the glass transition of general polymers, T_g changes about 7 °C for each order of magnitude of time scale.

5.3.2 Factors affecting glass transition of polymers

5.3.2.1 Effect of polymer chain compliance

In the chemical structure of polymer chains, the influence of chain flexibility on T_g is most important. The flexibility of polymer chains is caused by the internal rotation of a single bond in the main chain. The type of bonds in the main chain has a great influence on the internal rotation barrier. Ether bond $-O-$ is easier to rotate internally than $-C-C-$ bond, and $-Si-O-$ bond is easier to rotate internally. Therefore, polydimethylsiloxane has the lowest T_g (-123 °C) in common polymers. A double bond increases the internal rotation barrier. If there is a ring structure or trapezoidal polymer in the main chain, its internal rotation barrier is very large and T_g is very high. For example, the T_g of aromatic polyimide can reach above 300 °C. Large side groups, such as aromatic and tert-butyl groups, increase the steric hindrance of internal rotation and increase T_g, as shown in Tab. 5.3.

Tab. 5.3: The effect of substituent hindrance on T_g.

Polymer	Polyethylene	Polypropylene	Polystyrene	Poly (2,6-dichlorotoluene)	Polyvinyl carbazole
Lateral group	$-H$	$-CH_3$		Cl ... Cl	
T_g (°C)	-120	-13	81	167	208

When hydrogen on tertiary carbon atoms in the main chain is substituted by methyl or phenyl, the steric hindrance increases and T_g increases. For example, the T_g of α-methylstyrene increases to 180 °C, which is about 100 °C higher than that of polystyrene (PS). However, the increase of the length of aliphatic side groups will make the side base body more easily rotate inside, and increase the free volume of the polymer, thus improving the flexibility of the chain. For example, the influence of aliphatic ester group length on T_g in polymethacrylate polymers is shown in Tab. 5.4. Polytetrafluoroethylene is a rigid chain polymer because of the rigid helical conformation of the main chain caused by the substitution of fluorine atoms, so T_g is higher (126 °C).

Tab. 5.4: Effect of aliphatic ester group length on polymethacrylate T_g.

Ester type	T_g (°C)	Lateral base type	T_g (°C)
Methyl ester	105	N-Butyl ester	21
Ethyl ester	65	Hexyl ester	−5
Isopropyl ester	48	Octyl ester	−20
Propyl ester	35	N-Dodecyl ester	−65
Isobutyl ester	53		

5.3.2.2 Effect of intermolecular forces

Intermolecular forces can be expressed in terms of cohesive energy density. The higher the density of cohesion energy, the greater the intermolecular force and the higher the T_g. There are many factors affecting the intermolecular interaction force, such as the increase of molecular polarity, the increase of intermolecular interaction force and the increase of T_g. The side groups of polypropylene, polyvinyl chloride (PVC) and polyacrylonitrile are almost the same in size, but their polarity order is polypropylene < PVC < polyacrylonitrile. Therefore, T_g of polypropylene ($T_g = -18$ °C) < PVC ($T_g = 85$ °C) < polyacrylonitrile ($T_g = 101$ °C), respectively. Hydrogen bonding is a strong interaction between molecules, which makes T_g increase significantly. If polyacrylic acid has strong intermolecular hydrogen bond, its T_g (106 °C) is much higher than that of methyl polyacrylate (3 °C). Zinc polyacrylate has ionic bond and stronger intermolecular force, $T_g > 400$ °C.

The cohesive energy density of macromolecule is the sum of the contributions of each group. Therefore, an apparent T_{gi} can be determined for each group, and the T_g of polymer is:

$$T_g = \sum n_i T_{gi} \tag{5.18}$$

where n_i is the molar fraction of group i in polymer.

5.3.2.3 Effect of main chain symmetry
The symmetrical substitution of side groups tends to reduce the dipole moment, thus reducing the internal rotation barrier and decreasing the T_g. For example, the T_g of polypropylene is –20 °C, while that of polyisobutylene is – 70 °C, that of PVC is 87 °C and that of polyvinylidene chloride is – 19 °C.

5.3.2.4 Effect of relative molecular mass
T_g increased with the increase of relative molecular mass, but slowed down gradually, and finally reached the limit at the critical relative molecular mass. Expressed by formula as follows:

$$T_g = T_g^{\infty} - K/\overline{M}_n \tag{5.19}$$

K is the constant of the polymer, such as styrene $K \approx 1.75 \times 10^5$. The critical relative molecular mass is related to the flexibility of polymers, which can be understood as the relative molecular mass of chains. Polymers with high chain flexibility have low critical molecular weight. For example, the critical molecular weight of flexible polyisobutylene is about 1,000, equivalent to about 20 segments per chain. The critical molecular weight of rigid PS ranges from 12,000 to 40,000, which is equivalent to 100–400 segments per chain.

The effect of relative molecular weight on T_g can be explained by the increase of relative molecular weight and the decrease of the number of terminal groups per unit volume. Since the movement of the end group leaves free volume, the chain motion is prone to occur. Polymers with lower molecular weight contain relatively more end groups, so they can undergo glass transition at lower temperatures.

5.3.2.5 Effect of branching and cross-linking
The branching effect on the glass transition temperature of polymers has two aspects: after branching, the end groups of branched chains increase, the free volume increases, and the T_g decreases; however, the branching points restrict the movement of macromolecular chains, reduce the free volume and increase the T_g. Generally speaking, the former is dominant. However, when the polymer is highly branched, especially when star-shaped polymers are formed, the latter may dominate.

Cross-linking usually limits the movement of segments and increases T_g. When the cross-linking degree is low, T_g is only slightly higher than the polymer before cross-linking, such as vulcanized rubber. However, T_g of highly cross-linked materials such as epoxy resin and phenolic resin increased significantly. In addition, because the chemical composition of cross-linking agents is generally different from that of polymers, cross-linking also produces copolymerization effect, which may increase or decrease T_g.

Nielsen processed the literature data and obtained the empirical equation of the relationship between T_g and the number–average relative molecular mass of the cross-linking point:

$$T_g - T_{g0} \approx \frac{3.9 \times 10^4}{M_c} \tag{5.20}$$

where M_c is the number–average relative molecular mass between cross-linking points; T_{g0} is the glass transition temperature of polymer containing cross-linking agent but not yet cross-linked. Therefore, T_g-T_{g0} only reflects the effect of cross-linking, but does not consider the effect of copolymerization.

Dibenedetto summed up the empirical formula equation as follows:

$$\frac{T_g - T_{g0}}{T_{g0}} \approx \frac{KX_c}{1 - X_c} = \frac{2K}{n_c} \tag{5.21}$$

where X_c is the molar fraction of the monomer chain segments with cross-linking points, and n_c is the number of average atoms between cross-linking points. The constant K is between 1.0 and 1.2. It is a function of ratio of activity of the cross-linked front chain after cross-linking and is related to the cohesive energy density before and after cross-linking. $\frac{X_c}{1-X_c}$ can be used to characterize the cross-linking degree of polymers. For vinyl polymers, each monomer unit has two main chain carbon atoms:

$$\frac{X_c}{1 - X_c} \approx \frac{2}{n_c} = \frac{M_0}{M_c} \tag{5.22}$$

M_0 is the relative molecular weight of the monomer in the formula.

5.3.2.6 Effects of solvents and plasticizers

The T_g of polymers decreases with the addition of plasticizers, solvents or low molecular monomers. This phenomenon is called plasticization. Plasticizing can turn hard and brittle materials into soft and tough materials, which has great practical value. Nielsen proposed a formula for estimating volume fraction:

$$T_g = \varphi_p T_{gp} + \varphi_d T_{gd} \tag{5.23}$$

where p and d represent unplasticized polymers and plasticizers, respectively; φ is the volume fraction. T_{gd} generally ranges from −50 to −100 °C. An empirical constant K can also be introduced to obtain more accurate expressions:

$$-T_g = \frac{T_{gp} + (KT_{gd} - T_{gp})\varphi_d}{1 + (K-1)\varphi_d} \tag{5.24}$$

$$K \approx \Delta\alpha_d / \Delta\alpha_p \tag{5.25}$$

where $\Delta\alpha$ is the difference of volume expansion coefficients of pure components in liquid and glassy states. $\Delta\alpha \approx 4.8 \times 10^{-4}$ for most polymers. If $K = 1$, then eq. (5.24) becomes eq. (5.25).

5.3.2.7 Effect of crystallization and orientation

Crystallization and orientation generally increase the intermolecular force, which may lead to the increase of T_g. However, the effects of crystallization and orientation are complex, and some experiments show opposite effects. The crystalline region, like the breaking point, restricts the movement of the amorphous segments; the amorphous segments of the macromolecular chains passing through the two wafers are straightened and their mobility is reduced. These factors increase T_g. But outside the crystalline region, the components and impurities with low molecular weight tend to concentrate in the amorphous region, which plays a plasticizing role and reduces T_g.

T_g of polyethylene terephthalate, isotactic PS and isotactic polymethyl methacrylate (PMMA) increased with the increase of crystallinity; T_g of polypropylene and polytetrafluoroethylene did not change with the crystallinity but T_g decreased with the increase of crystallinity of poly-4-methyl-1-pentene. Some experiments show that orientation decreases T_g in the direction parallel to orientation, while orientation increases T_g in the direction perpendicular to orientation.

5.3.3 Glass transition of multicomponent polymers

For the relationship between glass transition temperature and composition of multicomponent polymers, there are some quantitative formulas which can be approximated. The most commonly used ones are

$$T_g = \varphi_A T_{gA} + \varphi_B T_{gB} \tag{5.26}$$

In formula φ_A and φ_B are the volume fractions of polymer components A and B, respectively. The T_g of compatible multicomponent polymers basically corresponds to the relationship of eq. (5.26) (line 1 in Fig. 5.1).

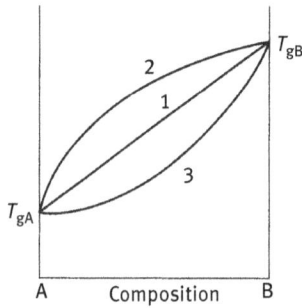

Fig. 5.1: T_g-composition relation of multicomponent polymers.

Such systems as cellulose nitrate/methyl polyacrylate and natural rubber/*cis*-butadiene rubber belong to this category. However, there are some compatible multicomponent polymers whose T_g-composition relationship does not conform to the linear relationship, but presents positive or negative deviations from the linear one, as shown in the curves 2 and 3 in Fig. 5.1. Generally, the negative deviation is more serious. Polymer plasticizers belong to this category. Their T_g composition can be expressed by the Fox equation:

$$1/T_g = W_A/T_{gA} + W_B/T_{gB} \tag{5.27}$$

where W is the mass fraction. The blends of PVC/NBR (nitrile butadiene rubber) and bisphenol A polyhydroxyether/polycaprolactam were in good agreement with eq. (5.27). Some compatible multicomponent polymers cannot be described by eqs. (5.26) and (5.27), but conform to the more complex Cordon–Taylor equation:

$$T_g = [W_A/T_{gA} + K(1 - W_B/T_{gB})]/[W_A + K(1 - W_A)] \tag{5.28}$$

where $K = \triangle \alpha_B / \triangle \alpha_A$. $\triangle \alpha$ is the difference of thermal expansion coefficients of polymers in glass and rubber states. Equation (5.27) is a special case of eq. (5.28) when $K = 1$. The relationship between T_g-composition of PMMA/polyvinyl acetate (PVAc) and styrene–butadiene rubber (SBR)/polybutylene (PB) systems is basically consistent with eq. (5.28).

When there is strong interaction between two polymer molecules in compatible multicomponent polymers, the T_g-composition relationship usually conforms to curve 2 in Fig. 5.1, such as polyvinyl nitrate/PVAc and polyvinyl nitrate/ethylene vinyl acetate systems. The T_g-composition relationship of such polymer alloys can be expressed by the following formula:

$$T_g = \varphi_A T_{gA} + \varphi_B T_{gB} + k\varphi_A\varphi_B \tag{5.29}$$

where K is the correction constant when miscibility is not ideal.

5.3.3.1 Characteristics of glass transition of multicomponent polymers

Figure 5.2 shows the glass transition characteristics of different types of multicomponent polymers. Single homopolymers and irregular copolymers usually have one or more secondary glass transitions, but usually only one primary glass transition. When two incompatible polymers are mixed, the glass transition of their homopolymer remains in both microdomains. Therefore, most of the blends, block copolymers, graft copolymers and IPN exhibit two primary glass transition. With the increase of compatibility between the two components, the glass transition peak area will become larger or the two transition temperatures will be closer or both will be larger and closer. At this time, there are two maximum values in the mechanical loss spectrum, and two turning points can be observed by dilatometer.

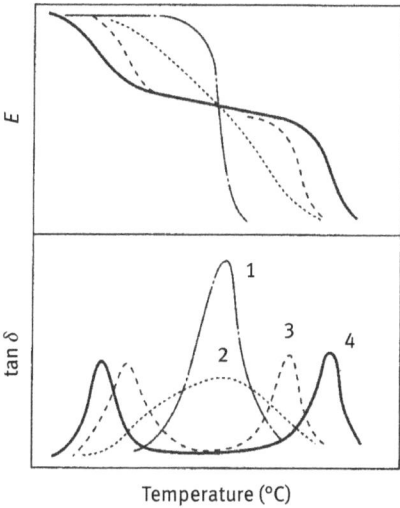

Fig. 5.2: Dynamic mechanical spectrum of polymer blends. 1, Completely compatible blends; 2, basic compatible blends; 3, partially compatible blends; 4, completely incompatible blends.

Qi Zongneng et al. studied the blends of polydimethylsiloxane (rubber) and polymethylphenylsiloxane block copolymers with the same composition by dynamic mechanical method. There are two primary glass transition in block copolymers and blends, and block copolymers or blends have little effect on T_g of rubber phase, but rubber components have more effect on T_g of plastic phase. This is because the T_g of the plastic phase is much higher than that of the rubber phase, and the molecular motion of the plastic component is frozen near the glass transition temperature of the rubber phase, while the molecular motion energy of the chain segment of the rubber component is affected near the glass transition temperature of the plastic phase.

By comparing the glass transition properties of blends and random copolymers, we can further understand their structural differences. For example, there are two transitions in the blends, and their exact location and width reflect the degree of miscibility. Random copolymers usually have only one transition. Figs. 5.3 and 5.4

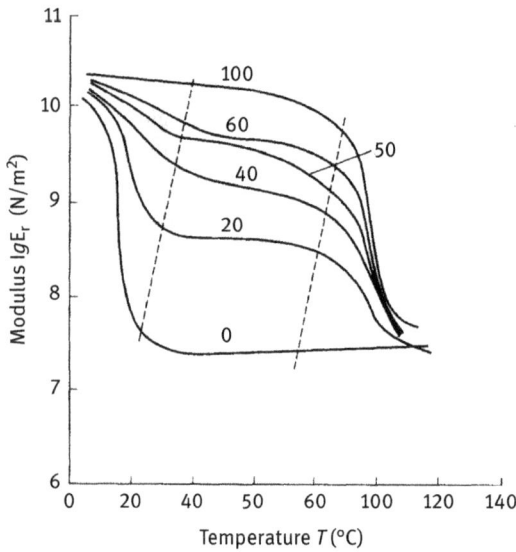

Fig. 5.3: Modulus–temperature relationship of polystyrene/styrene–butadiene rubber (*S*/*B* mass ratio = 30/70) blends. (The figure on the curve is the mass percentage of polystyrene.)

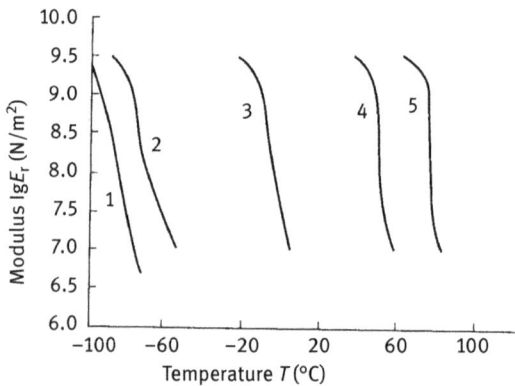

Fig. 5.4: Modulus–temperature relationship of random copolymer series of styrene, butadiene and styrene. 1, *S*/*B* = 0/100; 2, *S*/*B* = 20/80; 3, *S*/*B* = 50/50; 4, *S*/*B* = 80/20; 5, *S*/*B* = 100/0.

show the temperature-modulus characteristics of PS/SBR blends and random co-polymers of styrene and butadiene, respectively. The total *S*/*B* ratio in the two graphs is approximately the same. Figure 5.3 shows that both transitions are related to composition. As shown in the figure, the high temperature transition is not obvious in the blends with high SBR content, which indicates that the phase separation

between PS and SBR is not complete. On the contrary, a series of random copolymers in Fig. 5.4 behave more like homopolymers, with only one transition.

5.3.3.2 Free volume model for glass transition of multicomponent polymers

Takayanagi et al. proposed a free volume model for glass transition of polymer
blends based on the free volume theory of glass transition. The basic point is that
the chain motion of polymer is realized. Many small volume elements with free volume of several nanometers must be provided in the blends. For compatible polymer
blends, the free volume of the system is communal. All the small volume elements
are suitable for the chain motion of two polymers simultaneously; for process-
compatible systems or mechanically compatible systems, the free volume is partially communal, and only the local region (interfacial layer) of the free volume in
the blends is communal. Therefore, fully compatible systems have the same free
volume fraction (f), while process-compatible blends still have different free volume
fraction. Figure 5.5 is a schematic diagram of the free volume fraction distribution
function of various blends. The PM curve in the figure is a mechanically compatible
system, and the free volume fractions of the two components are f_A (P) and f_B (P),
respectively. The PFM curve is a completely compatible system with only one free
volume fraction (f), the S curve is a completely incompatible system, and the MH
curve is a compatible system. The compositions of all the curves in the figure are
equal, all are 50%.

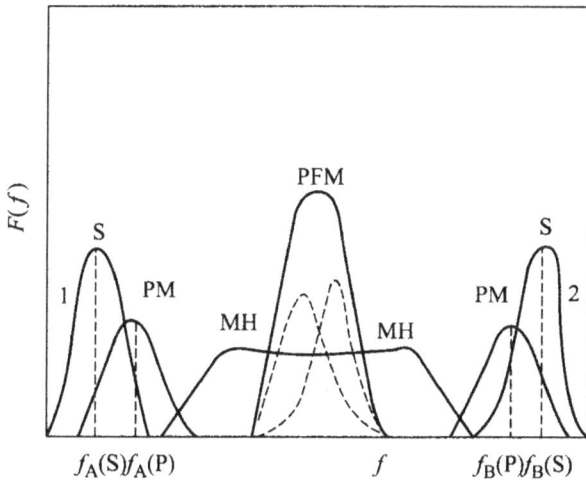

Fig. 5.5: $F(f)$ diagram of free volume fraction distribution function for various blends.

Using the theory of free volume model, various curves in the dynamic mechanics spectrum of Fig. 5.2 can be explained better. Because the two components of compatible

multicomponent polymer system have the same free volume fraction, the free volume fraction of glass transition is also the same. In this way, the two polymers will inevitably achieve chain motion or freezing at approximately the same temperature, and there will only be one loss peak in the dynamic mechanical spectrum.

For mechanically compatible multicomponent polymers, such as rubber-toughened plastics systems, because the two components are compatible to a certain extent, some molecules or segments of plastic phase permeate into the rubber phase, and some molecules or segments of rubber phase must also be mixed in the plastic phase. Since rubber is in a high elastic state, the free volume fraction of its macromolecule f_B (P) is much larger than that of the plastic phase f_A (P) in the glass state. Thus, in mechanically compatible blends, the rubber phase f_B (P) is naturally smaller than the free volume fraction f_B (S) of one component rubber. Similarly, because some rubber molecules are mixed into the plastic phase, the f_B (P) of the plastic phase is larger than the f_A (P) of the one-component plastics and the free volume fraction f_g of different polymers during glass transition are almost the same, that is to say, f_g is a fixed value. Therefore, in order to achieve the f_g value of the polymer component with lower free volume fraction, the glass transition temperature of the polymer must be higher than that of the original polymer. On the contrary, T_g decreased with the increase of f_A (P). This can explain that T_g of rubber phase in mechanical compatibility system occurs in higher temperature region than that of single component rubber, and the T_g of plastic phase in blends moves to lower temperature region than that of single component plastic. Moreover, as the compatibility of the two polymer components increases, the two T_g become closer and merge into one T_g when they are fully compatible.

5.4 Elastic modulus of polymer alloys

5.4.1 Modulus of elasticity and Poisson's ratio of polymer materials

The elasticity modulus of the material characterizes the ability of the material to resist deformation caused by external forces. It is the mechanical behavior of materials under short-term stress and small strain conditions. Generally, it is linear elastic behavior, that is, there is a linear relationship between stress and strain. Polymer blends also exhibit linear elastic behavior under short-term stress and small strain.

There are three basic types of elastic modulus of polymer materials, namely Young's modulus (tensile modulus) E, shear modulus G and bulk modulus B. G reflects the characteristics of changes in the shape of the material or product, B reflects the characteristics of changes in volume and E reflects the characteristics of changes in shape and size. In tension test, the ratio of transverse strain to longitudinal strain is expressed by v, which is called Poisson's ratio, and also reflects the volume change of the material during tension.

For isotropic materials, the relationships among the four parameters E, G, B and v are as follows:

$$E = 2G(1+v) = 3B(1-2v) \tag{5.30}$$

This shows that only two of the four parameters E, G, B and v are independent, that is to say, only two of them are known, which is sufficient to describe the elastic behavior of isotropic materials.

The bulk expansion occurs in most materials during tension, with v values ranging from 0.2 to 0.5. The v values of rubber and small molecule liquids are close to 0.5 (Tab. 5.3). However, some special polymer materials have negative Poisson's ratio. For example, in 1987, Lakes found that the Poisson's ratio of polyurethane foams with porous structure is −0.7.

For isotropic materials with Poisson's ratio of 0.5, the relationship between tensile modulus and shear modulus is as follows:

$$E = 3G \tag{5.31}$$

Tab. 5.5: Poisson's ratio of some materials.

Materials	Poisson's ratio	Materials	Poisson's ratio
Zinc	0.21	Glass	0.25
Steel	0.25–0.33	Stone material	0.16–0.34
Copper	0.31–0.34	Polystyrene	0.33
Aluminum	0.32–0.36	LDPE	0.38
Lead	0.45	Celluloid	0.39
Mercury	0.50	Rubber	0.49–0.50

5.4.2 Estimation of elastic modulus of polymer alloys

The simplest method for estimating the elastic modulus of blended modified plastics is according to the mixing rule:

$$M_c = \varphi_1 M_1 + \varphi_2 M_2 \tag{5.32}$$

$$\frac{1}{M_c} = \frac{\varphi_1}{M_1} + \frac{\varphi_2}{M_2} \tag{5.33}$$

where M_c, M_1 and M_2 are the modulus of elasticity of blends, component 1 and component 2, and φ_1 and φ_2 are the volume fractions of components 1 and 2.

Equation (5.32) gives the upper limit of elastic modulus of polymer alloys and formula eq. (5.33) gives the lower limit.

When the component with larger elastic modulus is continuous phase and the component with smaller elastic modulus is dispersed phase, the elastic modulus of the blend is close to eq. (5.32) and when the component with smaller elastic modulus is continuous phase and the component with larger elastic modulus is dispersed phase, the elastic modulus of the blend is close to eq. (5.33).

The elastic modulus of PS/polybutadiene alloy varies with composition as shown in Fig. 5.6.

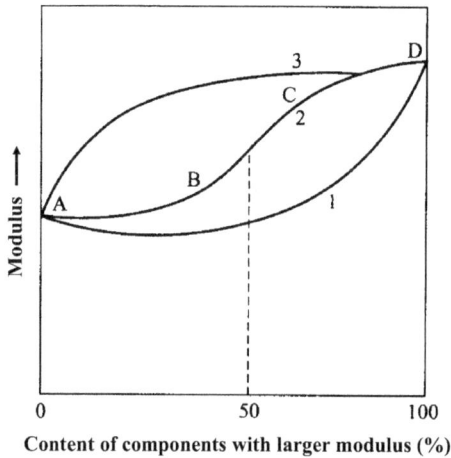

Fig. 5.6: Schematic diagram of elastic modulus of polystyrene/polybutadiene polymer alloy changing with composition.

Curves 1 and 3 show the lower and upper limits of elastic modulus of polymer alloys, respectively. Curve 2 shows the measured values of elastic modulus of polymer alloys. In AB region, the component with smaller elastic modulus is continuous phase, in CD region, the component with larger elastic modulus is continuous phase and in BC region, the phase inversion region of macromolecule alloys is represented. As shown in Fig. 5.6, the results of eqs. (5.32) and (5.33) are quite different from the actual ones, so many scholars have revised them.

Hashin proposed two approximate formulas for the upper and lower limits of elastic modulus:

$$G_c = G_2 + \varphi_1 \left[\frac{1}{G_2 - G_1} + \frac{6(B_2 + 2G_2)\varphi_2}{5G_2(3B_2 + 4G_2)} \right]^{-1} \tag{5.34}$$

$$G_c = G_1 + \varphi_2 \left[\frac{1}{G_2 - G_1} + \frac{6(B_1 + 2G_1)\varphi_2}{5G_1(3B_1 + 4G_1)} \right]^{-1} \tag{5.35}$$

For two consecutive blends, the relationship between the elastic modulus and the elastic modulus of the component polymer can be expressed by eq. (5.17) as described earlier:

$$G_0^{1/2} = \varphi_1 G_1^{1/2} + \varphi_2 G_2^{1/2} \tag{5.17}$$

5.4.3 Mechanical model of polymer blends

In order to quantitatively or semiquantitatively describe the elastic modulus and mechanical relaxation properties of polymer blends, Kawai Hiromichi et al. have developed mechanical models of two-phase polymer systems.

The mechanical model shown in Fig. 5.7 can be used to describe the blends of two incompatible polymers.

In the figure, P_1 and P_2 are components 1 and 2, and λ and φ are volume fractions of component 1 in parallel and series models, respectively. (1) represents the parallel model, which is an isomorphic system and (2) is a series model, which is an isostress system. If Young's modulus of components 1 and 2 are E_1 and E_2, respectively, Young's modulus E of the blends can be calculated according to the model.

For the parallel model:

$$E = (1 - \lambda)E_1 + \lambda E_2 \tag{5.36}$$

Similarly, for multicomponent parallel models:

$$E = \sum_{i=1}^{N} \lambda_i E_i, \quad \sum_{i=1}^{N} \lambda i = 1 \tag{5.37}$$

where N is the component of the blend.

For the series model

$$E = \left(\frac{1 - \phi}{E_1} + \frac{\phi}{E_2} \right)^{-1} \tag{5.38}$$

Series model for multicomponent blends:

$$E = \left(\sum_{i=1}^{N} \frac{\phi_i}{E_i} \right)^{-1}, \quad \sum_{i=1}^{N} \phi_i = 1 \tag{5.39}$$

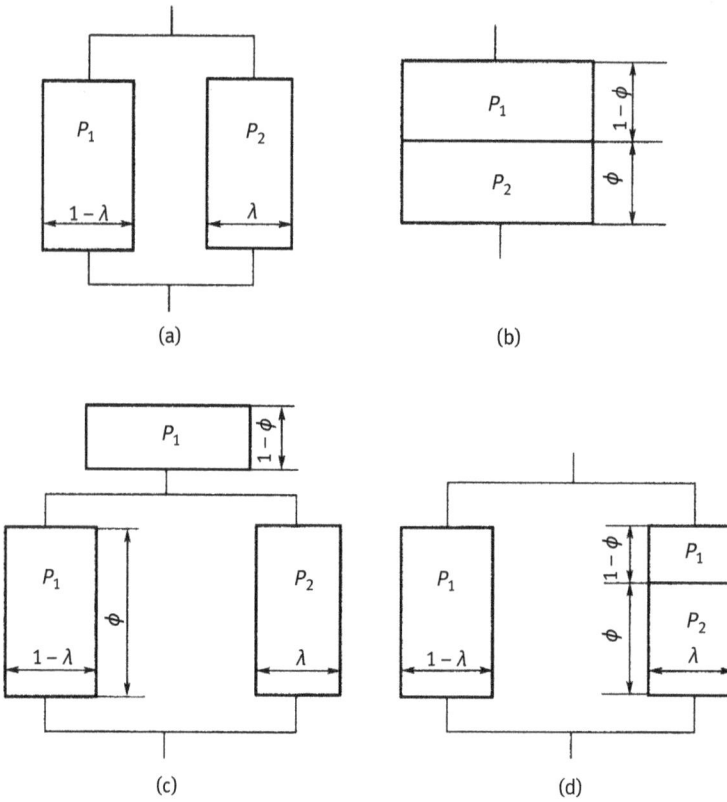

Fig. 5.7: Mechanical model of two phase polymer system.

Figure 5.7(a) and (b) shows two basic models. To further approximate the actual blends, two basic models can be further combined in different forms, and two combinations can be obtained as shown in Fig. 5.7(c) and (d). From (c) and (d) models, the Young's modulus of the blends can be obtained as follows:

$$E = \left[\frac{\phi}{\lambda E_1 + (1-\lambda)E_2} + \frac{1-\phi}{E_1} \right]^{-1} \tag{5.40}$$

and

$$E = \lambda \left(\frac{\phi}{E_1} + \frac{1-\phi}{E_2} \right)^{-1} + (1-\lambda)E_2 \tag{5.41}$$

In deriving these formulas, it is assumed that the polymer obeys Hooke's law, $\sigma = E\varepsilon$.

The above model assumes that there is no interaction between components, so it can only be applied to polymer blends with incompatible components. In fact, the components of the blend have mutual influence. When there is partial compatibility between two polymer components, there are concentration gradients of two polymer components in the blend, especially in the interface region. At this time, the blend is just like a system consisting of a series of copolymers with changing composition and properties. The mechanical properties of the blend can be described by the two models shown in Fig. 5.8.

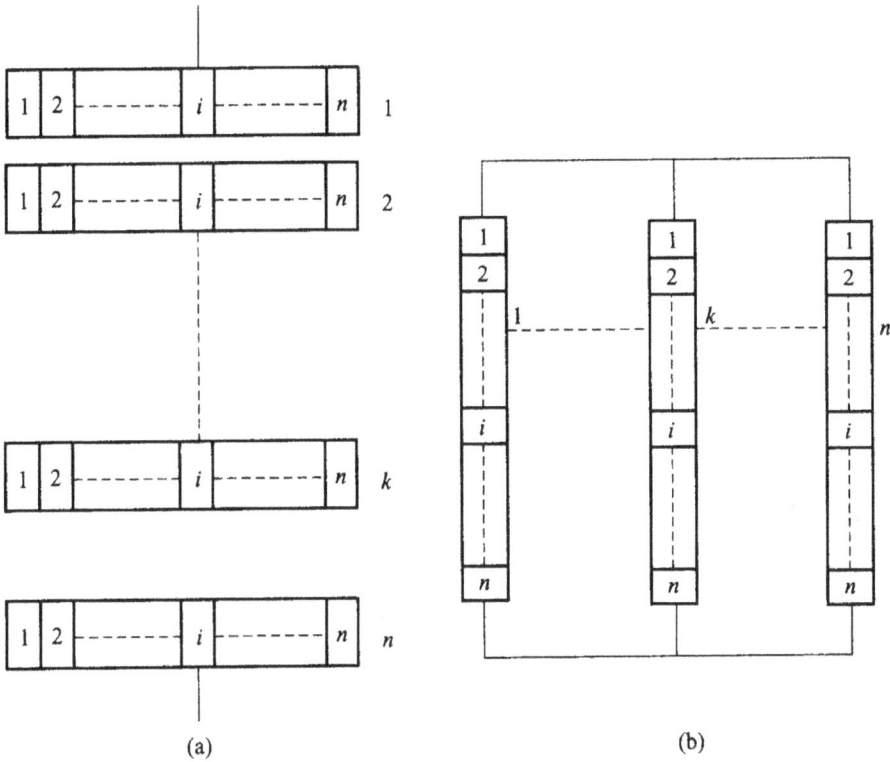

Fig. 5.8: Mechanical model of two-component partially compatible polymer blends: (a) series model and (b) parallel model.

For the series model, there are

$$\frac{1}{E} = \sum_{k=1}^{N} \lambda_k \left(\sum_{i=1}^{N} \lambda_i E_i \right)^{-1} \tag{5.42}$$

For parallel models

$$E = \sum_{k=1}^{N} \lambda_k \left(\sum_{i=1}^{N} \frac{\lambda_i}{E_i} \right)^{-1} \tag{5.43}$$

In the formula, λ_k is the weight of the contribution of each organizational unit k to the modulus E of the blend, and λ_i is the weight of the contribution of the i segment of each organizational unit to the modulus of the blend.

As such blends can be considered as heterogeneous systems of multicomponents, each component unit exhibits mechanical relaxation properties of a homogeneous polymer. Therefore, according to the above model, the mechanical relaxation properties of the blends can be expressed in Fig. 5.9. As can be seen from the graph, the mechanical relaxation time spectrum is greatly broadened.

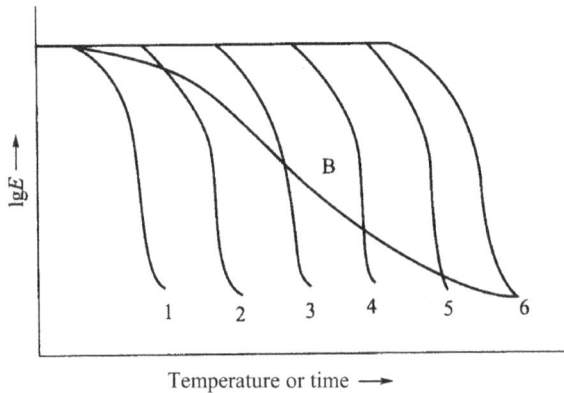

Fig. 5.9: Schematic diagram of modulus–temperature (time) relationship.
1–6: curves of six random copolymers with different compositions and B: copolymers consisting of 1–6 random copolymers.

According to the above model, the Williams-Landel-Ferry (WLF) equation of the blends can also be more clearly understood. For homogeneous polymers, the relaxation modulus $E(t)$ can be expressed by an improved Rouse function in the glass transition region.

$$E(t) = \frac{E_1 \tau_m^{1/2}}{\tau_m^{1/2} + t^{1/2}} + E_R \tag{5.44}$$

where t is time, τ_m is the minimum relaxation time of polymer, E_R is the modulus of polymer in rubber state (generally 4×10^{10} Pa) and E_1 is the modulus of polymer in glass state (generally 10^9–10^{10} Pa).

For blends, when the two components are partially compatible, they can be regarded as a heterogeneous system consisting of a series of different components, so eq. (5.44) can be improved as follows:

$$E(t) = E_1 \sum_{i=1}^{N} \frac{W_i \tau_{m \times i}^{1/2}}{\tau_{m \times i}^{1/2} + t^{1/2}} + E_R \qquad (5.44)$$

where W_i is the weight fraction of the part composed of i, $\tau_{m \times i}$ is the minimum relaxation time of the part composed of i and W_i meets the normalization condition:

$$\sum_{i=1}^{N} W_i = 1 \qquad (5.46)$$

The interpenetrating network system composed of 50/50 polyethyl acrylate/PMMA is basically in conformity with eq. (5.45). However, many blends do not conform to this formula.

Generally speaking, the time–temperature equivalent principle is not applicable to heterogeneous blends because the mobility factor α_T of the two polymers is different at the measured temperature. However, if the modulus of the two polymers varies greatly at the measured temperature, the time–temperature conversion principle can be approximated. For example, acrylonitrile–butadiene–styrene (ABS) resin, in which the T_g of rubber phase (polybutadiene) is very low, is in a highly stretched and elastic state at the measured temperature and has little contribution to stress relaxation. Therefore, the mechanical relaxation behavior of ABS is similar to that of acrylonitrile/styrene random copolymers, which basically conforms to WLF equation. Nevertheless, it can still be seen that the glass transition zone is obviously widened.

5.5 Stress relaxation of polymer alloys

5.5.1 Stress relaxation properties of blends

The stress relaxation of polymer refers to the phenomenon that the stress in polymer decreases with the increase of time at constant temperature and constant deformation, which is essentially the result of the movement of polymer molecules from unbalanced state to equilibrium state. In solid polymers, when a polymer chain is relaxed, it will affect other adjacent polymer chains to relax at the same time, and may have a synergistic effect. This in turn affects the relaxation characteristics properties of the polymer chain. Therefore, when the adjacent chains are the same, different or mixed chains, the stress relaxation characteristics may be different. Thus, the degree of mixing or separation of molecular chains of different blends can be revealed by stress relaxation test.

Figure 5.10 is a stress relaxation curve for a pair of incompatible polymers, while Fig. 5.11 is a stress relaxation curve for a pair of partially compatible polymers. From the stress relaxation curves of PVAc/PMMA blends in Fig. 5.10, two regions of rapid relaxation can be seen. The first one is near 35 °C, and the second

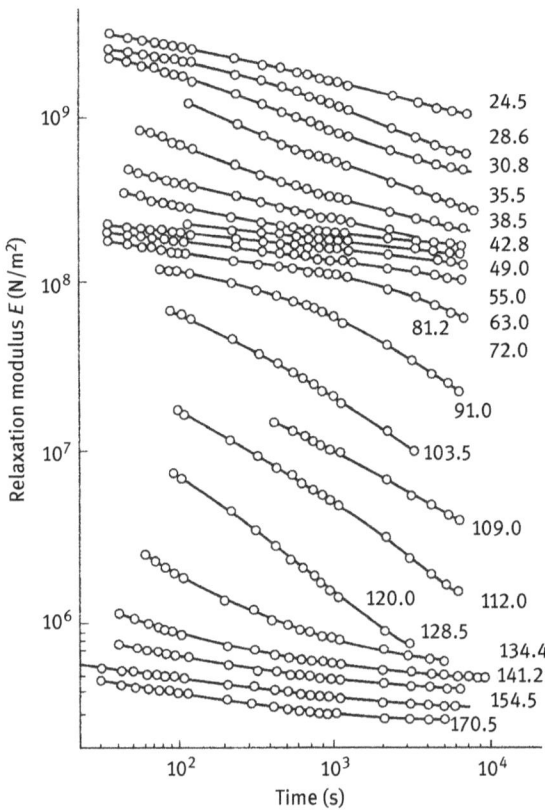

Fig. 5.10: Stress relaxation curves of PVAc/PMMA(50/50) blends. (The number on the right side is temperature, in units of °C.)

one is near 109 °C. These two relaxations correspond to the glass transition of PVAc and PMMA, respectively. Although PVAc and PMMA accounted for 50% of the blends, the low temperature transformation was not obvious. This indicates that the dominant or more continuous phase in the blend has a significant change. Importantly, the presence of two distinct transformations means that each polymer molecule is surrounded mainly by the same kind of molecule, thus proving that the two are incompatible.

Figure 5.11 shows that the stress relaxation curves of PVC/butadiene–acrylonitrile copolymer (PVC/NBR) blends are basically parallel lines. They have a wide relaxation time spectrum, indicating that there is no phase separation.

Experiments show that the relaxation characteristics of compatible systems can be predicted accurately by using WLF equation when the temperature is limited to one transition region. For incompatible blends with two transitions, satisfactory results can still be obtained if the WLF equation is applied to each transition.

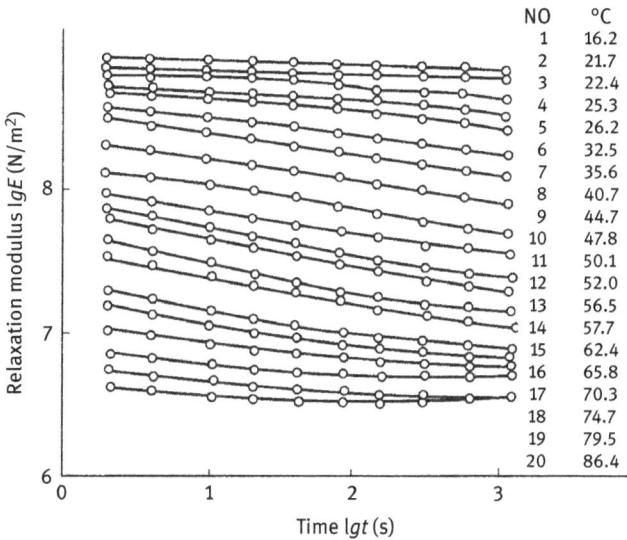

Fig. 5.11: Stress relaxation curves of PVC/NBR(77/23) blends. (The number on the right side is temperature, in units of °C.)

If the polymer blends have the same chemical compositions and different contents of the dispersed phase, the structure and relaxation characteristics of the corresponding microdomains are also different. The results show that the loss peak strength of the continuous phase component is high and that of the dispersed phase component is relatively low. The loss peak increases with the increase of dispersed phase content. Figure 5.12 is the dynamic mechanical spectrum of high-impact polystyrene (HIPS). Curves 3 and 4 in the graph are graft copolymerized HIPS. Curve 3 contains 5% PB and curve 4 contains 10% PB. The higher the PB component content, the higher the peak value of low temperature loss (curve 4 is higher than curve 3). Due to the high content of components, the energy dissipated by the relaxation of their molecular chains is correspondingly higher. Curves 4 and 2 contain 10% PB, but curve 2 represents mechanical blending HIPS and curve 4 represents graft copolymerized HIPS. The latter has honeycomb structure in the dispersed phase microregion. There is a certain amount of continuous phase component PS in the microregion. The volume fraction of microregion is much larger than that of mechanical blending with the same content. Therefore, it is more accurate to say that the higher the volume fraction of the dispersed phase, the higher the intensity of the loss peak, and the position of the peak also moves to high temperature.

Another important characteristic of mechanical relaxation properties of polymer blends is the broadening effect of mechanical relaxation time spectrum. Generally, the time range of glass transition zone of polymers on the time–temperature combination curve is about 109 s, while that of blends can reach 1,016 s. This broadening

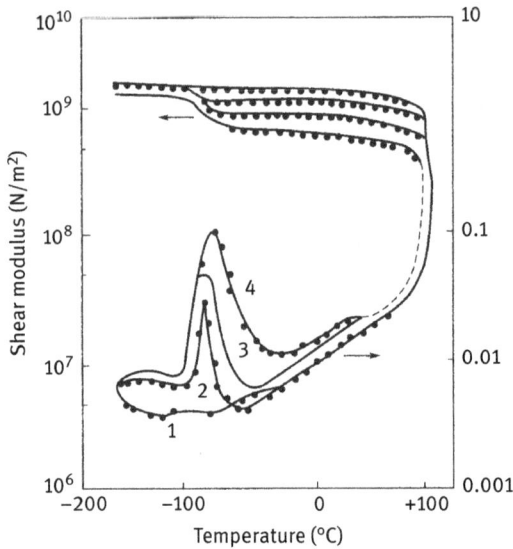

Fig. 5.12: Dynamic mechanical spectrum of HIPS.1, PS; 2, mechanical blending HIPS (containing l0%PB); 3, graft copolymerization of HIPS (containing 5% PB); 4, graft copolymerization of HIPS (containing 10% PB).

effect of mechanical relaxation time spectrum makes multicomponent polymers have better damping properties. For example, the damping effect of some IPN systems is very prominent, and they are very suitable for use as shock-proof and sound-proof materials. In fact, the broadening effect of mechanical relaxation reflected in the dynamic mechanical spectrum of multicomponent polymers (Fig. 5.12) is essentially consistent with the broadening effect of relaxation time spectrum. Two or more polymer molecules exist in the same system. Even though they are compatible to a great extent, they are polymers with different chain structures, so they have their own relaxation characteristics. The relaxation time spectrum and the widening of dynamic viscoelastic region are the results of two or more polymer molecules moving in synergy to a certain extent but not in complete synchronization.

5.5.2 Stress relaxation properties of grafted and block copolymers

In addition to the relaxation transition characteristics of general blends, graft copolymers or block copolymers often exhibit new and weaker relaxation transition corresponding to the relaxation transition of two homopolymers. These transformations are hard to see in mechanical blends, and they are characteristic reflections of the microphase separation structure of these copolymers.

Due to the diversity of molecular motion units, the relaxation mechanisms of polymers are also diverse. For amorphous linear polymers, there are at least four relaxation modes: local relaxation of molecular chains, relaxation caused by short and long chain segments and relaxation related to the whole molecular chain flow. Therefore, the relaxation principal curves of polymers can be decomposed into four relaxation moduli according to four relaxation mechanisms. Figs. 5.13 and 5.14 are the relaxation curves of one-component methyl polyacrylate (PMA) and PS corresponding to each relaxation mechanism, respectively. Figs. 5.15 and 5.16 are relaxation curves of PMA-g-PS copolymers with different degree of grafting. From this set of graphs, it is clear that there are two relaxation processes (GB and DF) which are not found in Figs. 5.12 and 5.13. GB is a weak relaxation between the T_g of two homopolymers in the graft copolymer. There is no such relaxation in the blends of PMA and PS. It is unique to copolymer systems with microphase separation structure, especially when the hard segment is in the dispersed phase and the soft segment is in the continuous phase. The occurrence of GB relaxation is related to the interfacial phase characteristics of microphase separation structures. DF relaxation is a relaxation process associated with the flow behavior of macromolecule chains. It is not caused by the real flow between molecules, but only by the structural flow of microregions composed of hard chains after "melting."

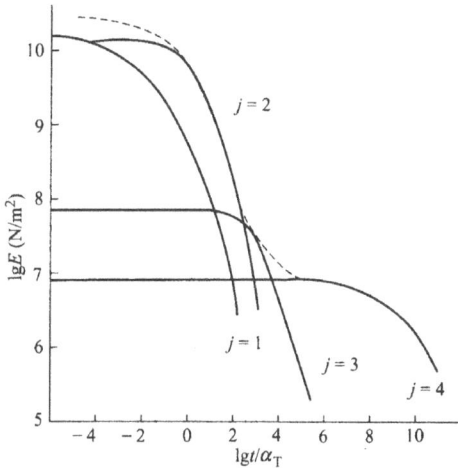

Fig. 5.13: Total apparent relaxation curve of methyl polyacrylate.

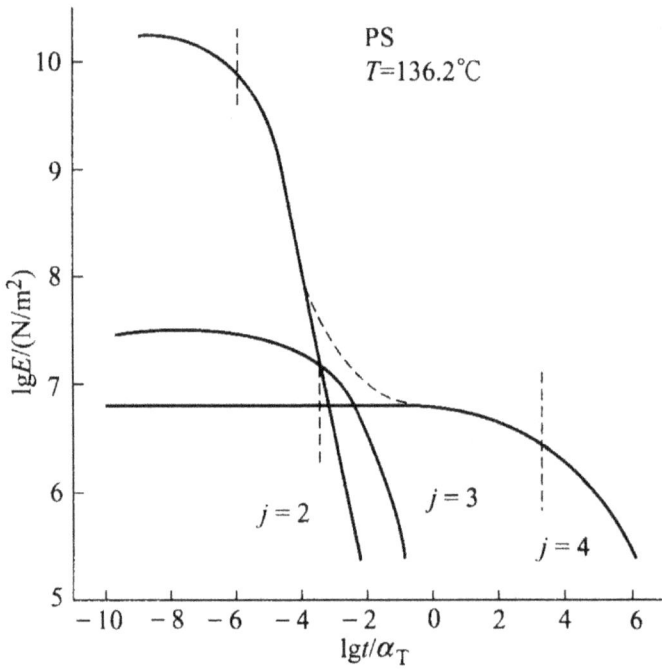

Fig. 5.14: Total apparent relaxation curve of polystyrene.

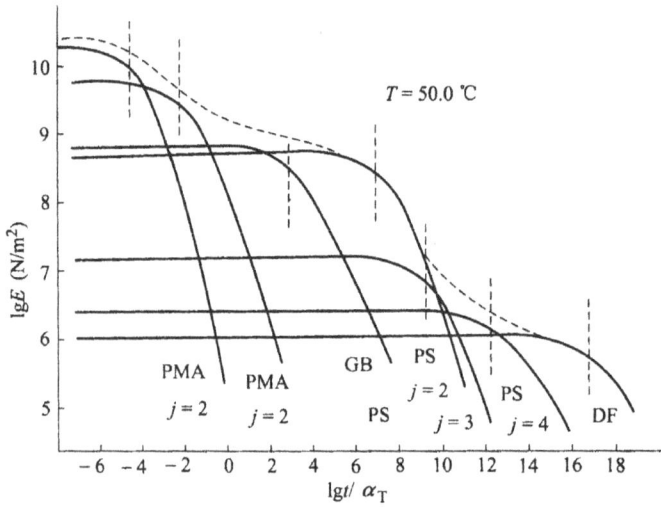

Fig. 5.15: Total apparent relaxation curve of PMA-g-PS.

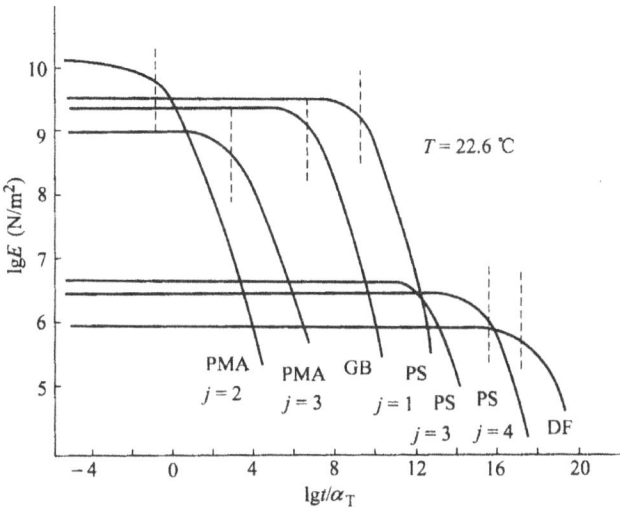

Fig. 5.16: Total apparent relaxation curve of PMA-g-PS decomposed into several relaxation modulus functions.

5.6 Deformation of polymer alloys

5.6.1 Mechanical strength and large deformation of polymer materials

The mechanical strength of polymer materials mainly includes tensile strength, tensile yield strength (yield stress), bending strength and elongation. Due to the viscoelastic properties of polymer materials, their mechanical strength depends not only on the structure of materials but also on the temperature and the speed of external forces. Sometimes, its mechanical properties will change greatly in a not too wide temperature range or a small external force acting speed range. The reason is that the polymer materials can present different mechanical states at different temperatures or external forces.

The mechanical strength of polymer alloys is related to the stress–strain relationship under large stress and large deformation. In this case, the linear relationship no longer exists, and there is a very complex relationship. The mechanical behavior is closely related to the mechanical state and transformation of polymer materials when the external force is large and the deformation is large.

In order to understand the mechanical properties of polymer alloys under large stress and strain, the mechanical behavior of pure polymers under large stress and strain is discussed first.

When the polymer sample is stretched under suitable conditions, it will yield to neck (local large deformation to form fine neck), that is to say, local large deformation will occur. A large number of experimental facts show that there are two mechanisms of polymer large deformation: shear deformation and crazing. In many cases, shear deformation and crazing coexist, and their respective proportions are related to the structure and field conditions of polymers.

The so-called shear deformation refers to the high orientation of macromolecule or macromolecule microaggregates in some planes under external forces, resulting in distortion deformation without obvious volume change. There are two kinds of shear deformation: diffuse shear deformation, that is, large-scale shear deformation occurring in the whole stress region, and shear band, that is, shear deformation occurring in the local regions.

Crazing refers to the void stripe-shaped deformation zone caused by stress concentration in some weak parts of polymer under tensile force, which is called craze, and this phenomenon is called crazing.

Crazes are different from cracks. The latter is empty and has no mass, while the former is not empty and has finer structure inside, so it shows a silver–white luster under light.

The shear deformation or craze of the specimen can be understood by the stress analysis of the polymer under uniaxial tension as shown in Fig. 5.17.

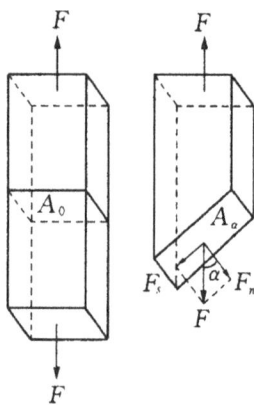

Fig. 5.17: Schematic diagram of stress analysis of polymer under uniaxial tension.

Assuming that a specimen with cross-sectional area of A_0 is subjected to uniaxial tensile force F, the σ_0 on the cross-sectional area is equal to F/A_0. If an inclined section with an angle of α between the cross-sectional surface and the specimen is taken, and its area $A = A_0/\cos\alpha$, the tensile force F acting on the A_α can be decomposed into two forces along the plane normal direction and along the plane tangent direction perpendicular to each other, and the two forces are recorded as F_n and F_s, respectively. Obviously, $F_n = F\cos\alpha$ and $F_s = F\sin\alpha$. Then the normal stress σ_{an} and the tangential stress σ_{as} on the oblique section are

$$\sigma_{an} = \frac{F_n}{A_\alpha} = \sigma_0 \cos^2\alpha \qquad (5.47)$$

$$\sigma_{as} = \frac{F_s}{A_\alpha} = \frac{\sigma_0 \sin^2\alpha}{2} \qquad (5.48)$$

It can be seen that the normal and tangential stresses in any section of the specimen are only related to the normal stress σ_0 and the inclination angle α of the section when the specimen is subjected to uniaxial tensile force. Once σ_0 is determined, σ_{an} and σ_{as} only change with the inclination angle of the cross section.

When $\alpha = 0°$, $\sigma_{an} = \sigma_0$, $\sigma_{as} = 0$;
When $\alpha = 45°$, $\sigma_{an} = \sigma_0/2$, $\sigma_{as} = \sigma_0/2$;
When $\alpha = 90°$, $\sigma_{an} = 0$, $\sigma_{as} = 0$

Figure 5.18 can be obtained by plotting α with σ_{an} and σ_{as}. It can be seen from the figure that the section with an inclination of 45° is the largest in terms of tangential stress. Similarly, the section with an inclination of 135° is the largest. Normal stress is the largest in cross section.

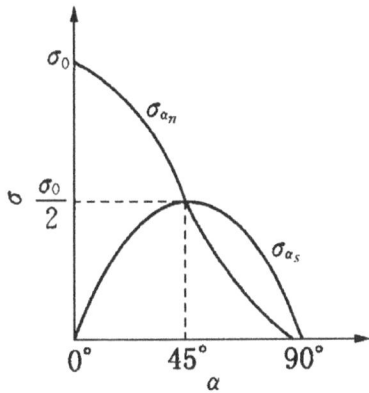

Fig. 5.18: Components of shear stress and normal stress in different sections under uniaxial tension.

When the polymer sample is subjected to uniaxial tensile force, if the shear yield strength of the polymer is first reached by the shear stress component on the oblique section with 45° and 135° cross section, the shear slip deformation bands (i.e. shear bands) with 45° and 135° cross section are first generated on the polymer sample, and a fine neck is formed. However, if the component of the radial stress reaches the failure strength of the polymer first, the fracture of the cross section will occur first. Usually ductile materials will yield to produce thin necks, while the brittle materials will fracture directly on the cross section (Fig. 5.19).

If the normal stress on the cross section reaches the yielding stress of the craze, the craze will occur.

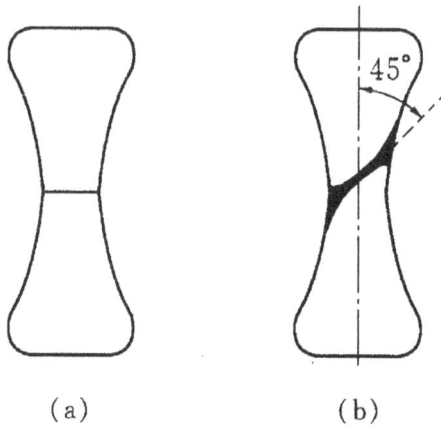

Fig. 5.19: Shear deformation band and fine neck of yielding specimens. (a) Specimen of tensile brittle fracture and (b) specimen of ductile materials at tensile yield.

There are two reasons for local yield necking. The first reason is pure geometry, that is, some fluctuation of the cross-sectional area of the specimen. When the cross-sectional area is smaller than the average value, the real stress is larger than the average stress, and the deformation value is larger than that of other places, which leads to the formation of fine neck. The second reason is that there are defects or heterogeneity in the structure, that is, there exists stress concentration substances. When an external force is applied, a stress concentration effect occurs, so the actual stress in some areas around the stress-concentrated substance is greater than the average stress, which results in large deformation and a narrow neck.

In addition to the geometric and stress concentration effects which cause local yield necking, strain softening and strain hardening can also occur.

Strain softening refers to the phenomenon that the resistance of material to strain decreases with the increase of strain once large deformation occurs. There are different explanations for strain softening. For example, the theory of thermal softening: when large deformation occurs, the rearrangement motion of some moving units overcomes internal friction, converts part of the work loss caused by external forces into heat and increases the temperature of the sample, which is conducive to the rearrangement movement of moving units. Stress activation theory: when large strain occurs, stress reduces the activation energy of the polymer motion unit rearrangement along the deformation direction, which is more conducive to the rearrangement movement of the motion unit along the deformation direction, and the deformation continues to increase. Argon's theory is that strain softening is due to the recombination of physical cross-linking points and the formation of supramolecular structures conducive to the development of deformation when large strains occur in polymers.

The so-called strain hardening refers to the orientation of polymer chains or segments or microaggregates along the direction of deformation, which decreases the conformational entropy and may increase the interaction between molecules and hinder the development of deformation. With the increase of strain, the orientation degree of these orientation units increases, and the resistance to deformation continues to increase. When the resistance to deformation is large enough to no longer increase, the deformation of the large deformation part ceases, which results in strain hardening.

Without strain softening, it is difficult to form a thin neck; without strain hardening, the fine neck becomes unstable, and the specimen will fracture rapidly at the fine neck. When the material is stretched, both strain softening and strain hardening occur, then a stable fine neck can be formed after the polymer specimen yields to neck. With the development of tension, the fine neck is no longer thinner, but the nonfine neck gradually becomes thin neck, forming a stable fine neck expansion; the specimen can undergo great deformation.

5.6.2 Large deformation of polymer alloys

In the polymer alloy system, the chemical structure and the aggregation structure of the polymers with different components may be different; under the same conditions, the polymers with different components may be in different mechanical states; the degree of difficulty, characteristics and development trend of the formation of shear bands and crazes of the polymers with different components may also be different. These factors complicate the stress–strain behavior of polymer alloys under stress, which not only has the same rules as common polymers but also has its own characteristics.

5.6.2.1 Stress concentration effect of elastomer dispersed phase

Elastomer-toughened plastics are generally single-connected and show continuous morphology. Under normal conditions, the continuous phase of resin presents glass or crystalline state, and the dispersed phase of the elastomer presents high elastic state, which belongs to typical heterogeneous structure system. The dispersed phase particles of elastomer play the role of stress concentrator. When toughened plastics are subjected to external forces, stress concentration effect occurs.

The magnitude of stress concentration effect can be expressed by stress concentration factor (the ratio of actual stress to average stress). The magnitude of stress concentration factor is related to the relative magnitude of elastic modulus of the stress concentration substance and the matrix, the shape of the stress concentration substance, the distance between the stress concentration substances and the direction of the external force.

The results show that the smaller the curvature radius ρ of crack tip is, the greater the stress concentration effect is. It is assumed that the elastomer particles are spherical, isotropic, with little content and no interference with each other; the elastomer particles are ideally bonded with the matrix resin. Under the action of tensile force, the stress concentration factor on the equatorial plane perpendicular to the direction of tensile force is the largest, which is 1.92. When there are resin inclusions in elastomer particles, such as ABS and HIPS, the elastomer dispersed phase particles contain a considerable amount of matrix resin, and the stress concentration factor decreases to about 1.54–1.89.

The stress concentration effect caused by dispersed phase particles is related to the relative elastic modulus of dispersed phase and continuous phase. For example, compared with pure elastomer, the stress concentration effect of ABS and HIPS is smaller. The basic rule is that the smaller the elastic modulus of dispersed phase is, the larger the stress concentration factor is. If the elastic modulus of the dispersed phase is the same as that of the matrix resin, no stress concentration will occur. The stress concentration factor decreases rapidly with the increase of the distance between the elastomer particles.

For practical elastomer toughened plastics, when the elastomer content is high, there is a strong interaction between the stress fields of elastomer particles, which may increase the stress concentration factor in the region between the particles.

5.6.2.2 Tensile properties of elastomer toughened plastics

In elastomer toughened plastics, besides the stress concentration effect of dispersed phase, the thermal expansion coefficients of dispersed phase and continuous phase are different. For example, in elastomer toughened plastics, the thermal expansion coefficient of elastomer is usually larger than that of matrix resin. When the blend system is cooled by melt, the thermal shrinkage stress is generated in the matrix resin around the elastomer particles. The thermal shrinkage stress is a kind of static tension, which can reduce the T_g of matrix resin around the elastomer and is beneficial to yield under external force.

Based on the earlier discussion, for rubber toughened plastics whose elastic modulus of dispersed phase is lower than that of matrix resin, under the action of tensile force, it is easy to cause a large number of crazes or shear bands in the matrix resin without excessive dispersion. Due to the stress concentration effect and thermal shrinkage stress, the average tensile stress is large, resulting in a reduction in yield stress, an increase in elongation at break, a decrease in tensile strength and elasticity and a decrease in elastic modulus.

However, the effects of shear band deformation and crazing on the tensile properties of materials are different. When a large number of crazes are formed, the elastic modulus of the material is greatly reduced due to the low modulus of the microfibers connecting the two crazes.

The shear deformation is different. The mechanical properties of the shear band are similar to those of the underformed polymer material without increasing the permeability of the material, and the degree of strain damage is small.

Yield deformation is an important characteristic of mechanical behavior of tough glassy polymers. For the rubber toughened plastics system, due to the low modulus of dispersed rubber particles, elongation deformation is more likely to occur under external force, and it becomes the center of stress concentration, especially on the equator of rubber particles, where the stress concentration is the greatest. A large number of crazes or shear bands around rubber particles cause local yield strain. The maximum stress concentration factor (the ratio of maximum principal stress to applied stress) near the equator of rubber particles can reach 1.92.

When the rubber particles contain resin inclusions, the stress concentration factor decreases. For example, the stress concentration factor near the equator of rubber particles in HIPS system is 1.54–1.89. As the distance from the particle surface increases, the stress concentration factor decreases rapidly (Fig. 5.20). When the distance between particles is small to a certain extent, the stress field of each particle will lead to superposition effect, which further increases the stress concentration factor. The increase of stress concentration factor is beneficial to yield deformation.

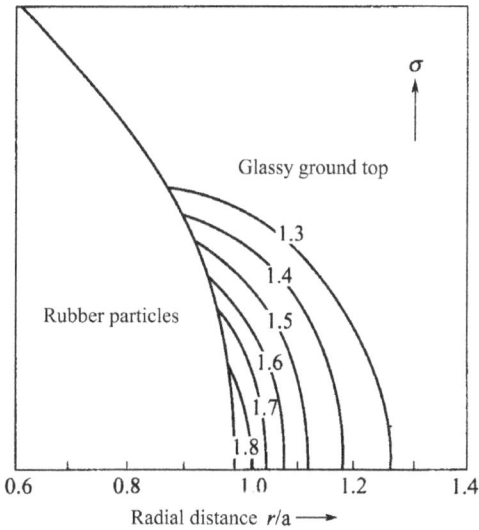

Fig. 5.20: Stress concentration factor near rubber particles in HIPS.

Another important reason for the yielding deformation of rubber toughened plastics system is the static tension of rubber particles on the surrounding plastic phase. The main reason for the static tension is that the thermal expansion coefficient of rubber is larger than that of plastic. When the toughened plastic is cooled from high temperature by melting to room temperature, the volume shrinkage of rubber phase is larger than that of plastic phase. As a result, the tensile stress (i.e. static tension) of rubber particles on the surrounding plastic phase matrix is produced. In addition, the static tension of toughened plastics will be further increased when they are subjected to external tension. This is because the Poisson's ratio of rubber (close to 0.5) is larger than that of plastic (about 0.3). When it is subjected to tensile stress, the transverse shrinkage of rubber is larger than that of plastics, which leads to the formation of new static tension. The formation of static tension increases the free volume in the plastic matrix, which leads to the decrease of glass transition temperature T_g of the plastic phase and is more conducive to yield deformation under external forces.

In a word, due to the existence of rubber particles in the blends, it is prone to yield deformation, which can change brittle polymers from brittle fracture to ductile fracture (Fig. 5.21). The deformation of PS without rubber toughening is very small. When the strain reaches 1.5%, craze can be observed. When it reaches 2%, the craze expands into a crack, which causes material to fracture. The initial stage of stress–strain curve of HIPS in Fig. 5.20 is similar to the linear relationship of PS. At this time, the craze has not yet formed, and the stress of material is lower than the critical stress of craze. When the stress increases to the critical stress, the material begins to whiten, which indicates that crazing occurs and the stress–strain behavior becomes nonlinear. When the stress continues to increase, the crazing rate accelerates rapidly until yielding occurs. After the yield point, due to the strain softening effect, the stress decreases to a certain value and remains constant, while

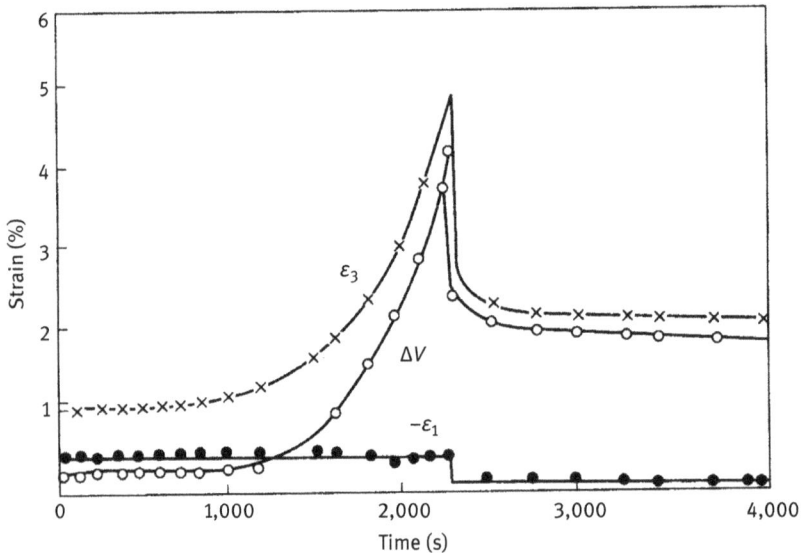

Fig. 5.21: Stress–strain curves of PS and HIPS (20 °C).

the strain continues to increase, which is equivalent to the creep process. At this time, the increase of deformation mainly comes from the increase of the number of crazes and the increase of the size of crazes. Ultimately, crazes are transformed into cracks under stress. At the time of fracture, the strain of the material has reached 40%, which is 20 times higher than that of PS.

5.6.2.3 Relationship between deformation mechanism and mechanical properties of polymer alloys

Different types of toughened plastics have different deformation mechanisms. HIPS is basically craze deformation and toughened PVC belongs to shear yield deformation, while ABS resin has both craze deformation and shear deformation. The mechanical behavior of blends with different deformation mechanisms is also different. The mechanism of glass deformation of blends is mainly related to the toughness of the matrix. The better the ductility, the better the deformation of shear band. For example, in HIPS/PS/polyphenylene oxide (PPO) blends, the PS content decreases with the increase of PPO content (Tab. 5.6), and the elongation at break increases significantly (D and E in Fig. 5.22). This is because that the toughness of the matrix increases (the matrix is composed of compatibilities formed by PPO and PS) with the increase of PPO content. At this time, the proportion of shear band deformation in the total deformation of the blends increases. The system with poor matrix toughness has smaller elongation at break.

Tab. 5.6: Composition of HIPS/PS/PPO System.

Blend system	HIPS (%)	PS (%)	PPO (%)
A	50	50	0
B	50	37.5	12.5
C	50	25	25
D	50	12.5	37.5
E	50	0	50

Under the tensile stress, the main deformation mechanism of the blends with high brittleness is the craze deformation, while for the blends with good toughness, the shear band deformation is the main deformation mechanism. The reason for this difference is related to the molecular chain network structure of the matrix. Recent studies have confirmed that when the molecular weight of glassy polymers reaches above the critical molecular weight, the molecular chains will form entangled network structures (Fig. 5.23).

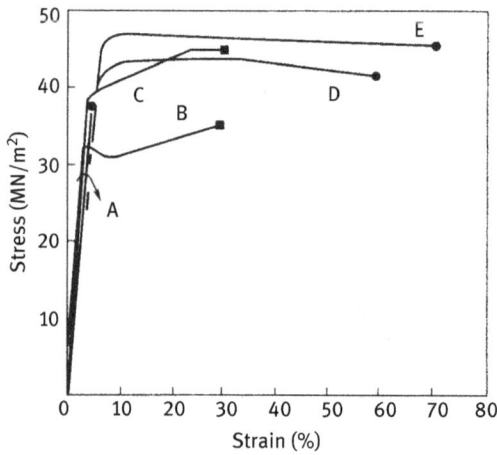

Fig. 5.22: Stress–strain curves of HIPS/PS/PPO ternary blends. (The composition of A, B, C, D and E can be seen in Tab. 5.6.)

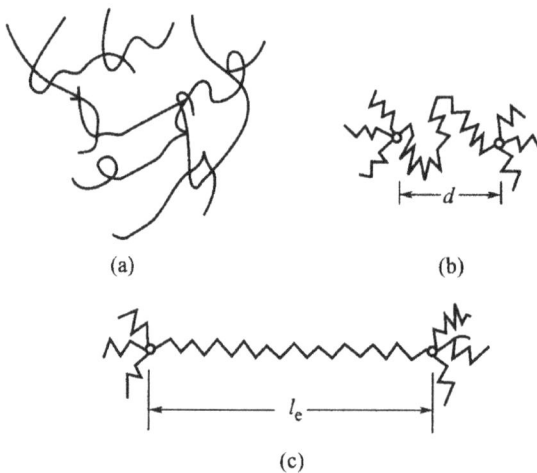

Fig. 5.23: Schematic diagram of entanglement and deformation of polymer chains: (a) polymer chain entanglement; (b) distance between undeformed entanglement points; and (c) distance l_e between two completely extended entanglements.

The parameters l_e and d are the basic parameters for describing the structural characteristics of entangled networks. As shown in Fig. 5.22, l_e is the full stretching length between two entanglement points. d is the size of the entanglement points when the entanglement network is not deformed. Under the action of tensile force, the maximum elongation ratio of network λ_{max} should be as follows:

$$\lambda_{max} = l_e/d \tag{5.49}$$

Equation 5.49 shows that the maximum elongation ratio of λ_{max} is closely related to l_e. When the density of entanglement point is low, that is, the length of l_e is large, the value of λ_{max} is large and the molecular chain can be fully extended. In this case, the molecular chains are easy to orientate to form microfibers and undergo cavitation, that is, crazing deformation. Polymers with poor matrix toughness are prone to crazing because the molecules are not easily entangled and the density of entanglement points is low. For molecular chains with high entanglement density, it is difficult to extend the entanglement chain, which will cause the strain hardening effect, so the local craze deformation will not fully develop at this time. When the stress on the weak part of the system exceeds the shear yield stress, the deformation of the blends is mainly contributed by the shear yield deformation. Polymers with good toughness, such as PPO and PC, have low λ_{max} and high entanglement density, so they are prone to shear yield deformation.

The effects of crazing and shear band on the properties of blends are different. The craze is porous, its modulus is much lower than that of the bulk, and it has great permeability to the liquid. At this time, if the material continues to be subjected to external forces, the craze is easy to develop into cracks and ultimately cause material damage, which is also called strain damage. Shear band deformation does not reduce the strength of the local strain zone too much nor does it increase the permeability of the polymer. The degree of strain damage caused by the shear band deformation is very small. In addition, the yielding deformation caused by crazes has a certain degree of restorability (high elastic deformation). Therefore, the cold forming method should not be used for polymers or blends which are mainly based on the craze deformation mechanism. However, the unrecoverable plastic deformation accounts for a large proportion in shear yield deformation, and there will be no large internal stress in the material. Therefore, the cold processing method can be used for polymers or blends that are primarily shear deformed.

Figure 5.24 shows the effect of crazing on the modulus of blends. The system with poor matrix toughness (HIPS/PS = 50/50) decreases very little with the development of strain modulus (Fig. 5.24(a) curve 1). However, when the specimen is stretched by 5% in creep test, and then the change curve of its modulus with deformation is measured, it can be found that the modulus decreases greatly(Fig. 5.24(a) curve 2). This is due to the formation of crazes in the specimens during creep tests, and the strength of crazes is much lower than that of bulk specimens. If the toughness of the matrix is changed, the situation will be different. Figure 5.24(b) shows a blend of HIPS/PPO (50/50). The toughness of matrix is much better than PS. The small changes in curves 1 and 2 indicate that after the creep test, the proportion of craze deformation decreases, the proportion of shear deformation increases and the influence of shear deformation on the modulus is relatively small.

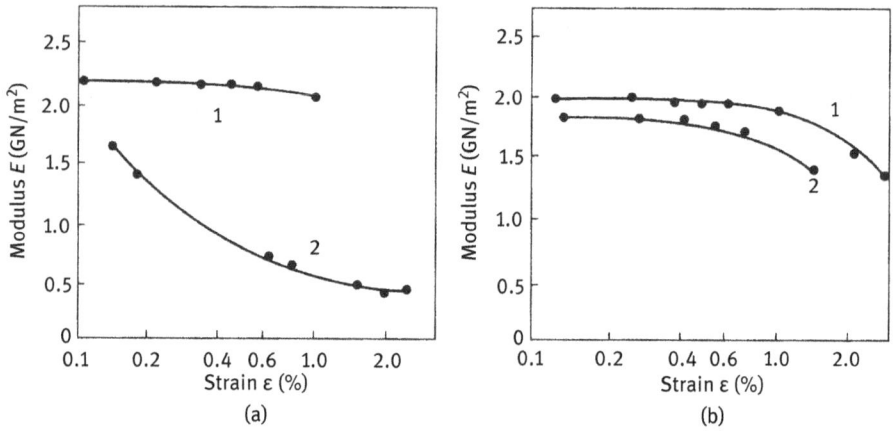

Fig. 5.24: Effect of craze deformation on modulus of blends. (a) Matrix with poor toughness and (b) matrix with good toughness. 1, First stretching; 2, second stretching.

5.6.2.4 Factors affecting deformation behavior of polymer alloys

1. Effect of resin matrix

As discussed earlier, the toughness of the matrix is the main factor determining the deformation mechanism of the blends. At the same time, the toughness of the matrix has a great influence on the creep behavior of the blends. Figures 5.25 and 5.26 show that due to the different toughness of the matrix, the creep curves of HIPS and toughened PVC under tensile stress are also different.

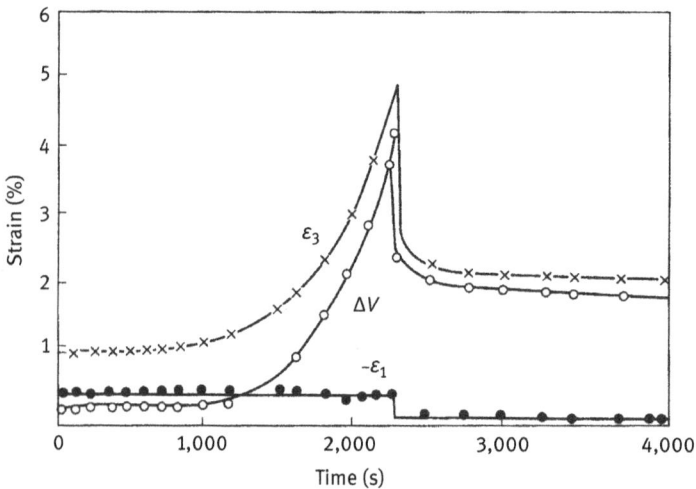

Fig. 5.25: Creep and recovery curves of HIPS under tension stress (19.7 MN/m^2) (20 °C).

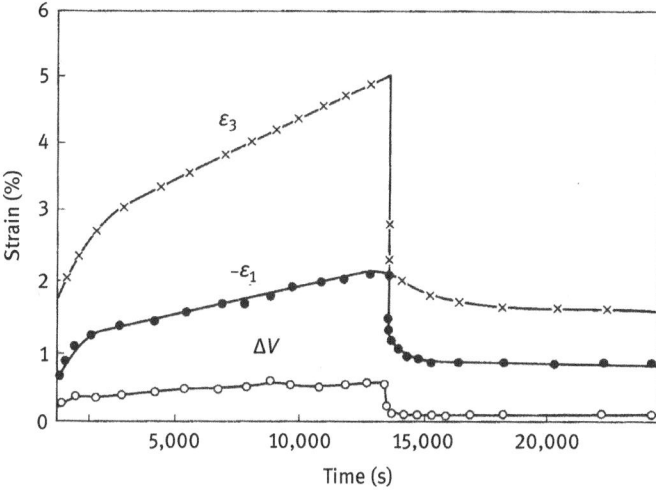

Fig. 5.26: Creep and recovery curves of toughened PVC under tensile stress (36.0 MN/m²) (20 °C).

2. Effect of stress and strain rate

The effect of stress on deformation is manifested in two aspects, which not only affects the deformation mechanism but also relates to the deformation speed.

The deformation mechanisms of the blends change with different stress types. HIPS undergoes a typical craze deformation under tension. However, the shear band is mainly formed under compressive stress. Because craze can only occur under tension stress, but shear band deformation can occur under shear stress, tension stress and compression stress.

For a system where two deformation mechanisms may occur at the same time, the magnitude of stress will also affect the deformation mechanism. Figure 5.27 is the creep curve of ABS. At low stress (26.5 MN/m²), the variation of V with time is very small (Fig. 5.27(a)), and the proportion of craze deformation is very low, which is mainly caused by shear deformation. However, at high stress (34.5 MN/m²), the shear deformation is still the main deformation in a short time (Fig. 5.27(b)). After 1,000 s, the crazing process increases significantly (the slope of ΔV curve increases). The stress increases from 26.5 to 34.5 MN/m², and the proportion of deformed crazing of ABS increases from 0 to 85% (Fig. 5.28). Both shear deformation and craze deformation increase with the increase of stress, which can be described by the Eyring equation:

$$\frac{\ln\varepsilon}{A} = \frac{V^*\sigma}{ykT} - \frac{\Delta H^*}{kT} \tag{5.50}$$

where ε is the strain rate, ΔH^* is the activation enthalpy, V^* is the activation volume, y is the stress concentration factor, A is the constant, k is the Boltzmann constant

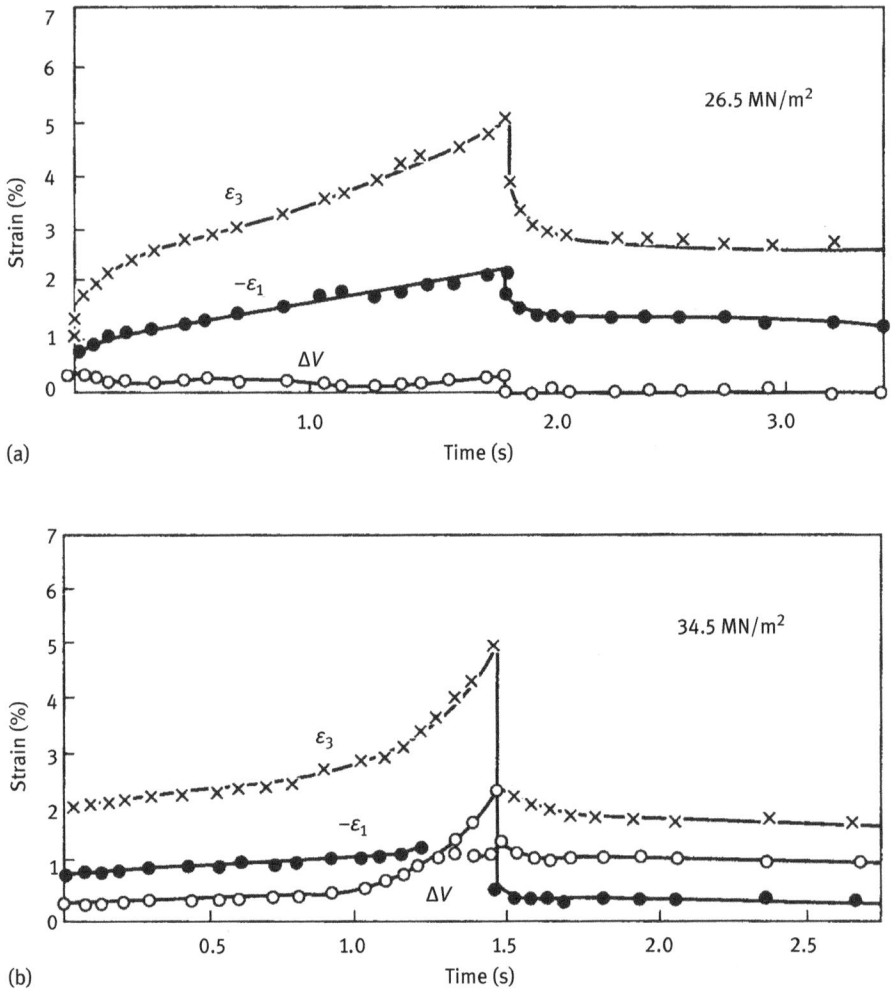

Fig. 5.27: Creep and recovery curves of ABS. (a) Under low stress and (b) under high stress.

and T is the absolute temperature. Eyring regards the deformation process as an activation process similar to the chemical reaction process. The motion unit of polymer needs to overcome a certain energy barrier in order to achieve movement and deformation. When the stress is applied, the increase of the stress decreases the enthalpy of motion activation of the moving element in the direction of deformation, thus increasing the strain rate.

The effect of strain rate on deformation mechanism is similar to that of stress. Stress whitening was observed in toughened PVC at high strain rate, indicating that craze deformation occurred. Because the strain rate is too high, the relaxation process of molecular motion cannot be completed, resulting in the formation of

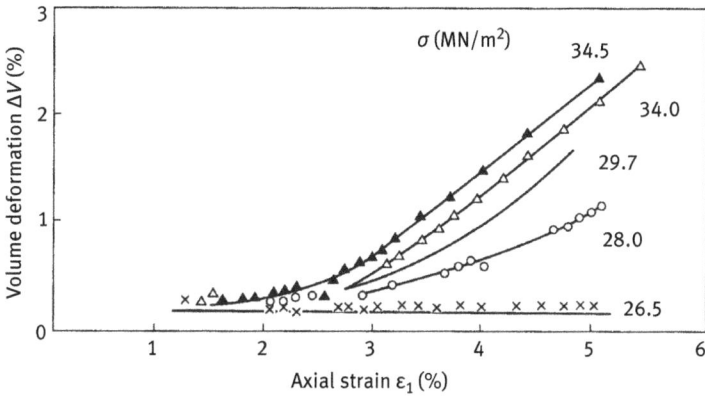

Fig. 5.28: Relation between volume deformation and elongation deformation of ABS (20 °C).

holes in the polymer, thus resulting in craze deformation. In slow creep tests, small volume changes were observed (Fig. 5.27).

3. Effect of temperature

The effect of temperature on deformation is also reflected in Eyring equation. The activation volume and deformation rate increase with the increase of temperature. As shown in Fig. 5.29, the slope of the creep curve of ABS increases with the increase of temperature. Take the creep compliance at 10^2 s for comparison, the creep compliance at 80 °C has an high order of magnitude than at 40 °C.

4. Effect of rubber content

With the increase of rubber content, the spacing between particles decreases and the stress field between particles overlaps. As a result, the stress concentration factor in the Eyring equation increases, and the deformation rate of the material is increased. The increase of rubber content also means that the number of rubber particles in the system increases, and the number of craze increases. However, when the number of crazes increases, the number of particles terminating the crazes also increases. The ultimate effect of these two aspects is not very obvious. That is to say, the increase of craze deformation rate depends mainly on the increase of stress concentration factor.

The rate of shear deformation increases with the increase of rubber content, but the effect is not as great as that of craze deformation. Therefore, the increase of rubber content results in the increase of the proportion of crazing deformation mechanism in the deformation process.

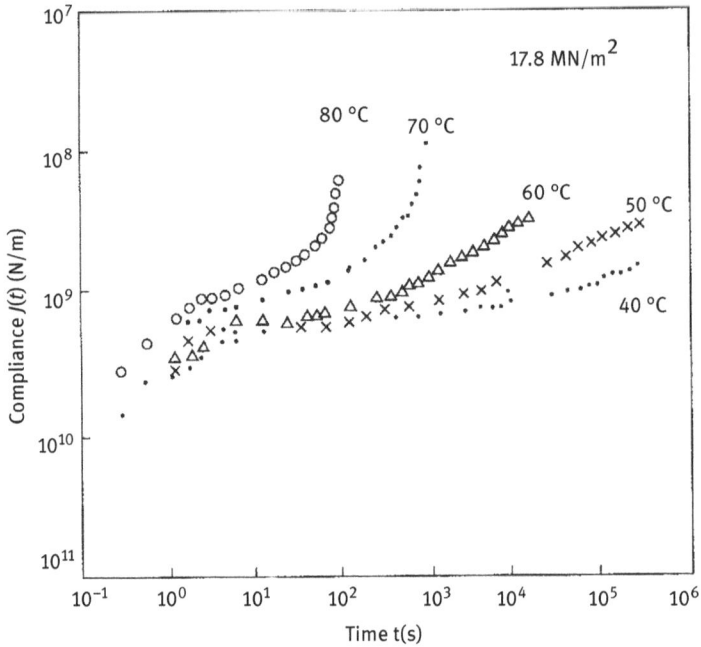

Fig. 5.29: Tensile creep curves of ABS at different temperatures.

5. Effect of tensile orientation

When toughened plastics are stretched, on the one hand, the molecular chains of the matrix are oriented. On the other hand, the dispersed rubber particles become elliptical, which reduces the stress concentration factor. Therefore, the strain rate of the stretched specimens decreases significantly. After orientation of ABS, the tensile test shows that the degree of stress whitening decreases, which indicates that the proportion of crazing deformation mechanism caused by tensile orientation is relatively reduced.

Exercises

1. What is the glass transition temperature of polymers? What are the factors affecting the glass transition temperature of polymers?

Answer:

Glass transition temperature (T_g) is an important index to measure their operating conditions. Many physical properties (such as thermal expansion coefficient, heat capacity, refractive index and electrical properties) have changed dramatically during the glass transition of polymers.

(1) Polymer chain compliance, (2) intermolecular forces, (3) main chain symmetry; (4) relative molecular mass, (5) branching and crosslinking, (6) solvents and plasticizers, and (7) crystallization and orientation.

2. What is stress relaxation in polymers?

Answer:
Stress relaxation of polymer refers to the phenomenon that the stress in polymer decreases with the increase of time at constant temperature and constant deformation, which is essentially the result of the movement of polymer molecules from unbalanced state to equilibrium state.

3. What are the mechanisms of large deformation of polymers?

Answer:
Many experimental facts show that there are two mechanisms of polymer large deformation: shear deformation and crazing. In many cases, shear deformation and crazing coexist, and their respective proportions are related to the structure and field conditions of polymers.

The so-called shear deformation refers to the high orientation of macromolecule or macromolecule microaggregates in some planes under the action of external forces, resulting in distortion deformation without obvious volume change. There are two kinds of shear deformation: diffuse shear deformation, that is, large-scale shear deformation occurring in the whole stress region, and shear band, that is, shear deformation occurring in the local zonal region.

Crayonization refers to the void streak-like deformation zone caused by stress concentration in some weak parts of polymer under the action of tensile force, and this phenomenon is called crazing. Cracks are different from crazes. The latter is empty and has no mass, while the former is not empty and has finer structure inside, so it shows a silver-white luster under the action of light.

Chapter 6
Rheological properties of multicomponent polymers

Polymer materials have value of direct use only when processed into products. In general, there are three basic steps from polymer materials to products: heating plasticization, flow molding and cooling and curing. In the process of melting and plasticizing into viscous fluids, all kinds of structures in polymer materials, except chemical structures, are almost destroyed. In the process of flow molding and cooling and curing, the aggregated structure and morphological structure of polymer are reformed. Therefore, the rheological properties of polymer melts are crucial to the structure and properties of products.

Due to the diversity of aggregated structure, the influence of rheological properties and product properties of multicomponent polymers is more complex. In order to comprehend the relationship of rheological properties and the structure and properties of multicomponent polymers, it is necessary to review the general flow characteristics of polymer materials.

6.1 Viscous flow of polymer melts

6.1.1 Characteristics of shear flow in polymer melts

The flow with the direction of velocity gradient perpendicular to the flow direction is called shear flow. Polymer shear flow has the following characteristics:

1. Through the coordinated rearrangement of chain segments along the flow direction.

There are many holes which are equivalent to the molecular size in the small molecule liquid. When subjected to external forces, the probability of molecules moving toward holes along external forces is greater than that in other directions. When the molecule transits forward, the original position of the molecule becomes a new hole and the subsequent molecule transits forward. In this way, molecules transit along the direction of external force through the holes between molecules, forming the macroscopic flow of liquid.

When a molecule transits to a hole, it must be resisted. This resistance is called flow resistance and is expressed by viscosity η. As the temperature rises, the thermal energy of the molecule increases, the holes in the liquid increase and expand

https://doi.org/10.1515/9783110596335-006

and the flow resistance, that is the viscosity, decreases. The relationship between η and temperature T can be expressed as follows:

$$\eta = Ae^{\Delta E_\eta/RT} \tag{6.1}$$

where A is a constant, R is the gas constant, and ΔE_η is viscous flow activation energy, which is the energy needed to overcome the interaction of the surrounding molecules when the molecular transits to the hole. Drawing $\ln\eta$ to map $1/T$, ΔE_η can be obtained from the slope of straight line.

The cavitation principle of small molecule liquid flow encounters difficulties in explaining the fluidity of polymer melts.

First, there is no hole in the polymer melt that can hold the whole macromolecule.

Second, if the whole macromolecule transits, the viscous flow activation energy of a long chain molecule containing 1,000 CH_2 chains is 2.1 MJ/mol, while the bond energy of C–C bond is only 3.4 kJ/mol. Macromolecules have been destroyed before the flow occurs.

Many experiments show that the viscous flow activation energy of polymers with different molecular weights has no relationship with their molecular weights. This fact shows that the flow of polymer melt is not a simple transition of the whole polymer, but a coordinated successive transition of the basic movement segments along the flow direction (coordinated rearrangement movement), thus realizing the relative displacement of the whole polymer. This process is similar to the creep of earthworms. Therefore, in the flow of polymer, the segment is flow unit.

2. Under general shear flow conditions, the flow of polymer melts does not conform to the law of Newton's fluid.

The viscosity of small molecules is a constant at a certain temperature, and the shear stress τ is proportional to the shear rate $\dot{\gamma}$:

$$\tau = \eta\gamma \tag{6.2}$$

Equation (6.2) is called Newton's law of flow. The viscosity does not vary with τ and $\dot{\gamma}$ and is constant at certain temperature. The fluid obeying Newton's law of flow is called Newton's fluid. The fluid that does not obey Newton's law of flow is called non-Newtonian fluid.

If only the relationship of viscosity and τ and $\dot{\gamma}$ is considered, and the relationship between rheological behavior and time is not considered, non-Newtonian fluids can be classified into three categories. Fluids that η decrease with the increasing of τ or $\dot{\gamma}$ are called pseudoplastic fluids, fluids that η increase with the increasing of τ or $\dot{\gamma}$ are called expansive plastic fluids and the non-Newtonian fluid that have yielding stresses is called Bingham fluid. Most polymer melts are pseudoplastic fluids under general shear flow conditions, also known as shear thinning fluids. Polymer suspensions, latex and polymer filler systems are generally expandable fluids. Mud, toothpaste, grease and paint are mostly Bingham fluid.

The viscosity of some non-Newtonian fluids is also related to time. The fluid whose η decreases with time at a constant $\dot{y}$ is called thixotropic fluid; fluid whose η increases with time is called fluidic fluid (or shaking fluid, when the concentration of potato starch is between 1.5% and 4.0%; the gelatinized fluid is shaking fluid).

Thixotropy of different degrees can also be observed in many polymer melts, and with the increase of molecular weight, the time required for viscosity to drop to a certain equilibrium value increases.

3. The viscosity of polymer melts is much larger than that of small molecules.

6.1.2 Apparent viscosity of polymer melts

Under certain temperature, the shear flow of polymer melt obeys the power formula.

$$\tau = K y^n \tag{6.3}$$

where K is called the consistency coefficient, n is the index representing the degree of deviation from Newtonian flow, called non-Newtonian index.

Under general shear flow conditions, the flow deformation of polymer melts is accompanied by an amount of high elastic deformation, and the viscosity is characterized by apparent viscosity η_a. It is defined as

$$\eta_a = \frac{\tau}{y} = K y^{n-1} \tag{6.4}$$

For pseudoplastic fluids, $n<1$, for expansive fluids, $n>1$ and Newtonian fluids can be regarded as a special case, $n=1$.

The n of polymer melt does not keep constant in a certain range of shear rate, but decreases with the increasing of y, that is, the pseudoplasticity increases with the increasing of y.

When the polymer melt is flowing at a very low shear rates, the high elastic deformation of the polymer melt disappears quickly, so the high elastic deformation is no longer accompanied by the flow deformation. The rheological behavior of the polymer melt conforms to Newton's law of flow and its viscosity is called zero shear viscosity, defined as:

$$\eta_0 = \lim_{y \to 0} \eta \tag{6.5}$$

Usually, the $\eta_a < \eta_0$ of polymer melts.

The fluidity of polymer melt can be characterized by the η_a of polymer melt: larger η_a means less fluidity, and smaller η_a means more fluidity. η_a of polymer melt is greatly affected by shear rate; η_a can decrease by 1–3 orders of magnitude with the increase of shear rate.

6.1.3 Factors affecting the apparent viscosity of polymer melts

6.1.3.1 Influence of polymer molecular structure

1. Influence of molecular weight

Although the molecular motion mechanism of polymer melt shear flow is the coordinated rearrangement of chain segments, the displacement of polymer along the flow direction and the mutual slip between chains are realized, the larger the molecular weight, the more the number of segments a polymer contains. In order to achieve the displacement of the center of gravity of whole polymer, the chain segment coordinated rearrangement movement needs to be completed. The more the number of times, the greater the resistance of the random thermal motion of the segment to the rearrangement motion, and the greater the apparent viscosity. Moreover, the slow increasing of molecular weight will result in a dramatic increase in the η_a (shown in Tab. 6.1).

Tab. 6.1: Relationship between apparent viscosity and molecular weight of high pressure polyethylene melt.

$\overline{M}_n \times 10^{-4}$	η_a (Pa· s) (190 °C)	$\overline{M}_n \times 10^{-4}$	η_a (Pa· s) (190 °C)
1.9	45	3.2	4,200
2.1	110	4.8	3.0×10^4
2.4	360	5.3	1.5×10^6
2.8	1,200		

2. Influence of molecular weight distribution

According to the earlier discussion, the effect of molecular weight distribution on the rheological properties of polymers has two aspects.
1) The contribution of the high molecular weight part to the apparent viscosity is much larger than that of the low molecular weight part.
2) The higher the molecular weight, the greater the sensitivity of viscosity to $\dot{\gamma}$, the greater the viscosity drop caused by shear and the lower the shear rate from the first Newton zone to the pseudoplastic zone, that is, the phenomenon that η decreases with the increase of $\dot{\gamma}$ at a lower $\dot{\gamma}$.

Therefore, there are three rules for the effect of molecular weight distribution on shear viscosity.

(1) As for same polymers with the same average molecular weight, the η_0 of polymer with a wider molecular weight distribution is higher than that of polymer with a narrower molecular weight distribution.
(2) The melt with a wide molecular weight distribution began to pseudoplastic flow at a lower shear rate value (see Fig. 6.1).
(3) The η_a in high shear rate region is lower than that of polymer melt with narrower distribution (see in Fig. 6.1).

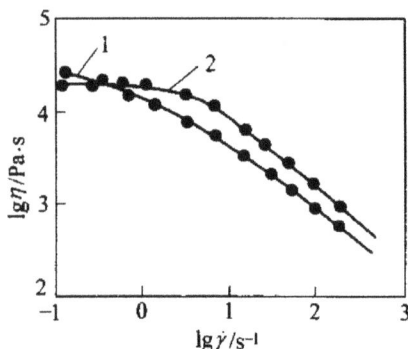

Fig. 6.1: Effect of molecular weight distribution on relationship between viscosity and shear rate (PS, 190 °C). (1) Wide distribution specimens and (2) narrow distribution specimens.

3. Influence of chain branching

Branching has little effect on melt viscosity when branching chain is not too long, and its η_0 is slightly lower than linear polymer with the same molecular weight. If the mean square radius of rotation is the same, the η_0 of the two is approximately equal. But if the branched chain is long enough to intertwine, the branching effect on the viscosity of the melt is significant.

From Fig. 6.2, we can understand the relationship between η_0 and mass average molecular weight, and the branching and linear polymer obey the same rule at lower molecular weight. However, when the molecular weight of the branched chain is more than twice the molecular weight between the entanglement points M_c, the viscosity of the branched polymer begins to rise rapidly and soon increases to more than 100 times that of the linear polymer.

The viscosity of long arm star-shaped polymers is more sensitive to shear rate. Compared with linear polymers of the same molecular weight, the viscosity deviation from Newtonian occurs in the lower shear rate region; in the higher shear rate region, the viscosity of star-shaped polymers is lower than that of linear polymers with the same molecular weight, as shown in Fig. 6.3.

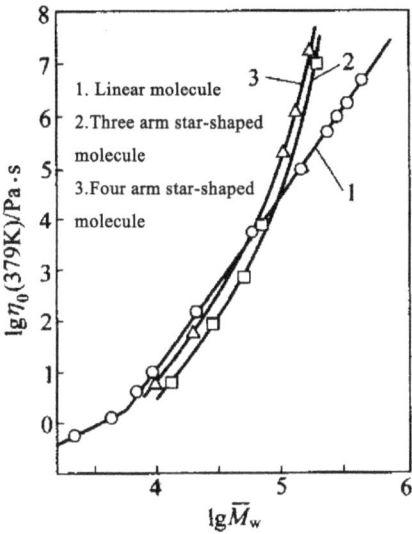

Fig. 6.2: Dependence of η_0 on the weight average molecular weight of branched polymers.

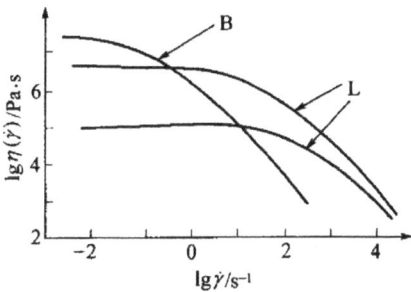

Fig. 6.3: Dependence of viscosity on shear rate of star-shaped polymer (B) and homogenous linear polymer (L).

4. Other structural factors

For polymers with the same molecular weight, the more stiff the molecular chain, the higher the viscosity of polymers. The greater the polarity of polymers, the greater the interaction force between polymers and the higher the viscosity.

6.1.3.2 Influence of processing conditions

1. Influence of temperature

Above the viscous flow temperature, the relationship between the viscosity of polymer melt and temperature is the same as that of small molecule liquids, which conforms to eq. 6.1. With the increase of temperature, the free volume of polymer melt increases, the chain activity increases, the intermolecular interaction weakens, the melt viscosity decreases exponentially with the increase of temperature and the fluidity increases.

The sensitivity of different polymers to temperature is different, that is, the viscous activation energy ΔE_η is different. The greater the molecular chain rigidity and the greater the ΔE_η value, the greater the sensitivity of the polymers to temperature. (see Tab. 6.2)

Tab. 6.2: ΔE_η of some polymers.

Polymer	ΔE_η (kJ·mol)	Polymer	ΔE_η (kJ·mol)
Polydimethylsiloxane	16.7	Styrene acrylonitrile copolymer	104.2–125
High-density polyethylene	26.3–29.2	Polycarbonate	108.3–125
Polypropylene	37.5–41.7	ABS (20% rubber)	108.3
Polybutadiene (cis)	19.6–33.3	ABS (30% rubber)	100.0
Polystyrene	94.6–104.2	ABS (40% rubber)	87.5
Polyvinyl chloride	147–168	Cellulose acetate	293.3

When the temperature drops below the viscous flow temperature, the viscosity and temperature of the polymer no longer conform to eq. 6.1, but conform to Williams-Landel-Ferry (WLF) equation.

$$\lg \frac{\eta(T)}{\eta(T_g)} = -\frac{17.44(T - T_g)}{51.6 + (T - T_g)} \tag{6.6}$$

Equation 6.6 can estimate viscosity of polymers at the range of $T_g < T < T_g + 100$ °C. For most noncrystalline polymers, viscosity is 1×10^{12} Pa·s at T_g.

2. Effect of shear rate and shear stress
The eq. 6.4 is transformed to logarithmic form.

$$\lg \eta_a = \lg K + (n - 1)\lg \gamma \tag{6.7}$$

Figure 6.4 is obtained by drawing A to B. It can be seen from the figure that under extremely low $\dot{\gamma}$ and fairly high $\dot{\gamma}$, the curve is two horizontal lines, indicating that η_a does not change with the change of $\dot{\gamma}$ and the shear flow of polymer melt obeys Newton's law of flow, so it is called respectively the first Newton region and the second Newton region. In the intermediate shear rate region, where η_a decreases with the increase of $\dot{\gamma}$, is pseudoplastic region.

In general, for the polymer with good flexibility, η_a is more sensitive to $\dot{\gamma}$. Therefore, for flexible polymers, increasing shear rate can increase the fluidity of polymer melt.

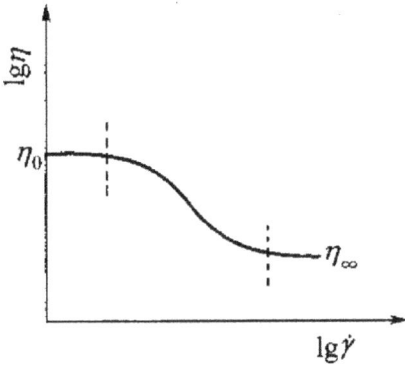

Fig. 6.4: Dependence of polymer melt viscosity to shear rate.

The effect of shear stress on the viscosity of polymer melt is basically the same as that of shear rate.

3. Influence of pressure

In the process of extrusion or injection molding, polymers often bear high hydrostatic pressure. Therefore, it is of practical significance to study the effect of pressure on shear viscosity of polymer melts.

Figure 6.5 shows the variation of shear viscosity of low-density polyethylene (LDPE) with shear stress under four pressure conditions at 210 °C. The shapes of the four curves are similar, but as the pressure increases, the curves move upward along the viscosity coordinates, the pressure increases from one atmospheric pressure (101.325 kPa) to 300 MPa, and η_a rises by nearly two orders of magnitude.

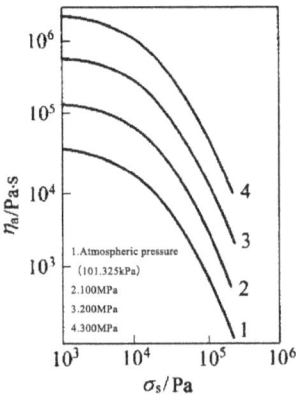

Fig. 6.5: Dependence of η_a of low-density polyethylene to pressure.

According to the concept of free volume, the free volume of polymer melt has a decisive effect on its viscosity. With the increase of pressure, the free volume decreases, the interaction between molecules strengthens, the flow resistance increases and the viscosity increases.

6.2 Viscous flow of multicomponent polymers

6.2.1 Dispersion of multicomponent polymer melts

The flow characteristics of multicomponent polymer melts are similar to those of general polymer melts, and they also belong to pseudoplastic fluids under normal shear flow conditions. However, the rheological properties of multicomponent polymer melts have their own characteristics because of the multiphase structure and the interaction between the phases.

A heterogeneous multicomponent polymer melt is formed if the polymer components that comprise the multicomponent polymer are still incompatible under melt conditions.

Heterogeneous polymer melts have two dispersive states: single phase dispersion and two-phase interlocking. In the former case, whether one component is in dispersed phase or continuous phase depends on the volume ratio, viscosity ratio, elasticity ratio, interfacial tension and blending conditions of the two polymers. The dispersed phase can be fibrous, ribbon or bead when flowing. For example, the blend melt of 10% polyethylene (PE) and 90% polystyrene (PS) has a zonal structure in the flow (see in Fig. 6.6), while during the flow of the blend melt of PE/PS=30/70,PS is dispersed in the PE continuous phase as beads and fibers, as shown in Fig. 6.7.

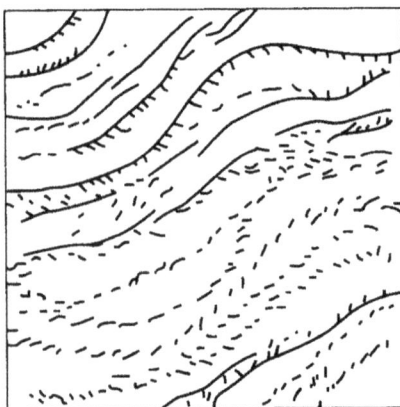

Fig. 6.6: Band zonal structure diagram of PE/PS (10/90) melt in the flow. (Dispersed state, parallel to the direction of flow. Black is PE.)

For the blend system composed of high-density PE (HDPE) and PS, when HDPE/PS=25/75 and the temperature is 200–220 °C, HDPE is a continuous phase and PS is a bead-like dispersed phase. When the temperature reaches 240 °C, PS becomes a continuous phase. When HDPE/PS=50/50, PS is dispersed phase at these temperatures. At HDPE/PS=75/25, the structure is interlocked at the above temperature, as shown in Fig. 6.8.

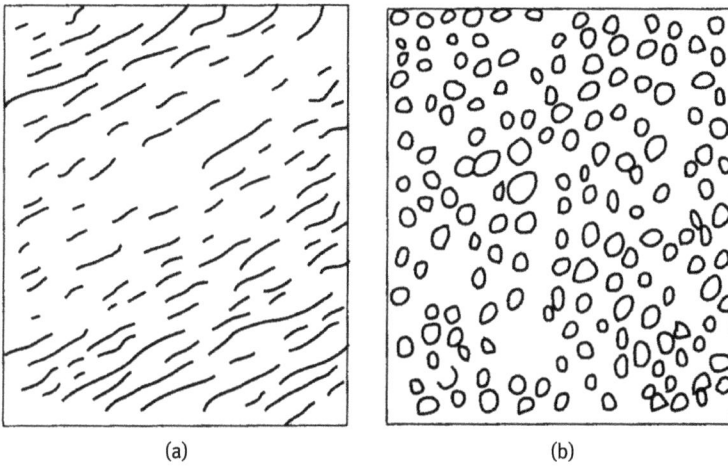

Fig. 6.7: Dispersion state of PE/PS (30/70) blends in melt flow. (a) Cross sections perpendicular to the flow direction; (b) Cross sections pparallel to the flow direction.

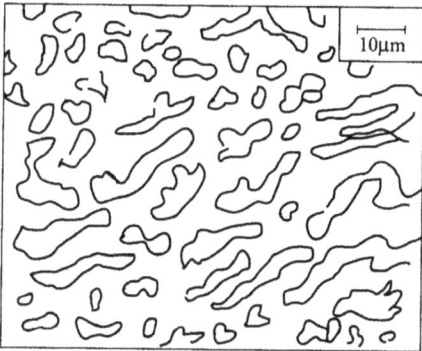

Fig. 6.8: Schematic diagram of extruded specimen cross section of HDPE/PS (75/25) blends.

6.2.2 Viscosity of multicomponent polymer melts

6.2.2.1 Relationship between viscosity and blend composition (blending ratio)

Compatible polymer blends, such as styrene-acrylonitrile/polycaprolactone (SAN/PCL) blends, have a linear relationship between viscosity and composition. With the increase of PCL content, the viscosity of the blends decreased, and the viscosity of the blends with different compositions was between SAN and PCL (viscosity of SAN was high, viscosity of PCL was low). Other compatible blends also had similar rules.

The melts of incompatible polymer blends belong to heterogeneous systems. Because of the complexity of rheological behavior of polymer blends, the flow mechanism of polymer blends is still not comprehended. Therefore, the quantitative description of the relationship between viscosity and composition of polymer blends is mostly empirical or semiempirical. Equation 6.9 is a more commonly used formula.

$$\log \eta = \varphi_1 \log \eta_1 + \varphi_2 \log \eta_2 \qquad 6.8$$

where η is the viscosity of blend, φ_1, φ_2, η_1 and η_2 is the volume fraction and viscosity of components 1 and 2.

The relationship between viscosity and composition of blends can be estimated by the mixture rule.

$$\eta = \eta_2 + \frac{\varphi_1}{\frac{1}{\eta_2 - \eta_1} + \frac{\varphi_2}{2\eta_2}} \qquad 6.9$$

$$\eta = \eta_1 + \frac{\varphi_1}{\frac{1}{\eta_2 - \eta_1} + \frac{\varphi_1}{2\eta_1}} \qquad 6.10$$

Equations (6.9) and (6.10) are upper and lower values, respectively. η is the viscosity of the blend, φ_1, φ_2, η_1 and η_2 is the volume fraction and viscosity of components 1 and 2, and $\eta_1 < \eta_2$.

There are many other improved formulas to describe the relationship between viscosity and composition, but they are all empirical formulas.

The viscosity and composition of the actual polymer alloy melt have the following relations:

1. The viscosity of the blend is between two pure components.

More kinds of blends belong to this situation such as PA6/ABS blends. At 250 °C, the viscosity of PA6 is low and that of ABS is high. With the increase of ABS content in the blend, the viscosity of the blend increases rapidly (see in Fig. 6.9). In the PA6/PS blend system, the viscosity of PA6 is higher and that of PS is lower. With the increase of PS content, the viscosity of the blend decreases. The melt viscosity of polypropylene (PP) and polyolefin elastomer (POE) blends is between the viscosity of each component polymers.

Some blend systems with liquid crystal polymers also belong to this type. For example, for polyetheretherketone (PEEK)/ liquid crystal polymer (LCP) blends system, the viscosity of PEEK is high and the viscosity of LCP is low. With the increase of the content of LCP in the blends, the apparent viscosity of the blends decreases obviously. The reasons are as follows: first, the orientation of melted LCP is easy under shear, which makes the slip between chains easier and reduces the viscosity of the blend; second, the existence of LCP hinders chain tangling, reduces the resistance to flow and also leads to the decrease of the viscosity (see in Fig. 6.10).

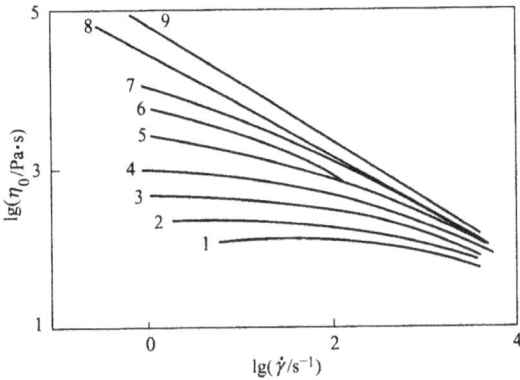

Fig. 6.9: Relationship between melt viscosity and blending ratio of PA_6/ABS blends (250 °C). (1) 100/0; (2) 90/10; (3) 80/20; (4) 70/30; (5) 60/40; (6) 50/50; (7) 40/60; (8) 30/70; and (9) 0/100.

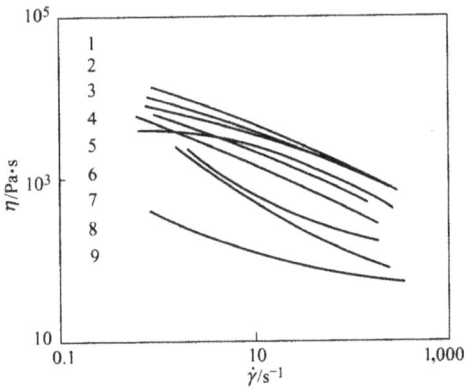

Fig. 6.10: Relationship between melt viscosity and shear rate of PEEK/LCP blend. (1) 97.5/2.5; (2) 100/0; (3) 95/5; (4) 75/25; (5) 50/50; (6) 90/10; (7) 25/75; (8) 10/90; and (9) 0/100.

2. The viscosity of the blend is less than that of the pure component.

Many incompatible blends belong to this situation. For example, the viscosity of PS/polymethyl methacrylate (PMMA) blends is lower than that of pure components at low shear rates, but close to that of PS at high shear rates (see in Fig. 6.11). The reason may be the deformation of dispersed phase under shear stress or the formation of phase slip.

The viscosity of all compositions of polyamide (PA) /LCP blends was significantly lower than that of pure compositions when flowing.

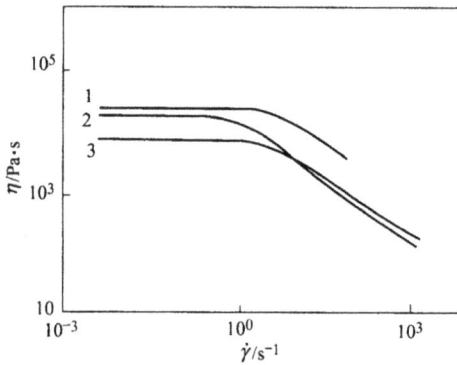

Fig. 6.11: Relationship between apparent viscosity and shear rate of blends at 210 °C. (1) PMMA; (2) PS; and (3) PMMA/PS=20/80.

3. Small proportion of blending produces a larger viscosity drop.

The relationship between the viscosity and composition of PP and copolymer of styrene and 2,2,6,6-tetramethylpiperidine methacrylate (PDS) blends was studied, and it was found that in the range of smaller PDS content (< 1%), the melt viscosity of PP–PDS blends decreased rapidly with the increase of PDS content. But when the content of PDS beyond 1%, the viscosity of the blend decreased slowly with the increase of PDS, as shown in Fig. 6.12.

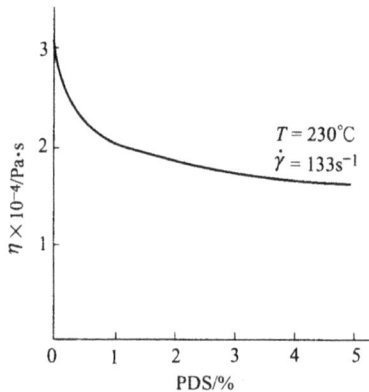

Fig. 6.12: Relationship between melt viscosity and PDS content in PP/PDS blends.

The explanation for this phenomenon is not consistent. One explanation is that this small proportion of blending results in a large viscosity drop which can be explained by the supermolecular structure of the melt. When a small amount of another incompatible polymer is contained in the polymer melt, the supramolecular structure of the melt can be greatly changed and the viscosity can be greatly reduced. As the content of the second polymer continues to increase, the supermolecular structure of the melt no longer changes significantly, so the viscosity changes slowly. It is also believed that a small amount of incompatible second polymer is deposited on the wall of the pipe, making the melt of the blend easy to slip along the wall of the pipe.

4. The viscosity of the blend has a maximum and a minimum.
PP/acrylonitrile-butadiene-styrene copolymer (ABS) blends belong to this type: when the ABS content is 10%, the viscosity of PP/ABS blends has a maximum, and when the ABS content is 20%, the viscosity of PP/ABS blends has a minimum. With the increase of shear stress, the difference between maximum and minimum decreases, as shown in Fig. 6.13.

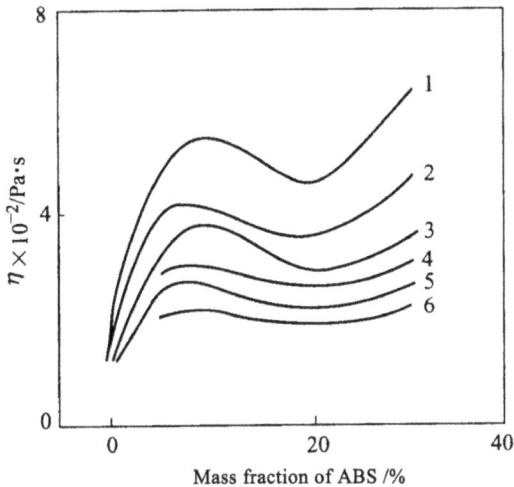

Fig. 6.13: Relationship between melt viscosity and composition of PP/ABS blends.
(1) $\tau = 5.77 \times 10^4$ N/m^2; (2) $\tau = 7.21 \times 10^4$ N/m^2; (3) $\tau = 8.65 \times 10^4$ N/m^2; (4) $\tau = 1.01 \times 10^5$ N/m^2; (5) $\tau = 1.15 \times 10^5$ N/m^2; and (6) $\tau = 1.29 \times 10^5$ N/m^2.

The relationship between the viscosity and composition of PS and polyethersulfone blends (PES) also belongs to this type: at 240 °C, the viscosity with 50% PES content reaches a maximum value, and at 30% and 70% PES content, the viscosity reaches a minimum value, respectively, as shown in Fig. 6.14.

There are chemical bonds between the two phases of block copolymers, and the bonding force between the two phases is very high. At a certain composition ratio, the melt viscosity of block copolymers reaches a maximum. Such as SBS, the relationship between viscosity and composition of melt is shown in Tab. 6.3. It can be seen that when the composition is about 50/50, the viscosity of the melt is the

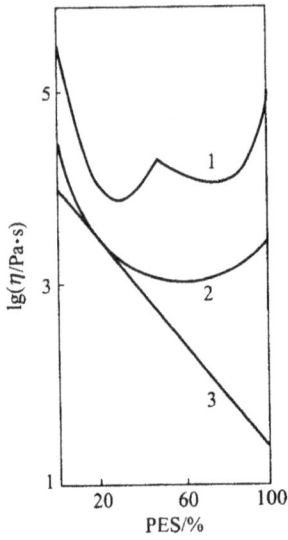

Fig. 6.14: Relationship between viscosity and composition of PS/PES blends at different temperatures. (1) 240 °C; (2) 260 °C; and (3) 280 °C.

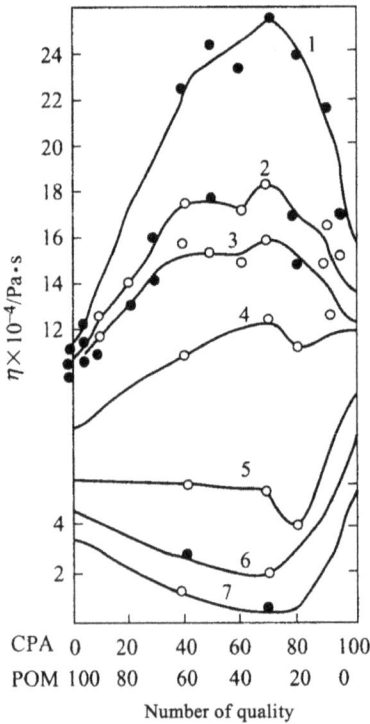

Fig. 6.15: Relationship between viscosity and composition and shear stress of CPA/POM blend melt. (1) $\tau = 1.27 \times 10^4$ Pa; (2) $\tau = 3.93 \times 10^4$ Pa; (3) $\tau = 5.44 \times 10^4$ Pa; (4) $\tau = 6.30 \times 10^4$ Pa; (5) $\tau = 12.59 \times 10^4$ Pa; (6) $\tau = 19.25 \times 10^4$ Pa; and (7) $\tau = 31.62 \times 10^4$ Pa.

Tab. 6.3: Relationship between viscosity and composition of SBS melt.

Polymer	10S-80B-10S	15S-70B-15S	25S-50B-25S	33S-34B-33S	40S-20B-40S
η_a (Pa·s)	3.2×10^2	2.9×10^3	1.18×10^4	2.8×10^3	5.5×10^2

largest. Because the flow of block copolymer melt destroys the two-phase regional structure, when the length of the block is similar, the energy consumed by the destruction of the regional structure is the largest, so the apparent viscosity is the largest.

6.2.2.2 Shear dependence of viscosity

The viscosity of multicomponent polymers melt is not only affected by the composition of the system, but also closely related to the shear conditions (τ and $\dot{\gamma}$) as the single polymer. The effects of τ and $\dot{\gamma}$ on the viscosity of multicomponent polymer melt are manifested in three aspects:

1. For polymer alloy with certain composition, the melt belongs to pseudoplastic fluid under general shear flow conditions, and the shear viscosity decreases with the increase of τ and $\dot{\gamma}$.

2. The relationship between viscosity and composition of polymer melt is affected by τ and $\dot{\gamma}$. For example, the relationship between the viscosity of melt and composition of POM (copolymer of formaldehyde and 2% 1,3-dioxane) and CPA (copolymer of 44% caprolactam, 37% hexanediol adipate and 19% ethylene glycol sebacate) blends is very sensitive to shear stress (see in Fig. 6.15).

3. The apparent viscosity of different polymers has different sensitivity to τ and $\dot{\gamma}$. Furthermore, the shear stress of the blend melt varies linearly in the radial direction when it flows in a capillary tube (or a general pipe), and the shear stress reaches maximum at the wall and reaches zero at the center of the pipe. Therefore, the radical change of shear stress will inevitably affect the viscosity ratio of the two polymer melts, and then affect the dispersion state of the blend melts, resulting in a significant impact on the apparent viscosity of the blend melts, and will complicate the relationship between the viscosity of the blend melts and the viscosity of the component polymer melts. Figure 6.16 shows the relationship between polymer viscosity and melt viscosity of polymer blends affected by $\dot{\gamma}$. Figure 6.16(a) are consistent with mixing laws, such as PP/HDFE and PC/PMMA blends Fig. 6.16(b) are examples of PS/LDPE and PS/PMMA blends and Fig. 6.16(c) are equivalent to cases where the blends have abnormally high viscosity, such as PS/PE (25/75). These types are not fixed and are related to the blending ratio and the range of shear rate. Various possible changes are shown in Fig. 6.16(d)–(g).

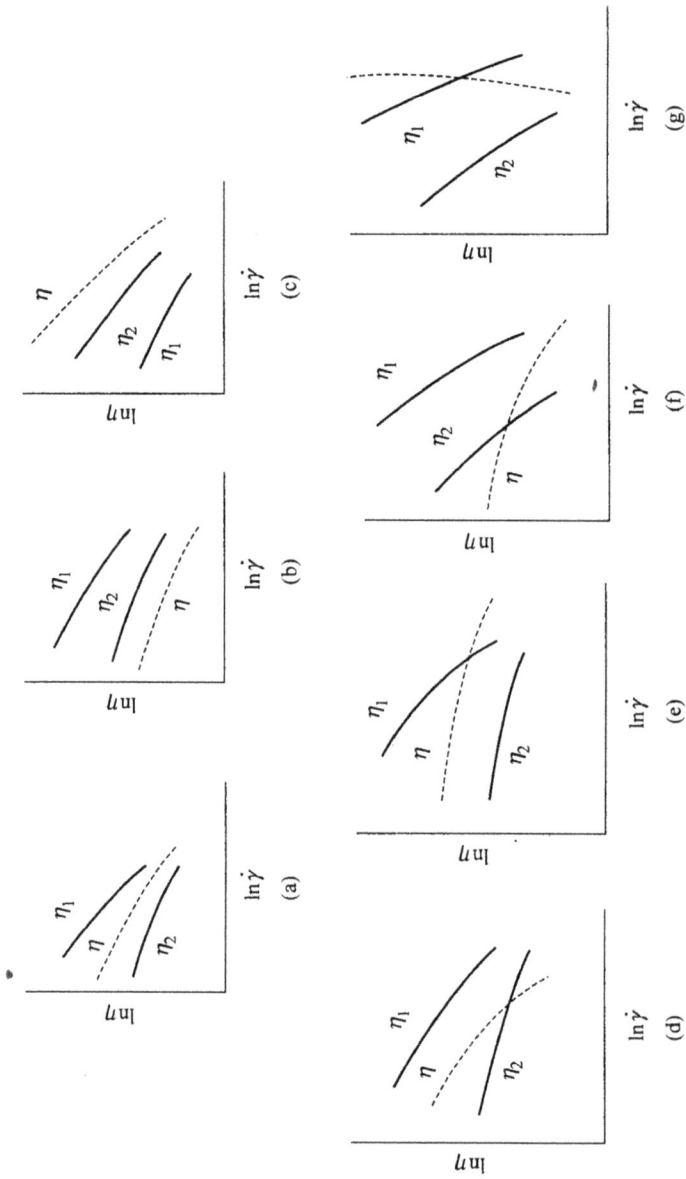

Fig. 6.16: Effect of shear rate on viscosity of blend melt.

6.2.2.3 Relationship between viscosity and temperature

The relationship between melt viscosity and temperature of multicomponent polymer is complex, which can be divided into the following types:

1. In a wide temperature range, the relationship between viscosity and temperature of multicomponent polymer melts conforms to the Arrhenius equation (see eq. 6.1). For example, HDPE/PS blend system.

The relationship between viscosity and temperature described by the Arrhenius equation is established when studying the relationship between viscosity and temperature in shear flow of small molecules. The definition of flow activation energy ΔE_η is very clear: it is the energy needed to overcome the effect of surrounding molecules when molecules transit to holes.

When studying the shear flow of polymer melt, it is considered that the flow is realized by the successive transition of chain segments, and the flow unit is not the whole macromolecule but the chain segments in the macromolecule. The size of the segment is not uniform. In addition, the shear flow of polymer melt has both real viscous flow and high elastic deformation. In this case, it is not reasonable to define the exponential ΔE_η as the flow activation energy. However, although the size and shape of the chain segment is not uniform, chain segment is still a kind of flow unit, and its physical significance is still clear.

But polymer alloy melt is a complex multicomponent multiphase system. The factors affecting the fluidity include the composition, the nature and size of the interfacial force, the size and dispersion of the dispersed phase, the morphology and association and coalescence of the dispersed phase and the difficulty of the deformation of the dispersed phase. There are more than one kind of flow units: not only chain segments of each component but also dispersed phase particles many times larger than the chain segments with different sizes and shapes. And other types of polymer microaggregates may exist. Therefore, the definition and physical meaning of E as flow activation energy must be unclear. For example, in the study of PP/HDPE/EPDM blends, it was found that although the composition of PP/HDPE/EPDM blends changed regularly, the measured ΔE_η was irregular. Therefore, for polymer blends, the ΔE_η obtained by experiment cannot be called flow activation energy, but only a parameter describing the temperature sensitivity of melt viscosity.

2. Influence of temperature on viscosity and composition of polymer blends

The viscosity of polyaryl compounds/LCP blends increases with the increase of LCP content at 280 °C, and decreases with the increase of LCP content at 300 °C. The reason is that LCP has not yet fully melted into an anisotropic liquid crystal state at 280 °C, and these solids act as rigid fillers. LCP melted completely at 300 °C turns into a liquid crystal state, and the orientable LCP at low shear rate decreased the viscosity of the blends.

Other liquid crystalline polymers have similar phenomena.

3. Temperature effects on relationship curves between viscosity and composition of polymer blends

The relationship between the viscosity and composition of PS/PES blends at 240, 260 and 280 °C is shown in Fig. 6.15. At 240 °C, when the PES content is 50%, the viscosity reaches its maximum. When PES content was 30% and 70%, the viscosity reaches their minimum, respectively. When the PES content was 50%, the viscosity reaches minimum at 260 °C, but at 280 °C, there was no maximum and minimum on the curve of the relationship between the viscosity and composition, showing an additive property.

The reason why the rheological behavior of blends varies greatly under different temperatures is melt of blends have different morphology at different temperatures: At 240 °C and 260 °C, the melt of the blend is a heterogeneous system, and the minimum value of the viscosity-composition curve is due to the formation of small flake structure of the dispersed phase in the polymer matrix. At 280 °C, blends mix well, so variation of viscosity with composition has additive property.

When the blend is below the viscous flow temperature and above the glass transition temperature, the relationship between the viscosity and the temperature is usually neither in accordance with the Arrhenius equation nor with the WLF equation. The main reason why it does not conform to the Arrhenius equation is that the cavitation condition of successive transitions of segments and other flow units in blends cannot be satisfied, and the flow is no longer a general activation process. The main reason for noncompliance with WLF equation is that the displacement factor α_T of polymers with different components in the blends is generally different.

6.3 Elastic effect of multicomponent polymer melts

6.3.1 Parameters for characterization of polymer melt elasticity

Polymeric alloy melts, like single polymer melts, flow with elastic deformation in general flow conditions. Because polymer alloys are multicomponent and multiphase systems, the elastic behavior of flow has its particularity and has an important impact on the quality of products processed.

The parameters of polymer melt and elasticity of polymer alloy melt are as follows: die swell ratio $B(d/D)$, first normal stress difference $N_1(\tau_{11}-\tau_{22})$, recoverable shear deformation S_R, apparent shear modulus G, and outlet pressure drop P_{out}. These parameters have the following relations:

$$\tau_{11} - \tau_{22} = 2\tau_w[2(d/D)^6 - 2]^{1/2} \tag{6.11}$$

$$S_R = (\tau_{11} - \tau_{22})/2\tau_w \tag{6.12}$$

$$G = \tau_w / S_R \tag{6.13}$$

$$\tau_{11} - \tau_{22} = P_{OUT} + \tau_w \frac{dP_{OUT}}{d\tau_w} \tag{6.14}$$

where d is the diameter of the effluent, D is the die diameter and τ_w is the shear stress at the tube wall.

6.3.2 Relationship between elasticity and composition of multicomponent polymer melts

The elastic effect of multicomponent polymer melt varies with the composition, and in some compositions, the elastic effect reaches its maximum or minimum. If using three kinds of PP and three kinds of PE with high melt flow rate (H), medium (M) and low (L) to form different types of PP/HDPE blends, the relationship between the large ratio of melt swells and composition under constant $\dot{\gamma}$ and τ conditions is shown in Figs. 6.17 and 6.18.

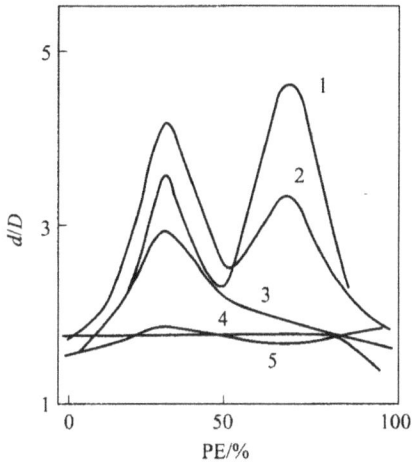

Fig. 6.17: Relationship between d/D and HDPE content in melt of PP/HDPE blends ($\dot{\gamma}$= 100 s^{-1}).
1. PPH/PEM; 2. PPM/PEM; 3. PPM/PEH;
4. PPM/PEL; and 5. PPL/PEM.

It can be seen that when the viscosity of the two components is high (the melt flow rate is low), such as PPM/PEL and PPL/PEM systems, the dilatation ratio of the blends is similar to that of the pure components; when the viscosity of the components is not high, the dilatation ratio of the blends has one or two peaks when the content of PE is 30% and 70% (mass). In fact, the maximum of elastic effect is the

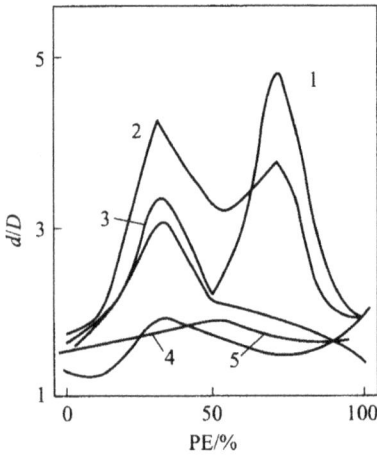

Fig. 6.18: Relationship between d/D and HDPE content in melt of PP/HDPE blends ($\tau = 3 \times 10^4$ Pa). (1) PPH/PEM; (2) PPM/PEM; (3) PPM/PEH; (4) PPM/PEL; and (5) PPL/PEM

minimum of viscosity, while the minimum of elastic effect corresponds to the maximum of viscosity. This is mainly determined by the dispersion state of the blends.

There are many blends with this rheological behavior, such as PE/PS blends, when the composition is 75/25, the elastic effect appears obviously minimum, but when the composition is 50/50 and 25/75, the elastic effect does not show obvious maximum.

6.3.3 Relationship between elasticity and shear of multicomponent polymer melts

Figure 6.19 is the relationship between the recoverable shear deformation S_R and the composition and shear stress τ of PP/ABS blend melt. For the same blend, the shear deformation of the melt increases with the increase of shear stress. When the content of ABS is 20%, the value of recoverable shear deformation reaches maximum, and reaches minimum when the content of ABS is 10%, which corresponds to the minimum value and maximum value of viscosity, respectively. The reason may be that when the content of ABS is 10%, the dispersed phase domains are small, difficult to deform, showing high viscosity and low elastic recovery ability; when the content of ABS is 20%, the dispersed phase domains are large, easy to deform, showing low viscosity and high elastic recovery ability.

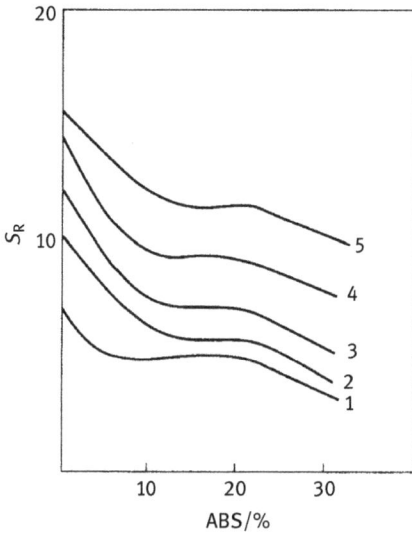

Fig. 6.19: Relationship between recoverable shear deformation and composition and shear stress of melt in PP/ABS blends (200 °C). (1) $\tau = 5.77 \times 10^4$ N/m^2; (2) $\tau = 7.21 \times 10^4$ N/m^2; (3) $\tau = 8.65 \times 10^4$ N/m^2; (4) $\tau = 1.01 \times 10^5$ N/m^2; and (5) $\tau = 1.15 \times 10^5$ N/m^2.

6.3.4 Elasticity of elastomer toughened plastic melt

For common elastomer toughening plastics, such as HIPS and ABS, the Die Swell Ratio d/D is much smaller than the corresponding homopolymer, and the value of d/D decreases with the increase of rubber content. There may be two reasons for this: first, the dispersed phase of elastomers in HIPS and ABS is moderately cross-linked, which contains a number of sausage-like structure of matrix resins and second, there are chemical bonds between elastomers and matrix resins, and there is a strong interaction and a large contact area, which makes the elastic deformation ability of the elastomer dispersed phase small, and a amount of energy is consumed at the interface energy, showing a higher viscosity and smaller elasticity. Block copolymers with chemical bonding between phases, such as SBS, have generally smaller elastic effects than corresponding homopolymers.

For blend modified plastics consisting of ABS containing elastomer, such as PP/ABS blends, the elastic behavior of the melt is shown in Fig. 6.20. The recoverable elastic deformation increases with the increase of shear stress τ_W, the recoverable elastic deformation decreases with the increase of ABS content, and the recoverable elastic deformation of pure PP is the largest. It shows that the elastic deformation of ABS droplets dispersed in PP matrix is very small. Moreover, these droplets may also absorb some deformation energy of the surrounding matrix and reduce the elasticity of the blends, so the addition of ABS can make the blends produce smooth surface at high shear rate.

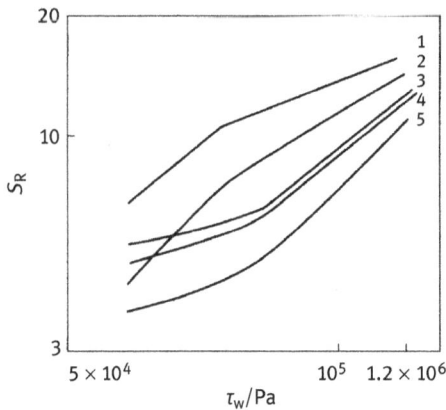

Fig. 6.20: Relationship between recoverable shear deformation and shear stress of melt in PP/ABS blends system (200 °C). Mass fraction of ABS %; 1,0; 2,5; 3,10; 4,20; and 5,30.

6.3.5 Influence of elastic effect of multicomponent polymer melts on molding process

6.3.5.1 Deformation, rotation and orientation of dispersed phase particles

The blended polymer melt is in a single continuous state containing deformable dispersed phase particles, such as elastomeric toughened plastics, when shear flow occurs, the dispersed phase particles are subjected to two kinds of stresses: tangential force and normal force, resulting in rotation, deformation and orientation. The rotation angle, orientation and deformation degree depend on the viscosity ratio, elastic ratio, interfacial tension, particle radius and shear rate of the two phases. The greater the particle radius and shear rate, the greater the degree of deformation and orientation.

For example, in injection molding, due to the rotation, deformation and orientation of the dispersed phase particles, the interior of HIPS injection products will present a three-layer structure, as shown in Fig. 6.21: surface layer, rubber particles

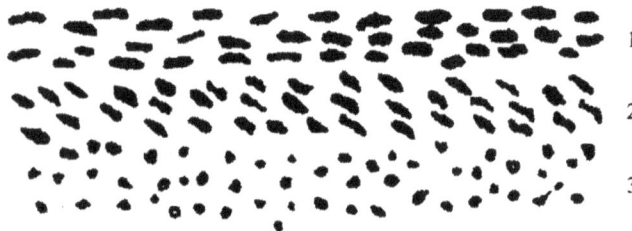

Fig. 6.21: Schematic diagram of rubber particle orientation in HIPS injection molded products. (1) The surface layer; (2) the shear layer; and (3) the central layer.

are ellipsoidal, the orientation of its long axis parallel to the flow direction, shear layer, rubber particles are ellipsoidal, its long axis is at an angle to the flow direction and center layer, rubber particles are basically spherical. The reason for this is that in injection molding, the material is stretched and deformed after entering the mold, and the rubber particles are deformed and oriented. After contacting with the cold die wall, the melt is cooled and solidified, and the deformed and oriented rubber particles have no time to recover, and their deformation and orientation state are frozen, thus the earlier surface layer is formed. The farther away from the die wall, the smaller the shear stress on the melt, and the smaller the deformation and orientation of the rubber particles. On the other hand, the slower the melt cooling rate is, the more time for the elastic deformation to relax and recover, resulting in the formation of surface layer, shear layer and center layer in turn.

Extrusion molding is different from injection molding. First, the shear rate of the melt flow in the extrusion molding process is lower; second, the temperature of the extrusion molding at the die orifice is higher than that of the injection mold, and the cooling rate of the product after leaving the mold is slower, so that the elastomer particles with a low degree of deformation and orientation may get sufficient time to relax the elastic deformation. This will have a negative effect: due to the elastic deformation and disorientation of the dispersed phase particles in the surface layer, the surface of the product will be wavy and lose its luster, as shown in Fig. 6.22. The surface distortion is related to the size of dispersed phase particles, the more the size of particles, the more heavy this phenomenon. This is the reason why the surface of HIPS extrusion products is coarser than the corresponding ABS products (HIPS have larger dispersed phase particles, while ABS have smaller dispersed phase particles).

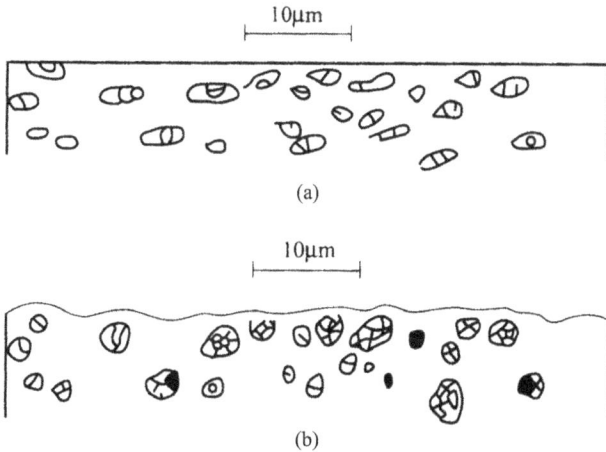

Fig. 6.22: Forming mechanism of wavy surface of HIPS extruded products. (a) Surface before relaxation and (b) A wavy surface after relaxation.

6.3.5.2 Transport phenomena of dispersed phase particles

In the process of flowing of elastomer toughened plastic melt, due to the different shear rates between the wall and the central axis area, as well as the deformation and orientation of elastomer particles disturbing the flow of surrounding materials, elastomer particles will radically migrate from the wall to the central axis, resulting in radial gradient of the concentration of elastomer particles. The larger the dispersed particles and the higher the shear rate, the more obvious the radial migration.

This radial migration, because of the reduction of the concentration of dispersed phase particles in the surface layer of the product, can improve the brightness of the product surface, but reduce the surface toughness of the product. If the radial migration degree of dispersed phase particles is too large, it may even result in delamination in the interior of product, thus affecting the strength of the product.

Exercises

1. Briefly describe the characteristics of shear flow of polymer solutions.

Answer:
(1) Through the coordinated rearrangement of chain segments along the flow direction.
(2) Under general shear flow conditions, the flow of polymer melts does not conform to the law of Newton's fluid.
(3) The viscosity of polymer melts is much larger than that of small molecules.

2. Please outline the factors affecting the apparent viscosity of polymer melts

Answer:
The influence of molecular structure of polymers: (1) molecular weight, (2) molecular weight distribution and (3) branching of molecular chains.
The influence of processing conditions: (1) temperature, (2) shear rate and shear stress and (3) pressure.

3. Briefly describe the relationship between melt viscosity and composition of polymer alloys.

Answer:
(1) The viscosity of the blend is between two pure components.
More kinds of blends belong to this situation such as PA6/ABS blends. Some blend systems with liquid crystal polymers also belong to this type.
(2) The viscosity of the blend is less than that of the pure component.

Many incompatible blends belong to this situation. For example, the viscosity of PS/PMMA blends is lower than that of pure components at low shear rates, but close to that of PS at high shear rates. The reason may be the deformation of dispersed phase under shear stress or the formation of phase slip.

(3)   Small proportion of blending produces a larger viscosity drop.

(4)   The viscosity of the blend has a maximum and a minimum.

PP/ABS blends belong to this type: when the ABS content is 10%, the viscosity of PP/ABS blends has a maximum, and when the ABS content is 20%, the viscosity of PP/ABS blends has a minimum.

4. What is the elastic effect of multicomponent polymer melts and what is its significance for molding?

Answer:

Polymeric alloy melts, like single polymer melts, flow with elastic deformation in general flow conditions. Because polymer alloys are multicomponent and multiphase systems, the elastic behavior of flow has its particularity and has an important impact on the quality of products processed.

1. Deformation, rotation and orientation of dispersed phase particles

The blended polymer melt is in a single continuous state containing deformable dispersed phase particles, such as elastomeric toughened plastics, when shear flow occurs, the dispersed phase particles are subjected to two kinds of stresses: tangential force and normal force, resulting in rotation, deformation and orientation. The rotation angle, orientation and deformation degree depend on the viscosity ratio, elastic ratio, interfacial tension, particle radius and shear rate of the two phases. The greater the particle radius and shear rate, the greater the degree of deformation and orientation.

2. Transport Phenomena of Dispersed Phase Particles

In the process of flowing of elastomer toughened plastic melt, due to the different shear rates between the wall and the central axis area, as well as the deformation and orientation of elastomer particles disturbing the flow of surrounding materials, elastomer particles will radically migrate from the wall to the central axis, resulting in radial gradient of the concentration of elastomer particles. The larger the dispersed particles and the higher the shear rate, the more obvious the radial migration.

Chapter 7
Other characteristics of multicomponent polymers

7.1 Permeability of multicomponent polymers

7.1.1 Introduction

When the pressure on both sides of the polymer film is unequal, the gas molecules diffuse from the side of high pressure to the side of low pressure, which is called the air permeability of the polymer. The two sides of the film are solutions of different concentrations; the molecules diffuse from the side of high concentration to the side of low concentration, which is called the permeability of the polymer.

The permeability and air permeability of polymers are of great practical significance. They have important applications in film packaging, purification, separation, sewage purification and medicine. Using external energy and chemical potential difference as driving force, the method of separating, purifying and enriching multicomponent solutes and solvents by polymer films is called membrane separation method. Due to the different pore size distribution and separation objects of the membrane, they can be reverse osmosis membranes (pore size 0.3–2 nm), ultrafiltration membranes (pore size 1.5–50 nm), microfiltration membranes (pore size 0.03–10 µm), electrodialysis membranes and gas separation membranes. In recent years, membrane separation technology has become a very attractive subject in the field of polymer science, and has a very broad application prospect. For example, ion exchange membranes have reached the level of industrial application in desalination and concentration of seawater. It has been widely used in the fields of purification, separation of substances, treatment of wastewater and electrolytic diaphragm. Some new applications are being developed, such as ultrafiltration, artificial kidney and artificial heart pacemaker, which have shown good results.

In addition to high mechanical strength, high selective permeability and high-penetration rate are the most important requirements for materials used as polymer separation membranes. Single-component polymer films are difficult to meet the comprehensive requirements of many aspects, and multicomponent polymers show more advantages than homopolymers in this separation technology. For example, the blend diaphragm made of cellulose triacetate and cellulose diacetate has high-selective permeability for desalination of seawater, and 99.9% salt can be removed by one-way treatment; the permeation rate of the diaphragm made of polyvinylpyridinone and polyurethane is twice as high as that of the ordinary cellophane diaphragm; and the ultrafiltration membrane made of sulfonated polystyrene and vinylidene fluoride can be used to concentrate biological proteins, treat sewage and purify pulp wastewater. Using blending technology to improve the permeability

https://doi.org/10.1515/9783110596335-007

and air permeability of polymer materials has become another important application field of multicomponent polymers. At the same time, the permeability of blends is closely related to the morphology and structure of blends, so the study of the permeability and air permeability of blends is also an important method to understand the morphology and structure of blends.

7.1.2 Air permeability of multicomponent polymers

The crystal region of polymer is impermeable, so the gas passes through the amorphous region of the polymer. For blends, gas penetration mainly occurs in the continuous phase, and the existence of dispersed phase microzone makes the path of gas molecules penetration become tortuous, thus increasing the distance length and reducing the air permeability.

The degree of tortuosity of gas penetration pathway is related to the morphology of dispersed phase. The factor that characterizes the degree of penetration tortuosity is τ, which is the ratio of the actual gas penetration path length to the thickness of sample. The predecessors have summarized many empirical formulas to describe the relationship between τ and the shape of dispersed phase. When the dispersed phase is spherical:

$$\tau = 1 + \frac{1}{2}\varphi_d \tag{7.1}$$

where φ_d is the volume fraction of the dispersed phase.

When the dispersed phase is flaky, and the flake is parallel to the surface of the sample, the following relationship exists:

$$\tau = 1 + (L/2b)\varphi_d \tag{7.2}$$

If the flake is perpendicular to the surface of the sample, then:

$$\tau = 1 + (b/2L)\varphi_d \tag{7.3}$$

In eqs. (7.2) and (7.3), L is the length of the flake and b is the thickness of the flake.

The relationship between diffusion coefficient P_c and τ of blends can be estimated as follows:

$$P_c = \frac{\varphi_m}{\tau} P_m \tag{7.4}$$

where P_m is the diffusion coefficient of polymer matrix and φ_m is the volume fraction of matrix. By substituting eqs. (7.2) and (7.3) into eq. (7.4), the upper and lower limits of diffusion coefficients of blends can be obtained:

$$P_c = \frac{\varphi_m}{[1+(L/2b)\varphi_d]}P_m \qquad (7.5)$$

$$P_c = \frac{\varphi_m}{[1+(b/2L)\varphi_d]}P_m \qquad (7.6)$$

If $L \gg b$, from eq. (7.6):

$$P_c = \varphi_m P_m \qquad (7.7)$$

It shows that the gas penetration path is the shortest, and the diffusion coefficient of the blend is only related to the volume fraction of the matrix material.

When the dispersed phase is spherical:

$$P_c = \frac{\varphi_m}{[1+1/2\varphi_d]}P_m \qquad (7.8)$$

The contribution of dispersion to the air permeability of relative polymer blends cannot be neglected, sometimes even very large. At this time, the P_c of polymer blends can be estimated by series combination model (eq. (7.9)) or parallel combination model (eq. (7.10)):

$$\frac{1}{P_c} = \frac{\varphi_d}{P_d} + \frac{\varphi_m}{P_m} \qquad (7.9)$$

$$P_c = \varphi_d P_d + \varphi_m P_m \qquad (7.10)$$

Because continuous phase plays a dominant role in the air permeability of blends, when the component with larger diffusion coefficient is in continuous phase, the P_c of blends is close to the calculated value of eq. (7.10), and when the component with smaller diffusion coefficient is in continuous phase, the P_c is close to the calculated value of eq. (7.9). When the two components are completely miscible to form a homogeneous system, the diffusion coefficients of the blends generally conform to the following equation:

$$\ln P_c = \varphi_A \ln p_A + \varphi_B \ln p_B \qquad (7.11)$$

The specific applications of polymer permeability are mainly in film packaging and gas purification and separation, in which oxygen-enriched membranes are the most attractive research and application in gas separation. Oxygen-enriched membranes refer to the permeable gas that passes through the membranes under pressure and gets higher oxygen concentration in the air (oxygen-enriched air) on the other side of the membranes.

Polymer blends can be used as oxygen-enriched membrane materials to obtain the most favorable combination of materials with high permeability and poor selectivity or materials with poor permeability and high selectivity. For example, polydimethylsiloxane (PDMS) has high air permeability to ordinary gases, but its selectivity and strength is not ideal. Block copolymers with PDMS as soft segment and polysulfone (PSF) or polycarbonate (PC) as hard segment can be prepared by copolymerization, and the air permeation rate and separation effect of O_2 could be considered through controlling the ratio of two blocks. In addition, fluorine-containing polymers have good selectivity for oxygen, but the air permeability coefficient is too small. Fluorine and silicone copolymers can be prepared by plasma polymerization of organosilicon and fluorocarbon polymers, and the air permeability of composite membranes is significantly improved. It has been proved that the selective oxygen permeability of PDMS in graft copolymers, block copolymers or blends will not change if the content of PDMS is above 60% (in continuous phase). Therefore, if PDMS is blended with highly separation coefficient materials, it is possible to obtain oxygen-enriched membrane materials with better performance.

7.1.3 Permeability of multicomponent polymers

The permeation of liquid to polymer blends is similar to that of gas, but when vapor and liquid pass through the film, swelling often occurs, which changes the relaxation properties of the blends. Therefore, the permeability coefficient of the copolymer is highly dependent on the concentration of the liquid.

As the liquid separation membrane material, blends show excellent separation effect. At present, most of the membrane materials used for separating organic solvents such as absolute ethanol is blend membranes, such as polyvinylpyridine ketone/polyacrylonitrile blend membranes and polyvinyl alcohol/PSF blend membranes. They are used for separating water/ethanol, water/tetrahydrofuran, water/dioxane and other systems, and have good separation effect. The permeability coefficient of the blend consisting of polyvinyl alcohol and polyacrylamide-sodium acrylate copolymer (for H_2O/C_2H_5OH system) is higher than that of polyvinyl alcohol, and increases with the increase of acrylamide content in the copolymer. This is because the copolymer is more hydrophilic than polyvinyl alcohol, and the solubility of water increases, leading to the increase of permeability coefficient. At the same time, after high-energy radiation, the separation coefficient of the blend used N,N-methylacrylamide as cross-linking agent is higher than that of single component. In general, the separation coefficient will decrease with the increase of permeability coefficient. This is because the copolymer is easy to cross-link with N,N-methacrylamide under radiation conditions, which leads to the increase of separation coefficient.

Cellulose acetate and aromatic polyamide (PA) can selectively absorb water and have high absorptivity while salt is selectively rejected. They can be blended or grafted with other polymers to make excellent membrane for desalination separation.

7.2 Barrier properties of multicomponent polymers

Polymer materials play an increasingly important role in packaging field. Compared with traditional packaging materials such as metal and glass, they have the advantages of light weight, easy forming and processing, and unbreakable. However, their barrier property to gases and organic solvents is very poor, which greatly limits the application of polymer packaging materials. By using polymer mixing technology and improving barrier properties of different polymers and their structural characteristics, polymer can be widely used in the field of packaging materials.

7.2.1 Barrier properties of polymers

The barrier property of polymer refers to the shielding ability of polymer products to small molecule gases, liquids, vapors, spices and medicinal flavors. The transmission coefficient is used to characterize the barrier ability of polymer. The smaller the transmission coefficient of polymers, the higher the barrier capacity. The barrier of materials to small molecules such as H_2O, CO_2, O_2 and gasoline is often considered in practical application. The permeation process of small molecules into polymers can be divided into four stages: (1) small molecules adsorbed on the surface of polymers; (2) small molecules dissolved in polymers; (3) small molecules passing through polymers at a certain concentration gradient and (4) small molecules desorbed on the other surface of polymers. Anything that can delay any one of these processes is conducive to improving the barrier properties of polymers.

The barrier ability of polymers to small molecules of gases and liquids under certain external conditions depends on their molecular structure and aggregate structure. The permeation of small molecules with different polarities in different polar polymers is quite different. For example, water molecules can easily pass through PA and ethylene-vinyl alcohol copolymer (EVOH) with hydrophilic segments, while nonpolar benzene and gasoline can hardly pass through these two polymers. Conversely, benzene and gasoline are easy to penetrate through nonpolar polymers of polyolefins, while water molecules are difficult to penetrate through such polymers. In addition, the more compact and orderly the polymer aggregates are, the better its barrier properties are. Generally, crystallization and orientation are beneficial for improving the barrier properties of polymers. PA, polyethylene naphthalate (PEN), random copolymer of EVOH and polyvinylidene chloride (PVDC) can be used as barrier resins because of their special chain structure. Here is a brief introduction.

1. PA: PA is a kind of barrier polymer. But the barrier properties of PA of different varieties are quite different. In recent years, various kinds of high-barrier PA have appeared one after another. In 1976, Toyobo Company of Japan successfully developed high-barrier PA resin MXD 6. In the 1980s, DuPont Company also developed modified PA6/66 copolymer (brand Selar RB) and amorphous copolyamide Selar PA with excellent barrier properties. In addition, Umo Japan and Allied Chemical Company have developed various high-barrier PA resins.

High-barrier PA has excellent performance at low temperature and high temperature, good mechanical properties, better barrier performance than EVOH and PVDC at high temperature and humidity, and lower price. Therefore, it has been widely used as a high-barrier material.

2. EVOH: EVOH is a random copolymer of ethylene and vinyl alcohol. It has good crystallinity and is the best barrier material at present. The existence of polar ethylene alcohol chain in EVOH resin makes it have good barrier to nonpolar solvents such as hydrocarbons. The existence of nonpolar-ethylene chain end helps to improve its barrier to polar solvents such as water. Therefore, it can effectively block the penetration of oxygen, carbon dioxide, water vapor and other gases, ensure the taste and flavor of food, while significantly prolonging the price life and storage period of food, and has a wide range of applications in food packaging. EVOH resin can also be used in packaging of chemical solvents, pharmaceutical products, cosmetics and electronic products and has good barrier to organic solvents. EVOH can be blended and compounded with other resins to produce packaging materials with better properties. PE, PA, PET, PP and so on are commonly used for blending and compounding with EVOH.

3. PEN: PEN is the polyester obtained by polycondensation of naphthalic acid and ethylene glycol. Its structure is similar to that of PET. However, naphthalene ring is a larger conjugated molecular structure than benzene ring, so the molecular chain is more rigid and more plane distribution. The application of PEN in container packaging has the following advantages: (1) good gas barrier (about 5 times that of PET); (2) absorption of ultraviolet radiation below 383 nm, which can avoid food deterioration caused by ultraviolet radiation; (3) high glass transition temperature, good heat resistance; (4) good chemical resistance. PEN is an ideal packaging material for beverages, food, condiments, cosmetics and pharmaceuticals due to its excellent characteristics.

Polypropylene naphthalate (PTN) with similar structure is a new type of polyester developed by Shell Chemical Company. It is said that the carbon dioxide barrier of this PTN resin is 18 times higher than that of PET and 3.5 times higher than that of PEN. The barrier to oxygen is 9 times higher than that of PET and 1 time higher than that of PEN. PTN resin also has good thermal stability. Plastic bottles processed with this material will not crack when placed under pasteurization conditions (60 °C, 30 min).

Futura Polyester Company of India made plastic bottles by blending PET and PTN. Its CO_2 barrier is twice as high as that of PET bottles, and its processing cost is comparable to that of glass bottles.

4. PVDC: PVDC has excellent barrier performance, and has the characteristics of flame retardant, transparent, oil resistant and chemical resistant. Composite materials with PVDC as barrier resin have excellent properties in flavor insulation, moisture-proof, oil-proof and antipermeability, and prolonging shelf life of food. For example, the coating of ham sausage is mainly prepared by PVDC. However, there are some problems in PVDC, such as high cost, monomer residue and material recovery.

7.2.2 Barrier properties of blended polymers

DuPont Company, USA, first proposed the barrier technology of blended polymer. They developed Selar RB barrier resin in the early 1980s. Later, they also obtained high-density polyethylene (HDPE)/PA blends with layered structure by morphology control, which had good barrier properties.

Most two-phase polymer blends are thermodynamically incompatible systems, and their two-phase dispersion structure can be divided into: (1) island structure, in which the dispersed phase is dispersed in continuous phase in granular, rod or ellipsoid form; (2) dispersed phase is dispersed in continuous phase in the form of a large number of microfibers; (3) dispersed phase is dispersed in continuous phase in the form of two-dimensional thin layers. The aspect ratio of dispersed phases of the two structures ((2) and (3)) is much larger than that of the first (1). When the aspect ratio reaches a certain level, the single continuous structure becomes an approximate two continuous structure, which is called the primary continuous phase and the secondary continuous phase respectively. The barrier polymer (PA) can form a subcontinuous layer structure in the matrix resin continuous phase, which can improve the barrier of the matrix resin to small gas–liquid molecules.

The interfacial morphology of polymer binary blends is related to the viscoelasticity of the melt in the flow field. Viscosity plays a major role in determining the equilibrium shape of the interfacial between the two fluids. According to rheological theory, two kinds of polymer melts with different viscosities always move according to the "soft–hard" rule, that is, the fluid with low viscosities should automatically coat the fluid with high viscosities, which requires the least energy, that is, the principle of minimum energy. In blending, the viscosity of dispersed phase must be greater than that of continuous phase in order to form normal laminar flow of two phases, so that the dispersed phase cannot be coated by continuous phase. When the viscosity of the dispersed phase is greater than that of the continuous phase, it will not disperse into droplets to break up the melt under the shear force, but will be easily elongated and oriented under the shear force, which is conducive to the formation of layered dispersion structure.

From the barrier mechanism, the layered structure improves the barrier properties of materials by extending the diffusion path of permeable molecules in the

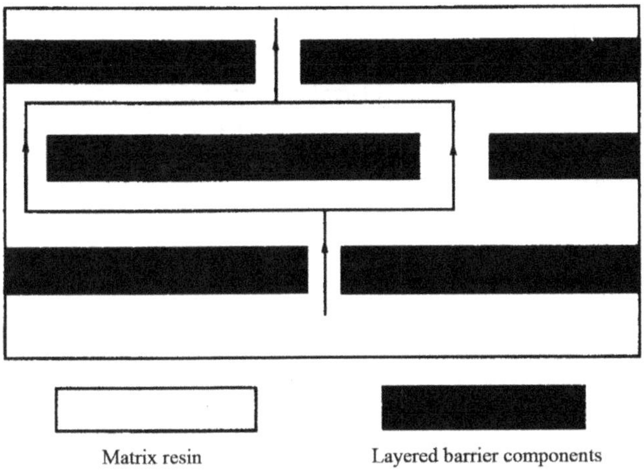

Fig. 7.1: Schematic diagram for improving barrier property of plastics with layered structure.

matrix. As shown in Fig. 7.1, if white represents HDPE and black represents PA, small molecules such as gasoline must bypass barrier layer PA to continue to penetrate into the material. The permeability path of small molecules is greatly prolonged, so the permeability of small molecules decreases in a limited time. In terms of kinetics, Chengmin Lian et al. proved that the mass transfer process of hydrocarbon solvents through HDPE/PA layered blends conformed to the dissolution–diffusion model.

$$D/D_0 = a_1 e^{b1/\varphi} \tag{7.12}$$

where D_0 is the solvent diffusion coefficient of HDPE, D is the solvent diffusion coefficient of HDPE/PA laminar blends, a_1 and b_1 are dimensionless coefficients and φ is the volume fraction of PA sheets. The larger the size of the barrier subcontinuous phase, the better the barrier effect.

7.2.3 Influencing factors of barrier property of blend polymer

The barrier properties of blended polymers are controlled by the composition of materials and processing conditions. Figure 7.2 illustrates the preparation process and control factors of layered blends with polyolefin as an example. According to the rheological principle of laminar blending, in order to satisfy the viscoelastic conditions during processing, the relative molecular mass and its distribution of raw materials must be properly selected. In general, the main continuous phase component should select the raw materials with higher melt flow rate, while the secondary continuous phase component requires less melt flow rate. HDPE/PA is the most widely studied

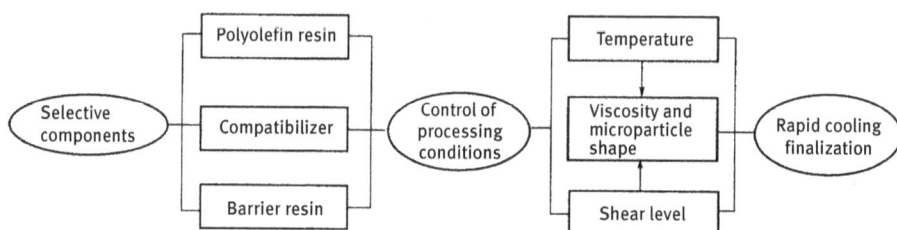

Fig. 7.2: Preparation process and control elements of polyolefin layered blends.

barrier system of layered blends. This may be because the melt flow rate values of HDPE and PA match well and the processing conditions can be easily controlled.

7.2.3.1 The effect of compatibilizer on barrier property

Resins with good barrier to nonpolar hydrocarbons usually have larger polarity and poor compatibility with nonpolar polyolefins. Therefore, suitable compatibilizer and its dosage are indispensable factors for obtaining layered blends. The two phases of the blend have good cohesion and the layered dispersion form is stabilized, only under the strong compatibilizing effect of compatibilizer.

Compatibilizers generally include ionic copolymers and modified polyolefin graft copolymers. The amount of compatibilizer should not be too large; otherwise, the layer structure will be eroded and dispersed to form "island" structure, which cannot improve the barrier property of materials. A barrier vessel with xylene permeability of only 0.3% was prepared by HDPE/PA layered blending technology. They tried to use noncompatibilizer and use chemically modified PE, chemically modified PP and ionomer as compatibilizers respectively. The permeability of dimethylbenzene for 28 days at room temperature was 2.35%, 0.64%, 0.91% and 0.75%, respectively. The results indicate that the effect of matrix resin graft as compatibilizer is the best. EVOH/HDPE layered blends were prepared by using Zn^{2+} ionomer as compatibilizer. The obtained materials have good mechanical properties and barrier properties in high humidity environment. In addition, HDPE/EVOH high-barrier materials were prepared by using HDPE-g-MAH as compatibilizer, confirming that the compatibilizer could largely change the melt viscosity ratio of EVOH and HDPE, and improve the elasticity of EVOH, making it easier to disperse in HDPE and form layered distribution. When there are too much compatibilizer in the barrier masterbatch, it is beneficial to adjust the melt viscosity ratio of the dispersed phase to the matrix phase to be close to 1, but at the same time, it will cause the melt elasticity of the dispersed phase to be too large, which is not conducive to the formation of layered morphology, and on the contrary, it will affect the rheological properties of the blend system.

HDPE/PET layered blends were prepared by using ethylene-acrylic acid copolymer and ethylene-vinyl acetate copolymer as compatibilizers to research the relationship between submicroscopic phase and compatibility of layered blends. It was concluded that the trend of PET layering increased with the increase of compatibility of the system, but when the content of compatibilizer exceeded a certain value, layered dispersed phase states begin to be destroyed from the middle. Then the concept of stratified compatible zone is proposed.

7.2.3.2 Effect of viscosity ratio of two-phase melt on barrier property

For blended HDPE/PA6 layered barrier materials, PA6 is uniformly distributed in the polyethylene matrix with layered structure, which is conducive to the improvement of barrier properties. The melt viscosity between PA6 dispersed phase and HDPE continuous phase is close, which is beneficial to the adhesion of the interface. The good bonding between the two phases determines the formation of uniform PA6 layered morphology.

However, some researchers believe that the viscosity of dispersed phase should be higher than that of continuous phase in order to make the dispersed phase distribute in layers. The results show that the greater the difference of melt viscosity between PA6-dispersed phase and HDPE continuous phase, the more favorable the formation of uniform-PA6-layered morphology. Therefore, in order to obtain large deformation of dispersed phase droplets, it is necessary to make the two-phase viscosity ratio sufficiently large.

The main reason for the divergence of these results is that in the blending process, the formation of layered structure of dispersed phase is more complex, and there are many influencing factors, so the results are not consistent.

7.2.3.3 Effect of forming processing conditions on barrier property

(1) Effect of processing temperature

The viscosity of polymer melt is not only affected by shear, but also strongly depends on temperature. Therefore, in order to make the melt viscosity of the dispersed phase higher than that of the continuous phase, besides choosing suitable blending components, reasonable processing temperature is the key to prepare layered blend products.

Reasonable processing temperature is conducive to the dispersal of barrier resins in matrix resins with layered structure, and the determination of processing temperature is mainly related to the types of barrier resins. If PA6 is chosen as barrier resin, the general processing temperature is about 235 °C. Some researchers believe the viscosity of PA6 dispersed phase is higher than that of PE at a reasonable processing temperature (about 240 °C), which is conducive to the formation of the layered structure of the dispersed phase. When preparing HDPE/PA6 hollow container, the temperature is over 235 °C, forming is difficult, and the layered structure of PA6 is difficult to form, and the barrier performance decreases.

(2) The influence of shear force

In two-phase flow of melt blends, spherical particles of dispersed phase need to be sheared properly before they can be stretched into lamellar structure. The shear force is too small to make the droplets of dispersed phase oriented, deformed and stretched. The shear force is too large to break the droplets of dispersed phase into small and uniform droplets. When the shear force is too large, unstable turbulence will occur in the two phases. It is generally believed that too high screw speed makes the shear rate of the blend larger, the dispersed phase easier to refine, and the barrier performance decreased. The screw speed is too low and the shear rate is too small to make the dispersed phase droplets deform and orient, and the barrier performance will not be high. The results show that with the increase of the screw speed of extruder, the barrier performance of HDPE/PA6 hollow container first decreases and then increases. There is an optimal speed, and the optimal speed varies with the volume fraction of dispersed phase.

(3) The influence of mixing method

Different mixing methods also have great influence on the barrier properties of the blends. When HDPE, PA6 and compatibilizer are mechanically mixed and added into extruder for extrusion blow molding, the barrier property of the product is better. However, after PA6 and compatibilizer were first fused and mixed to produce masterbatch, and then extruded and blow molded together with HDPE, the permeability of the product to hydrocarbon solvent was 5 times as high as that of the product prepared by the former mixture method.

In addition, because HDPE and PA both are semicrystalline polymers, the change of crystallization behavior after blending will have a greater impact on the barrier properties and other properties of products.

7.3 Transparency of multicomponent polymers

7.3.1 Transparency of polymer blends

Most amorphous polymers are clear and transparent when they do not contain impurities, fillers and crazes. But most incompatible blends do not have optical transmittance. For example, PS, PC and PMMA are transparent polymers. When they are blended with other polymers, they usually become less transparent or opaque materials. This is due to the reflection and refraction of light at the interface of multicomponent polymers. However, if the particle size of the dispersed phase is small or the refractive index of the two polymers is close, the blend can still be transparent. Therefore, it is possible to prepare polymer alloys with high-optical transmittance and excellent mechanical properties.

When two polymers with similar refractive index form blends, although micro-areas of dispersed phase exist, the light from the continuous phase matrix into the dispersed phase area will not cause a large amount of scattering of light, and the blends still show good transmittance. In the blend system of rubber-toughened plastics, the rubber particle size is smaller than the wavelength of visible light. Due to the phenomenon of light diffraction, the toughened plastics have a certain trans-mittance. The smaller the size of rubber particles, the better the transmittance. For example, SBS resin with polybutadiene as disperse phase and PS as continuous phase contains 20% rubber, but its particle size is very small (0.040–0.050 μm), so it still has good light transmittance and can be used as food packaging materials. But for general toughened plastics such as HIPS, in order to meet the requirements of toughening effect, the size of rubber particles cannot be too small, and the refrac-tive index of rubber and plastics is quite different, so it is impossible to be transpar-ent. When the content of rubber is constant and the refractive index of the two polymers is constant, how big is the rubber particle to meet the requirements of transmittance? How can the refractive index of the two components of polymer match to be transparent when the size of rubber particles is fixed? The following is to discuss the relationship between them around these problems and provide the basis for the development of transparent blend products.

7.3.2 Transparency of rubber toughened plastic blends

7.3.2.1 Relationship between rubber content, rubber particle size, refractive index and transparency

In order to study the relationship between rubber content, particle size, refractive index and light transmittance of blends, several hypotheses should be made:
(1) The dispersed phase particles are spherical particles with the same size.
(2) Each phase is isotropic and does not absorb visible light.
(3) Only the light scattering of rubber particles is considered, and the interference effect between scattered light is not considered.
(4) The scattering of other impurities is not considered.

The transmittance of materials is usually expressed by the ratio of transmittance (I) to incident intensity (I_0):

$$\text{Transmittance} = \frac{I}{I_0} = e^{-\tau x} \tag{7.13}$$

where x is the thickness of the sample and τ is the turbidity. It can be seen from the formula that the smaller τ, the better light transmission. τ is mainly determined by the size of dispersed phase size and scattering coefficient K. If the number of colloidal particles is N and the radius of particles is r, then:

$$\tau = KN\pi r^2 \tag{7.14}$$

where K is related to the refractive index and particle size of the two components in the system. When the particle size is less than 1/10 of the incident wavelength, it has the following relationship:

$$K = \frac{8\alpha^4 (m^2 - 1)^2}{3(m^2 + 2)^2} \tag{7.15}$$

where m is the ratio of refractive index of dispersed phase to continuous phase, that is, $m = n/n_o$, α is the ratio of particle circumference to wavelength (λ) of light in medium, that is, $\alpha = \pi d/\lambda$ (d is particle diameter).

From the above relations, if the K value is controlled within an appropriate range, the material can have a certain degree of transparency. When the thickness of the sample is constant, the rubber content and the refractive index of the two components are known, the size of the rubber that achieves the required transmittance can be predicted. If the rubber content and particle size are known, the difference between the two components of the required transmittance can be calculated. Table 7.1 lists the matching data between the transmittance calculated from the above relationship and the volume fraction of rubber phase φ_B, relative deviation of refractive index ($m-1$) and particle size. The data in the table show that the smaller the particle size, the better the transmittance of the material. The smaller the value of $|M-1|$ is, the larger the allowable particle size will be when the transmittance reaches the same level.

Tab. 7.1: The relationship between transparency, volume fraction, particle size and relative refractive index of rubber phase[1].

φ_B	Particle size $m-1$	I/I_0			
		0.10	0.50	0.80	0.90
0.10	0.050	0.068	0.045	0.031	0.024
	0.025	0.11	0.072	0.049	0.038
	0.010	0.33	0.14	0.091	0.071
	0.005	1.3	0.40	0.14	0.12

Tab. 7.1 (continued)

φ_B	Particle size $m-1$	I/I_0			
		0.10	0.50	0.80	0.90
	0.050	0.054	0.036	0.025	0.019
0.20	0.025	0.085	0.057	0.039	0.030
	0.010	0.17	0.11	0.072	0.056
	0.005	0.66	0.20	0.12	0.090
	0.050	0.047	0.031	0.022	0.017
0.30	0.025	0.075	0.050	0.034	0.027
	0.010	0.14	0.092	0.062	0.049
	0.005	0.44	0.13	0.10	0.078

$^1\lambda = 0.5893$ μm, $n_0 = 1.60$, $x = 3.3$ mm.

In addition, the transmittance of PVC/MBS blends is related to the content of components. The relationship between MBS dosage and transmittance and turbidity is shown in Tab. 7.2. With the increase of MBS dosage, the transmittance of the blends decreased while the turbidity increased. When MBS dosage exceeded 11 phr, the transmittance decreased significantly. With the increase of MBS dosage, the turbidity of blends increases. Besides the above factors, the turbidity of MBS (6%) is higher than that of PVC (2.65%) and the interaction between them is also related.

Tab. 7.2: Relationship between MBS dosage and transmittance and turbidity.

MBS phr	0	3	7	11	15
Transmittance(%)	89.3	87.1	86.6	85.5	84.1
Turbidity (%)	2.65	4.21	5.56	6.44	8.36

7.3.2.2 Effect of temperature on transparency

Because the refractive index of different polymers is affected by temperature, the transmittance is also affected by temperature. For example, in the system of rubber-toughened plastics, the refractive index of rubber phase is much more sensitive to temperature than that of plastics, as shown in Fig. 7.3. Therefore, the refractive index of the two components varies greatly at one temperature, and the blend is opaque, whereas at another temperature it is possible that the refractive index becomes similar and transparent.

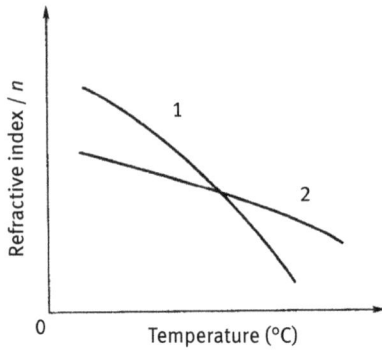

Fig. 7.3: Effect of temperature on refractive index of polymer.

7.3.2.3 Effect of particle size on transparency

For systems with different particle sizes, the influence of temperature on transmittance is different. Figure 7.4 shows the transparency of the blends of polyacrylonitrile and polyisoprene with two particle sizes at different temperatures. It can be seen when the particle size is 0.1 μm, the blends have good transparency in a wide temperature range. When the dispersed phase particles are 1 μm, the transparency zone is greatly reduced, only the blends under nearly 30 °C are transparent.

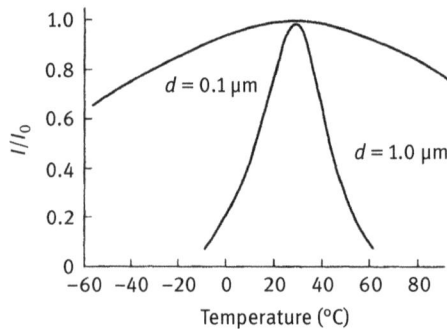

Fig. 7.4: The relationship between transparency and particle size.

7.3.2.4 Effect of complexes on transparency

For blends with various complexes, the selection of complexes has a great influence on the transmittance. In general, nonpolluting, colorless or light-colored liquid additives should be used. The particle size of the powder additive should be as small as possible, and the refractive index should be close to the polymer component. For example, most of the fillers for transparent rubber products are transparent white carbon black, which is a kind of highly active fine particle silica. Magnesium carbonate is sometimes used as a transparent rubber filler, and its refractive index is close to that of rubber.

7.3.3 Refractive index and transmittance of copolymers

Both rigid and elastic block copolymers have much better light transmittance than general blends, which is determined by the structure characteristics of microphase separation of block copolymers. Unless the molecular weight is very high, the microsize is generally much smaller than the wavelength of visible light (400 nm). For example, SBS thermoplastic elastomer microzone size is about 20–30 nm, with better transparency. However, with the increase of molecular weight of PS in hard segment, the size of microregion increases accordingly. When the PS content (mass) exceeds 23%, slight turbidity begins to appear, and the PS content reaches 47%, material becomes translucent.

Some crystalline block copolymers are also transparent. For example, a block copolymer composed of polyester (polybutylene terephthalate) block and polyurethane block (butylene glycol-diphenylmethane diisocyanate) has good transparency, which may be due to the low crystallinity and the small size of the microzone in the system.

Copolymers are often used as components of transparent blends because the refractive index of copolymers has similar additive property. In this way, the refractive index of the copolymer can be adjusted in a larger range. If the refractive index difference between rubber phase and plastic phase is controlled below 0.005 and the rubber particle size is about 0.2 μm, transparent toughening plastics can be prepared. Figure 7.5 shows the relationship between refractive index and monomer content of styrene-methyl methacrylate copolymer (MS).

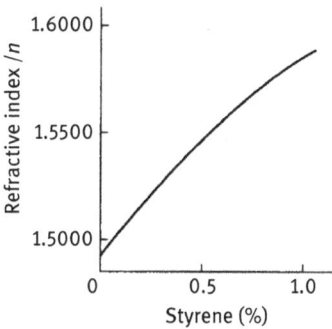

Fig. 7.5: Relationship between refractive index and monomer content of styrene-methyl methacrylate copolymer.

As shown in Fig. 7.5, the refractive index of the copolymer can be adjusted from 1.5000 to 1.6000. Therefore, the possibility of MS blending with other polymers to become transparent products is greatly increased. The refractive index of the polymers listed in Tab. 7.3 is basically within this range. For example, MBS, a blend of MS and polybutadiene, is transparent, while ABS is opaque. Because, the refractive index of *cis*-butadiene rubber containing polystyrene (PS) is 1.526, while that of

Tab. 7.3: Refractive index of common polymers.

Polymer	Refractive index (25 °C)	Polymer	Refractive index (25 °C)
Polyvinyl chloride	1.54 (1.52–1.55)	Polyacrylonitrile	1.520
Polycarbonate	1.585	Butadiene–acrylonitrile rubber	1.520
Polystyrene	1.593 (1.59–1.60)	Styrene–acrylonitrile (AN24%)	1.575
Poly(α-methylstyrene)	1.56	Polybutadiene	1.515
Polymethyl methacrylate	1.496 (1.49–1.50)	cis-Polybutadiene rubber	1.526
Polybutylacrylate	1.463 (20 °C)	Butadiene–styrene rubber (St23.5%)	1.535
2-Ethylhexyl polyacrylate	1.503		

SAN copolymer is about 1.575 (24% AN). The difference of refractive index between them is 0.05, and the particle size of rubber cannot be less than 0.1 μm, so ABS cannot have transparency. The refractive index of MS can be adjusted to 1.520 to 1.530, so that the refractive index difference between the two is less than 0.005, so the MBS with particle size of about 0.2 μm is transparent plastic.

PVC toughened with MBS or MABS can also be a transparent blend. The refractive index of PVC is 1.54. The refractive index of MBS can be adjusted to close to this value. Usually using following methods, MMA and St are grafted with styrene-butadiene latex (30% St), or add a small amount of AN, the gum content is controlled between 40% and 60%. By blending this copolymer with PVC, the toughened PVC with transparent, high-impact strength can be obtained. Because the toughening mechanism of PVC plastics is shear band effect, the particle size of latex is controlled below 0.2 μm, which can meet the requirements of transparency without affecting the toughening effect.

PC is an engineering plastic with high-impact strength. In order to improve its stress cracking and processing properties, elastomers are commonly used for blending modification. If the blend product is required to have a certain transmittance at the same time, blending ABS with PC, which has smaller particle size, lower gum content and similar refractive index, can prepare the PC/ABS alloy with better transparency and high-impact strength.

7.4 Dielectric properties of multicomponent polymers

With the development of electronics and power industry, it is very important to research and produce polymer materials with low cost, high dielectric constant and low-dielectric loss to produce capacitors with small volume and large capacitance. For a capacitor of a certain shape, the stored energy of the capacitor is determined by the dielectric constant of the dielectric material used at the same operating voltage ($C = \varepsilon A/t$, $W = 1/2CU^2$). Therefore, the volume and weight of capacitors can be greatly reduced by using dielectric materials with high-dielectric constant. At the same time, the heat dissipation capacity of capacitors is also an important performance index. When the voltage U, frequency f and capacitance C are the same, the heat generation of the capacitor depends on the dielectric loss tan δ ($P = 2\pi f\ CU^2$ tan δ). Therefore, capacitor materials are required to have high-dielectric constant and low-dielectric loss.

Polymer materials, such as polypropylene (PP) and PS, have been widely used in capacitors of power electronic devices due to their low dielectric loss, good stability, high volume resistivity, low water absorption and low thermal shrinkage. But its dielectric constant is usually low (between 2.2 and 3.0), which restricts its application to a certain extent. Multicomponent polymers prepared by means of blending and copolymerization can make up for the shortcomings of dielectric and mechanical properties of single polymers, and make polymer materials widely used in this field.

7.4.1 Relationship between dielectric properties and composition of multicomponent polymers

PP has lower dielectric loss but lower dielectric constant ($\varepsilon = 2.0–3.0$). Polyvinylidene fluoride (PVDF) is a semicrystalline polycrystalline thermoplastic polymer. It crystallizes into four phases: α, β, γ and δ under different conditions. The strong polar β phase has piezoelectric and ferroelectric properties, and its dielectric constant is about 8, which is much higher than that of PP. Therefore, blending of PVDF and PP can greatly improve the dielectric constant of the material. The experimental results are shown in Fig. 7.6. It can be seen from Fig. 7.6(a) that the dielectric constant of the composites increases with the increase of the content of PVDF. When the frequency is lower than 10^6 Hz, the dielectric constant of the composites changes gently with the increase of frequency, which is basically a straight line. However, when the frequency is higher than 10^6 Hz, the dielectric constant of the composites decreases obviously. Corresponding to Fig. 7.6(b), the orientation frequency of polar groups in composites is far behind the change frequency of electric field, and the dielectric loss is aggravated. Figure 7.6(b) shows that when the frequency is lower than 10^5 Hz, the dielectric loss of the composites is lower than 0.025, which has great practical application value.

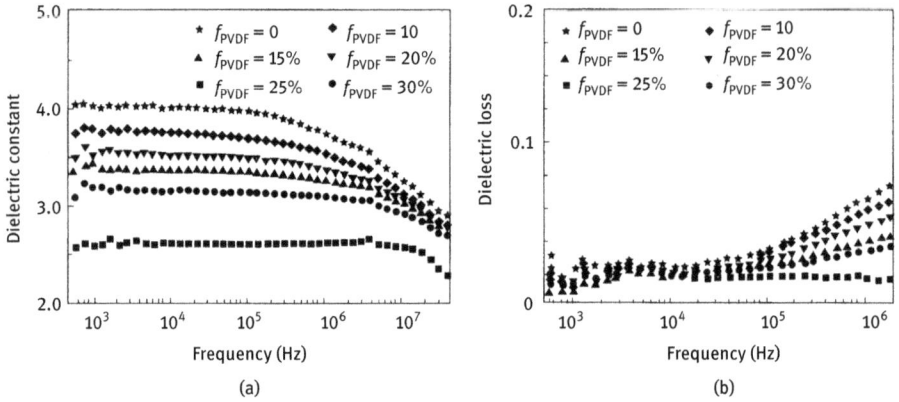

Fig. 7.6: Dielectric properties of PP/PVDF composites with different PVDF volume contents: (a) dielectric constant and (b) dielectric loss.

The dielectric properties and compositions of PA6/PP blends are shown in Figs. 7.7 and 7.8. The dielectric constant and loss of PP/PA6 increase with the increase of PA6 content at the same temperature. This is because the change of PA6 content makes the morphology and structure of the system change. PP as the matrix has a certain shielding effect on PA6 particles, which weakens the water absorption and dielectric properties of PA6 phase. When the content of PA6 phase increases, the shielding effect of PP decreases and the compatibility of the two phases becomes worse. Therefore, the influence of PA6 on the dielectric properties of the system increases, and the dielectric constant and dielectric loss also increase.

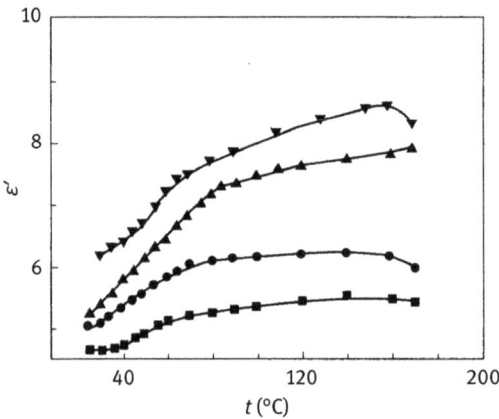

Fig. 7.7: The relationship between the blend ratio and the dielectric constant of PP/PA blends. ■ PP/PA6 (80/20); ● PP/PA6 (70/30); ▲ PP/PA6 (60/40); ▼ PP/PA6 (50/50).

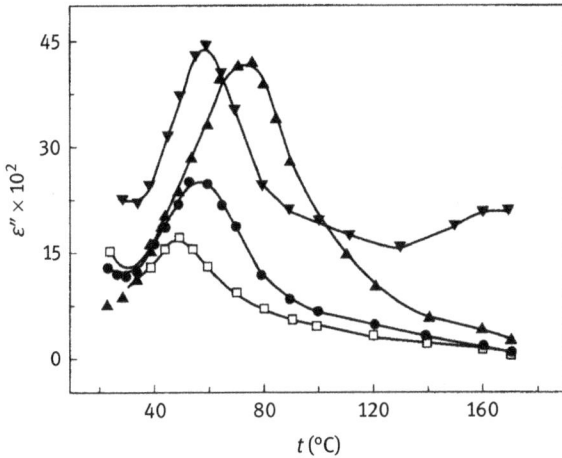

Fig. 7.8: The relationship between blend ratio and dielectric loss of PP/PA blends.
□ PP/PA6 (80/20); ● PP/PA6 (70/30); ▲ PP/PA6 (60/40); ▼ PP/PA6 (50/50).

Figure 7.9 is a scanning electron microscope photograph of the above blends. When the content of PA6 is 20%, the shape of PA6 particles is similar, most of them are regular spherical and the size distribution is uniform, about 1–2 μm. The interfacial layer between PA6 and PP is not obvious, indicating that the bonding degree between PA6 and PP is better. However, when the content of nylon is 50%, not only is the particle size greatly increased, but also some parts of the structure area have become two continuous structures, which have reversed. At the same time, there is a large gap between the PA6 spherical protrusion on the fracture surface and the adjacent matrix, so the bonding force is very poor. Therefore, the smaller the particle size

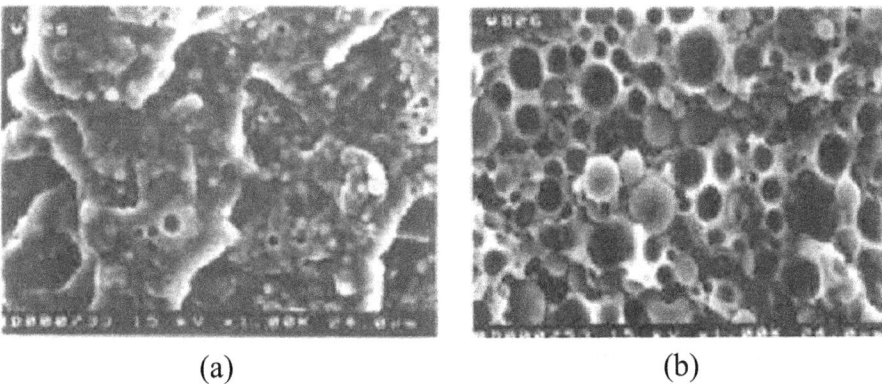

(a)   (b)

Fig. 7.9: SEM graphs of PP/PA6 fracture: (a) PP/PA6 = 80/20 and (b) PP/PA6 = 50/50.

of PA6 dispersed phase, the more uniform the dispersion in PP continuous phase, and the stronger the cohesive force between the two phases, that is, the better the compatibility of the two phases, the smaller the dielectric constant and dielectric loss of the blend system.

In the study of dielectric properties of styrene–acrylic acid substituted ethyl ester copolymers, the dielectric constants and dielectric losses of the copolymers at 1,000 Hz were measured, as shown in Figs. 7.10 and 7.11. The curves show a downward trend with the increase of styrene content.

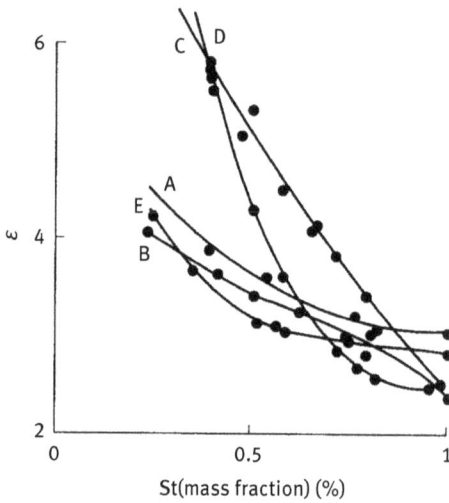

Fig. 7.10: Relationship between dielectric constant and composition of styrene–acrylic acid substituted ethyl ester copolymer.
A, $CH_2=CHCOOCH_2CH_2Cl$; B, $CH_2=CHCOOCH_2CH_2Br$; C, $CH_2=CHCOOCH_2CH_2CN$; D, $CH_2=CHCOOCH_2OCH_3$; E, $CH_2=CHCOOCH_2CH_3$.

PS is a kind of nonpolar polymer. It has no orientational polarization under applied electric field. The main factors that affect dielectric constant and dielectric loss are electronic polarization and atomic polarization, so its dielectric constant and dielectric loss are low ($\varepsilon = 2.4$–3.1, $\tan = 10^{-4}$). From the composition curve of dielectric properties, when the content of styrene reaches 100%, the dielectric constant and loss of pure PS correspond to the curve, but in fact, the two parameters are within a certain range. This is because the dielectric properties of polymers are related to the molecular weight and distribution of polymers, the morphology and thermal history of materials and so on. Different preparation methods will lead to different dielectric properties. Acrylic acid substituted ethyl ester polymer is flexible polar polymer, dipole orientation polarization makes it have much larger dielectric constant and dielectric loss than PS ($\varepsilon > 8$, $\tan = 10^{-2}$). The copolymerization of styrene and

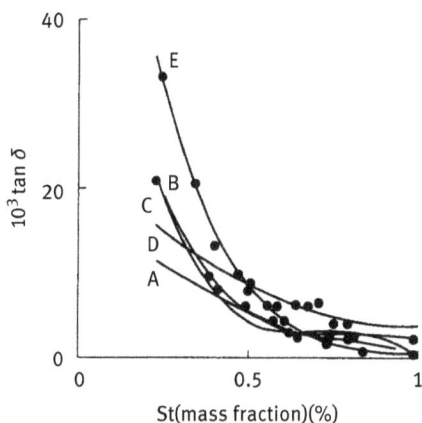

Fig. 7.11: Relationship between dielectric loss and composition of styrene-acrylic acid substituted ethyl ester copolymer.
A, $CH_2=CHCOOCH_2CH_2F$; B, $CH_2=CHCOOCH_2CH_2Cl$; C, $CH_2=CHCOOCH_2CH_2Br$;
D, $CH_2=CHCOOCH_2CH_2CN$; E, $CH_2=CHCOOCH_2CH_3$.

acrylic acid substituted ethyl ester makes the copolymer chain contain a certain number of random styrene and acrylic acid substituted ethyl ester units, which play a role in the macrodielectric properties of the copolymer. It is not difficult to understand that the number of acrylate units carrying polar side groups decreases with the increase of styrene units in the molecular chain. When applied with an external electric field, the orientational polarization intensity caused by a permanent dipole decreases, and the dielectric constant of the copolymer decreases. Similarly, as the density of polar groups in the chain decreases, the van der Waals force between chains decreases accordingly, and the internal friction force to be overcome by the movement of molecular chains or segments decreases undoubtedly, and the dielectric loss also decreases.

7.4.2 The relationship between dielectric properties of multicomponent polymers and compatibilizers

Compatibilizer can improve the compatibility of blends, increase the interaction and dispersion of polymer components, and improve the dielectric properties of the blends.

The effect of compatibilizer in PP/PVDF blend system is shown in Fig. 7.12. Figure 7.12(a) shows that when the volume content of compatibilizer increases from 10% to 15%, the dielectric constant of the composite increases from 3.5 to 3.9, and then with the content of compatibilizer increasing to 20% and 25%, the dielectric constant decreases to 3.4 and 3.6 respectively. This is because when the content of

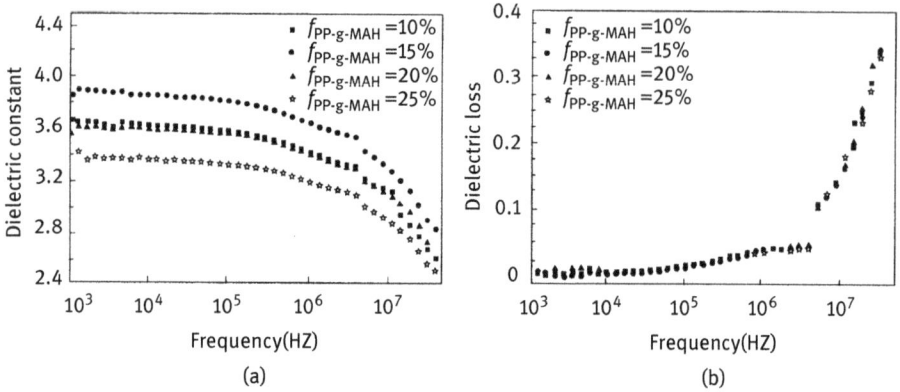

Fig. 7.12: Effect of compatibilizer on dielectric properties of PP/PVDF blends: (a) dielectric constant and (b) dielectric loss.

compatibilizer is higher than 15%, the effect of compatibilizer on the crystallization of PVDF limits the formation of β-phase PVDF and leads to the decrease of dielectric constant of the composites. However, in Fig. 7.12(b), the dielectric loss of composites with different compatibilizer content is almost the same, concentrating below 0.05. Only when the frequency increases to 10^7 Hz does it suddenly rise.

Figure 7.13 is the temperature spectrum of dielectric constant and dielectric loss of PP/PA6 blends at 330 Hz with 5 phr, 10 phr and 25 phr, respectively, maleic anhydride grafted polypropylene (PP-g-MAH) as compatibilizer. Comparing the dielectric spectra and morphology of the three systems, it is found that the dielectric loss and permittivity of the system are the largest when the content of compatibilizer is 2.5 phr at the same temperature. At this time, the dispersion of PA6 particles in PP

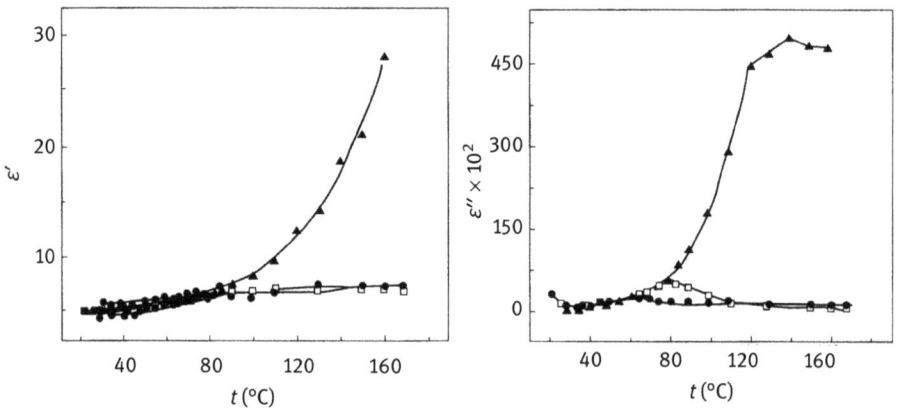

Fig. 7.13: Relation between dielectric properties and compatibilizer content of PP/PA6.
□ PP/PA6 (5 phr); ● PP/PA6 (10 phr); ▲ PP/PA6 (25 phr).

matrix is obviously inhomogeneous, there is a clear gap between PP and PA6 phase, and the material is easy to crack, so its dielectric constant and dielectric loss are larger than the other two. When the content of compatibilizer is 5 phr, the average particle size of PA6 particles has been reduced, the dispersion in PP phase has become uniform, and the degree of adhesion with PP matrix is better, so the dielectric constant and dielectric loss of the system are significantly lower than those of PP/PA6 = 25 phr. When the content of compatibilizer is 10 phr, the compatibility of the system is very good. PA6 particles can hardly be distinguished from the continuous phase of PP, indicating that the cohesive force between the two phases in the system has been significantly improved compared with the two systems mentioned earlier. However, the dielectric constant and dielectric loss of the system are similar to those of PP/PA6 at 5 phr. This is because the compatibilizer PP-g-MAH itself is a polar molecule, and orientational polarization can occur in alternating electric field. If it is introduced into PP/PA6 blend system, the dielectric constant and dielectric loss of the system will increase. For PP/PA6 system with 10 phr, although increasing the content of compatibilizer improved the phase structure of PP/PA6 system, and reduced its dielectric constant and dielectric loss. At the same time, it increases the polarity of the whole system, resulting in the increase of dielectric constant and dielectric loss. Considering these two factors, the dielectric properties of the system have no obvious change compared with the system with 5 phr content of compatibilizer.

7.3.3 Relationship between dielectric properties and processing conditions of multicomponent polymers

The processing conditions of the blend polymer will affect the aggregate structure of the system, and then the dielectric properties.

When PP/PA6 blends were prepared by melt blending, the rotational speed of screw extruder (proportional to the shear stress and shear rate of polymer melt) directly affected the rheological properties of components, and thus affected the dispersion degree of dispersed phase in the blends melt. Figure 7.14 shows the dielectric spectra of PP and PA6 at 330 Hz for three systems with screw speed of 20, 30 and 40 r/min, respectively, but with same composition, compatibilizer content and blending temperature. It can be seen from the figure that when the screw speed is 30 r/min, the dielectric constant value is small, which indicates that PA6 particles disperse most evenly in the continuous phase and the particle size is the smallest. However, the dielectric loss of the screw at 30 r/min is larger than that of the 20 r/min system, which is related to the poor interface bonding between PP phase and PA6 phase. If the bonding degree of the interface between the two phases is not good, the response time of the two phases to the electric field variations is different, which results in the

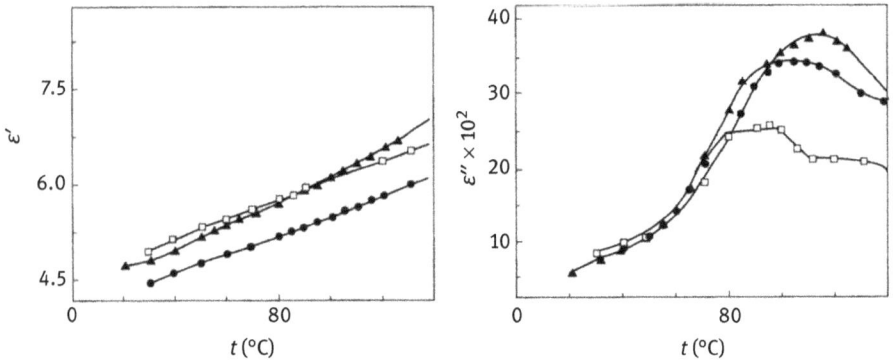

Fig. 7.14: Effect of screw speed on dielectric properties of PP/PA6 blends.
□ 20 r/min; ● 30 r/min; ▲ 40 r/min.

dielectric loss of the system larger than that of the system with tightly integrated phases and molecules with uniform response to electric field changes.

Blending temperature is one of the important preparation parameters. Figure 7.15 is the temperature spectrum of dielectric constant and dielectric loss at 330 Hz of three PP/PA6 blends prepared at different temperatures (230, 235 and 240 °C). The experimental results show that when the blending temperature is 240 °C, the PA6 particles in the blend disperse well in the PP continuous phase, and the interface layer can hardly be observed, that means the bonding degree between PP and PA6 in the blend is very good. As for dielectric properties, it is also the best in the three systems.

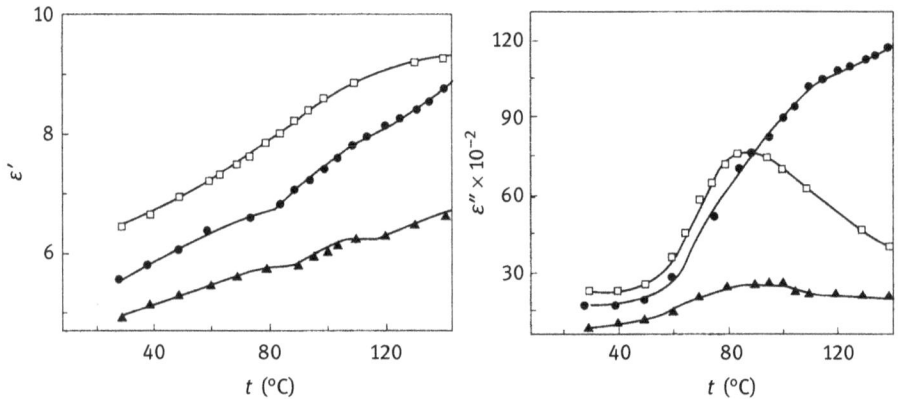

Fig. 7.15: Dielectric spectra of PP/PA6 blends at different temperatures.
□ 230 ℃; ● 235 ℃; ▲ 240 ℃.

Figure 7.16 shows the dielectric constant and dielectric loss of PP/PVDF blends with different compatibilizer content before and after stretching. The dielectric constant diagram shows that the dielectric constant of the specimens with 10% compatibilizer and 15% compatibilizer increases from 2.7 and 3.2 to 4.7 and 5.5, respectively, after stretching, which is about one time higher than that before stretching. This is because the dielectric properties of the composites are mainly supplied by the PVDF of the β phase. Tensile can promote the transformation of the PVDF of the α phase to the β phase in the composites, thus increasing the content of the PVDF of the β phase, so the dielectric constant of the blends has been improved obviously. The dielectric loss of the blends before and after stretching did not change much, remaining a low value.

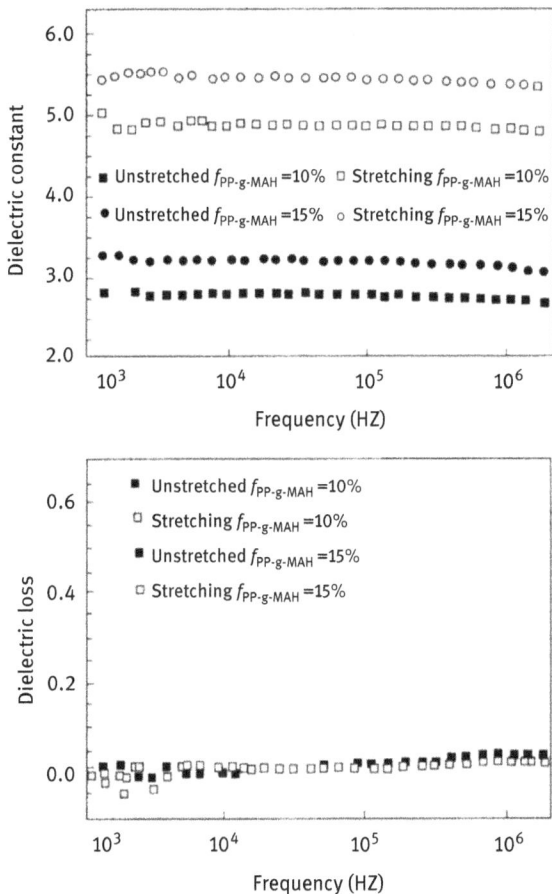

Fig. 7.16: Dielectric properties of PP/PVDF blends before and after stretching.

Exercises

1. Please illustrate which part of the multicomponent polymer plays a dominant role in its gas permeability.

Answer:
The crystal region of polymer is impermeable, so the gas passes through the amorphous region of the polymer. For blends, gas penetration mainly occurs in the continuous phase, and the existence of dispersed phase microzone makes the path of gas molecules penetration become tortuous, thus increasing the distance length and reducing the air permeability.

2. The barrier properties of multicomponent polymers are very important. Please outline the barrier mechanism and influencing factors of multicomponent polymers. Then try to design a multicomponent polymer with good barrier properties and its application.

Answer:
The barrier property of polymer refers to the shielding ability of polymer products to small molecule gases, liquids, vapors, spices and medicinal flavors. The permeation process of small molecules into polymers can be divided into four stages: (1) small molecules adsorbed on the surface of polymers; (2) small molecules dissolved in polymers; (3) small molecules passing through polymers at a certain concentration gradient; and (4) small molecules desorbed on the other surface of polymers. Anything that can delay any one of these processes is conducive to improving the barrier properties of polymers. Autonomous design is omitted.

3. What are the factors affecting the transmittance of rubber-toughened plastic blend system?

Answer:
(1) Rubber particle size: The smaller the particle size, the better the transmittance of the material.
(2) Refractive index: When the refractive index of the two components is within a certain range, the material can be transparent.
(3) Temperature: Because the refractive index of different polymers is affected by temperature, the transmittance is also affected by temperature.
(4) Complexes: For blends with various complexes, the selection of complexes has a great influence on the transmittance. In general, nonpolluting, colorless or light-colored liquid additives should be used. The particle size of the powder additive should be as small as possible, and the refractive index should be close to the polymer component.

4. Why do compatibilizers have a great influence on the dielectric properties of multicomponent polymers? What other factors affect dielectric properties?

Answer:
Compatibilizers can improve the compatibility of blends, increase the interaction and dispersion of polymer components and improve the dielectric properties of the blends.

Other factors include composition and processing conditions.

5. What is the oxygen index (OI) and what performance does it characterize?

Answer:
OI refers to the lowest oxygen concentration required for flame combustion of materials in oxygen–nitrogen mixture under specified conditions. It is expressed in terms of the volume percentage of oxygen. High OI means that the material is not easy to burn, and low OI means that the material is easy to burn. Under the specified test conditions, the minimum oxygen concentration of the material at room temperature in the mixture of O_2 and N_2, which just maintains flaming combustion, is expressed as a percentage of volume. The higher the OI, the better the flame retardancy.

Chapter 8
Graft copolymers

8.1 Concepts of graft copolymers

8.1.1 Definition and development of graft copolymers

Graft copolymers are copolymers formed by chemical linking of two or more macro-molecules with different structures and properties. These polymers have unique properties and play an extremely important role in the preparation and application of multicomponent polymers.

In the field of polymer chemistry, the study of graft copolymers has a history of half a century. In fact, examples of graft copolymers can be traced back to earlier ages.

In 1912, Matthews applied for a patent, introducing the polymerization of sty-rene (St) monomers in the presence of natural rubber. In 1930, Starck et al. carried out vinyl acetate emulsion polymerization in the presence of polyvinyl alcohol to prepare adhesives, and it was found that some polyvinyl acetate grafted on polyvi-nyl alcohol. In 1935, Roedel, DuPont of the USA, carried out graft copolymerization of acrylonitrile (AN) onto natural rubber latex. In 1946, Stoops et al. blended St, po-tassium persulfate, 16 alkyl mercaptan and sodium oleate into St–butadiene rubber (BR) latex and heated at 50 °C for 2 h, and the St obtained had excellent impact resistance.

Although there was no concept of graft copolymers at that time, the above poly-mers were obviously all graft copolymers.

In September 1951, at the annual meeting of the International Union of Pure and Applied Chemistry held in New York, Smets et al. reported that vinyl acetate was polymerized in benzene solution containing polymethyl methacrylate (PMMA) and 0.5% benzoyl peroxide at 75 °C. The products obtained included both graft co-polymer of polyvinyl acetate on PMMA and homopolymer. The mixed system can be dissolved in hot ethanol, but the graft copolymer can be precipitated at lower temperature. Therefore, the graft copolymer can be separated by changing the tem-perature. Later, Roland et al. successfully grafted ethylene onto polyvinyl acetate using peroxide as initiator. Therefore, the concept of graft copolymer was put for-ward clearly and systematically.

Because the main and side chains of graft copolymers consist of polymers with different structures and properties, they have attractive special properties and at-tracted much attention, initiating active research on the synthesis and properties of

https://doi.org/10.1515/9783110596335-008

graft copolymer. Since the 1980s, many new methods of polymer synthesis have been found, which provides a powerful means for the synthesis of graft copolymers. Many new graft copolymers have been developed, which greatly enriches the research and application of graft copolymers.

So far, graft copolymerization has become an important means and basic method of polymer modification. The object of graft copolymerization also involves the whole polymer field, such as natural polymers (starch, wool, cellulose, natural silk and natural rubber) and synthetic polymers (synthetic rubber, synthetic fibers and plastics). In recent years, some inorganic materials, such as carbon black, silica gel and metal oxides, have also been modified by graft copolymerization. Since the 1990s, the rise of nanotechnology has also been applied to the graft copolymerization technology. People use the method of grafting modification to modify the surface of nanomaterials, which greatly improves the application level of nanomaterials. In this respect, successful examples include nanocalcium carbonate, nanosilica, nanotitanium dioxide, fullerene, carbon nanotubes (CNTs) and graphene grafting modification.

8.1.2 Representation of graft copolymers

In general, a copolymer in which segment composed of A monomer and segment composed of B monomer are connected head to tail is called block copolymer. The structure of block copolymers is shown in Fig. 8.1(a). The copolymers consisting of A monomer as the main chain and B monomer as the branch chain are called graft copolymers, and the structure of the copolymers is shown in Fig. 8.1(b). In fact, the main and branch chains of the graft copolymers themselves can also be copolymers, as shown in Fig. 8.1(c).

The nomenclature of graft copolymers generally follows Ceresa's rule proposed in 1962:

1. When naming homopolymer, just put "poly" before the monomer name such as polyethylene (PE), polyvinyl chloride (PVC) and PMMA.
2. When naming copolymers, the names of two or more copolymers are linked by "-" and then preceded by the word "poly." Usually, the copolymers contain more components in the first place and less in the second place, such as PVC–vinyl acetate and acrylonitrile–butadiene–polystyrene (ABS).
3. In order to point out the arrangement of monomer units in copolymer structure, the letter "g" (graft) denotes grafting and "b" (block) denotes block. For the graft copolymer, the component before "g" represents the main chain, and the component after "g" represents the branched chain. For example, PMMA-g-PS represents a PS graft copolymer with PMMA as the main chain.

AAAAAAAA BBBBBBBBB AAAAAAAA

(a)

AAAAAAAA AAAAAAAA
B   B
B   B
B   B
B   B
B   B
                 B

(b)

AAAAAAAA AAAAAAAA   ABAABBABA BBBAABB
B   B   A   A
A   A   A   A
B   A   A   A
B   B   A   A
A   B   A   A
B   A   A   A
                 B   A

(c)

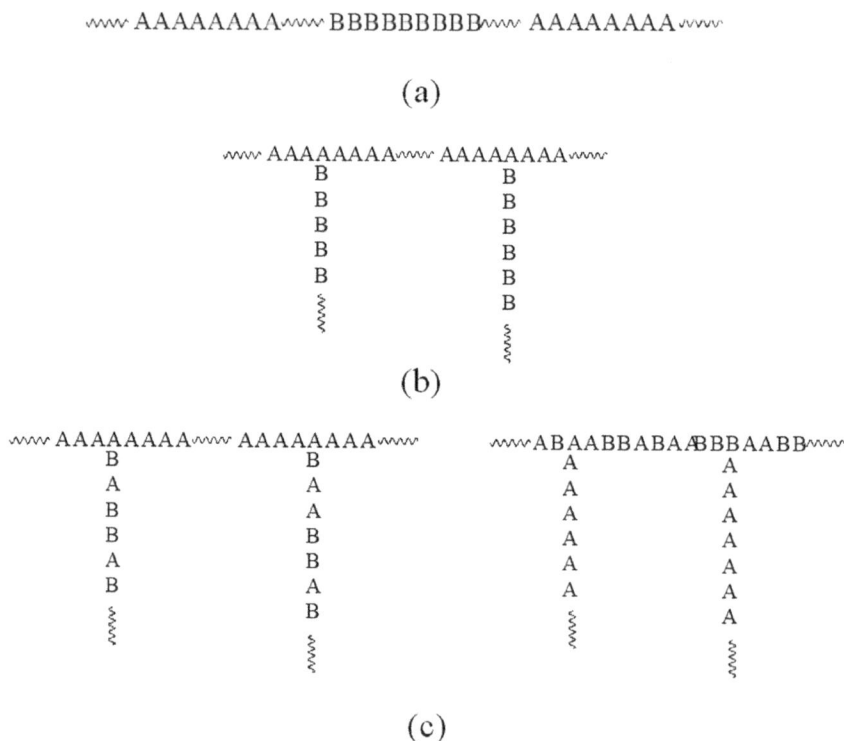

Fig. 8.1: Structural schematic diagram of block copolymers and graft copolymers.

8.2 Synthesis of graft copolymers

8.2.1 Basic methods for preparing graft copolymers

Different from random copolymers and alternating copolymers, graft copolymers cannot be prepared by ordinary copolymerization. According to the branched chain structure of the grafted polymer, the synthesis of the grafted polymer can usually be achieved through the following ways:

1. In the main chain of a polymer, introducing active group, through initiation and polymerization of another monomer, the branched chain of another polymer is formed.
2. A polymer molecular chain is linked to another polymer molecular chain as a branched chain by grafting (chemical bonding).
3. In the preparation of one polymer, another polymer is formed at the same time and is linked to the former as a branched chain.

Since graft copolymerization involves the polymerization of monomers on the main chain of the polymer skeleton, in principle, various methods could induce the polymerization reaction. For example, vinyl monomers can be used to synthesize grafted polymers by radical or ion-initiated addition polymerization or ring-opening polymerization of cyclic monomers. Several main synthetic routes are briefly described further.

8.2.2 Chain transfer method

Chain transfer method is an old but commonly used method to prepare grafted polymers.

According to the kinetics, the polymerization process of vinyl monomers can be regarded as composed of chain initiation, chain growth, chain transfer and chain termination. Chain transfer method is to prepare grafted polymers by chain transfer reaction in free radical polymerization. Through chain transfer reaction, the activity of free radicals can be transferred to monomers, solvent molecules, transfer agent molecules (such as mercaptan) and polymer molecules. If the polymer with chain transfer has the same composition as the monomer, then the branched polymer is obtained by polymerization. If the component of the polymer with chain transfer is different from that of the monomer, then the grafted polymer is obtained by polymerization. The activity of the graft chain can be terminated by coupling or disproportionation. Therefore, the simplest method is to polymerize liquid vinyl monomer A, which is soluble in homopolymer or copolymer. The reaction mode is shown in Fig. 8.2.

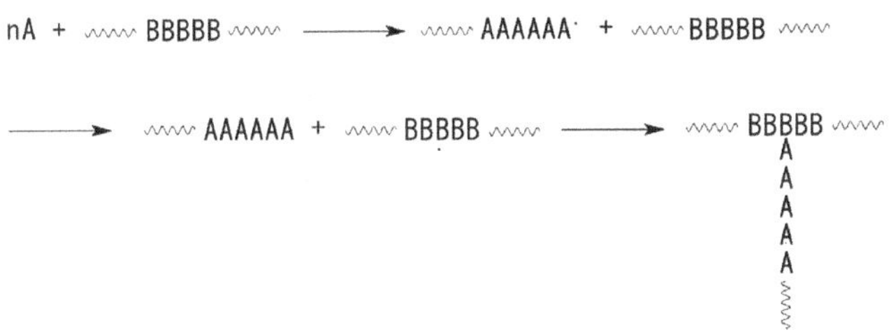

nA + ⌇⌇ BBBBB ⌇⌇ ⟶ ⌇⌇ AAAAAA· + ⌇⌇ BBBBB ⌇⌇

⟶ ⌇⌇ AAAAAA + ⌇⌇ BBBBB· ⌇⌇ ⟶ ⌇⌇ BBBBB ⌇⌇
                                                 A
                                                 A
                                                 A
                                                 A
                                                 A
                                                 ⌇

Fig. 8.2: Graft copolymer prepared by chain transfer method.

In general, pure grafted polymers cannot be obtained by this method, and the products usually contain two kinds of homopolymers. However, for most industrial products of polymer blends, there is no need for pure grafted polymers. If the product contains 2–5% of the graft copolymer components, it can play a compatible role, which effectively reduces the tendency of phase separation between the two homopolymers, and is conducive to improving some properties of the system.

The structure and grafting rate of the grafted polymer depend on the chain transfer rate, temperature, concentration, reaction ability of monomers and free radicals, and the activity of active atoms on the macromolecular chain. In addition, the activity of initiator or initiator system is also an important factor. Increasing monomer concentration, polymer concentration and introducing active groups (e.g., -SH) into the polymer could improve the yield of the grafted polymer.

Generally, all kinds of polymerization processes can be used to carry out graft copolymerization, for example, suspension polymerization, emulsion polymerization, bulk polymerization and solution polymerization. The initiation methods can be thermal initiation, photoinitiation, peroxide initiation, redox reaction system initiation and radiation initiation.

The high impact PS (HIPS) used as a commodity in the early stage was a graft copolymer prepared by the chain transfer method. For example, polybutadiene is dissolved in St monomer, and then graft copolymerization is initiated by heat or peroxide.

Another typical polymer modified by chain transfer is ABS copolymer. The two most commonly used methods are suspension polymerization of butadiene-eye rubber in St monomer solution, swelling of polybutadiene latex by solution of AN and St mixed monomer, and then grafting polymerization. The initiator used may be organic peroxide or redox initiator system.

The synthesis of grafted polymers by chain transfer method is not limited to using polymers with higher molecular weight as backbone chains, but also using polymers with lower molecular weight as backbone chains. For example, epoxy resin, polyethylene oxide, polypropylene oxide, copolyether and shellac are easily soluble in monomers such as methyl methacrylate (MMA), ethyl acrylate and St. In these solutions, the addition of peroxide initiator makes the polymerization to take place, and the grafted polymer products with different properties can be obtained. The relative molecular mass of the polymer used as the backbone main chain in the above-mentioned grafted polymers is not high (usually 500–2,500), while the relative molecular mass of the branched chains on the grafted polymers can be as high as tens of thousands.

Halogen atoms can easily be removed from polymer molecules as the backbone of the main chain by chain transfer, so halogen atoms are potential active sites for grafting. In the presence of other polymerizable monomers, halogen-containing polymers such as PVC, chloroprene rubber and chlorinated PE (CPE) can form graft polymers by chain transfer. For example, in the presence of benzoyl peroxide, the graft polymer of PS onto PVC can be obtained by heating PVC in St monomer. If there is no oxygen at this time, the graft copolymer with higher grafting rate can be obtained.

Grafting by chain transfer is not only related to the properties of potential grafting sites on the skeletal polymer chains, but also to the activity of free radicals formed by monomers. A typical example is that MMA can be easily grafted onto the PS main chain, but vinyl acetate cannot be grafted onto the PS main chain under the same conditions. This is because the free radicals of PS are relatively stable, vinyl acetate monomers are also very stable and stable free radicals are not easy to initiate stable monomer polymerization; hence, the above phenomenon occurs.

8.2.3 Active group method

Some reactive monomers can be introduced into the polymer chain as units through chemical treatment of the polymer in the skeleton main chain. For example, a pere-ster can be formed by treating acrylic polymers or copolymers with phosphorus pentachloride and reacting with hydrogen peroxide. The active sites for grafting can be obtained by decomposition. Sometimes peroxide or hydrogen peroxide can also be used for direct peroxidation treatment. A simple example is, when air goes through the natural rubber latex system, ammonia is used as a stabilizer. At this time, rubber molecules are more prone to peroxidation, thus introducing hydrogen peroxide groups. Usually, rubber latex contains some hydrogen peroxide groups that can be used for grafting. For relatively stable latex systems, monomers are usu-ally used to swell them, followed by tetravinyl pentamine or ferrous sulfate solu-tions, in order to initiate (redox) polymerization at room temperature. The results showed that if the monomers in the reaction system are St or MMA, there will be no homopolymer in the reaction products. Because in these reaction systems, only the active part of the rubber skeleton can initiate the polymerization reaction.

Ozone reacts with a variety of polymers because of its high activity, so the range of synthesis of grafted polymers by active group method is expanded. Some typical ozonation-based grafting systems are shown in Tab. 8.1.

Tab. 8.1: Graft polymerization system using ozone reaction.

Backbone polymer	Graft monomer	Backbone polymer	Graft monomer
Starch	Styrene	Polyvinyl chloride	Acrylonitrile
Cellulose (cotton)	Acrylonitrile	Polyvinyl chloride	Methyl methacrylate
Polyester	Styrene	Polytrifluorochloroethylene	Acrylonitrile
Polyethylene	Acrylic acid	Wool	Acrylonitrile
Polyvinyl chloride	Vinyl acetate	Silk	Acrylonitrile
Polyvinyl chloride	Styrene		

By ozonation reaction, hydrogen peroxide group can also be introduced into the main chain of the skeleton, and then redox initiator can be added to form redox initiator system, which can initiate graft polymerization. For example, MMA and acrylamide can be grafted onto polybutadiene in this way. In addition, MMA and St can be grafted onto polypropylene in this way.

In industrial production, it is very difficult to make the solid material ozonate uniformly, and the monomer is not easy to diffuse to the active part of the material. So, this method is only suitable for the grafting of thin film materials.

Graft copolymerization by active group method has an important application in industry, which is to treat fibers. For example, grafting polyacrylic acid onto PE fibers can improve the dyeability of PE.

8.2.4 Radiation methods

Ultraviolet radiation can activate polymers and form active sites for grafting. When irradiated by ultraviolet light, the absorbed energy is low, so it is suitable for surface grafting. The yield of grafted polymers can be greatly increased by activation of dyes or other photosensitive initiators that are sensitive to ultraviolet light. Ultraviolet radiation can also be carried out in the presence of air, thus introducing hydrogen peroxide groups into skeleton polymers. Then, the graft copolymerization was initiated by oxidation–reduction method. Photosensitive groups can also be part of the skeleton polymer chain. For example, chlorine atoms at the side groups can be photolyzed to obtain active sites for grafting.

In practical applications, high-energy radiation technology is more widely used. For example, the γ-ray radiation technology using Co^{60} as radiation source has been studied in depth and has been widely used in industrial production. There are many methods to synthesize grafted polymers by γ-ray radiation. The simplest method is to radiate the monomer–polymer system directly in liquid or gas phase. If the polymer is incompatible with the monomer, or the polymer cannot be swelled by the monomer, then the grafting reaction often occurs only on the surface of the polymer. However, when the radiation dose is high, the degradation of the graft chain or the main chain will occur, which makes the reaction process more complex. When the polymer is swollen by monomers, the graft polymerization usually occurs only in the swollen region.

When the polymer is irradiated in the presence of oxygen, peroxide and hydrogen peroxide groups can be formed on the polymer. These groups are usually attached to polymer molecules as side groups. Then the monomer in liquid or gas phase contacts with the peroxide polymer to form the graft polymer. Due to hydrogen peroxide groups, graft copolymerization can be initiated by redox initiation. In this way, graft copolymerization can be carried out at a low temperature, and the homopolymer ratio in the product can be reduced. For example, using ferrous salts and diphenylacetone of acetylacetone at room temperature, monomers such as St and AN can be irradiated onto PE and polypropylene.

In some cases, irradiation leads to degradation and chain breakage, resulting in block polymers rather than grafted polymers.

The essence of radiation grafting of polymers dissolved in monomers is bulk polymerization. Homopolymerization is not easy to exclude when the conversion is high. In order to reduce homopolymerization, polymer latex swollen by monomer can be grafted. This method has been successfully applied to the grafting of natural rubber with PMMA and polypropylene with polyvinyl acetate.

Radiation of halogenated polymers, such as PVC and polytetrafluoroethylene (PTFE), makes it easy to obtain the active sites for grafting by losing halogen. Using radiation, swelling and oxidation techniques, monomers such as St, methacrylate and vinyl acetate have been successfully grafted onto PTFE.

Radiation grafting can also be carried out on inorganic materials. For example, in the presence of MMA, acrylic acid or vinyl chloride, the grafting system can be obtained by irradiating inorganic materials such as carbon black, silica gel and magnesium oxide with high-energy β-rays. It is reported that the grafting reaction takes place on the lattice surface of inorganic materials.

8.2.5 Additive polymerization and ring-opening polymerization

Macromolecular free radicals can undergo additive reactions when they encounter additional macromolecules containing unsaturated bonds, and then react with the monomer, eventually forming a graft polymer. For example, polyacrylonitrile (PAN)-g-PMMA (PAN-g-PMMA) is obtained by adding PAN macromolecular free radicals to the double bonds at the end of the molecule, as shown in Fig. 8.3.

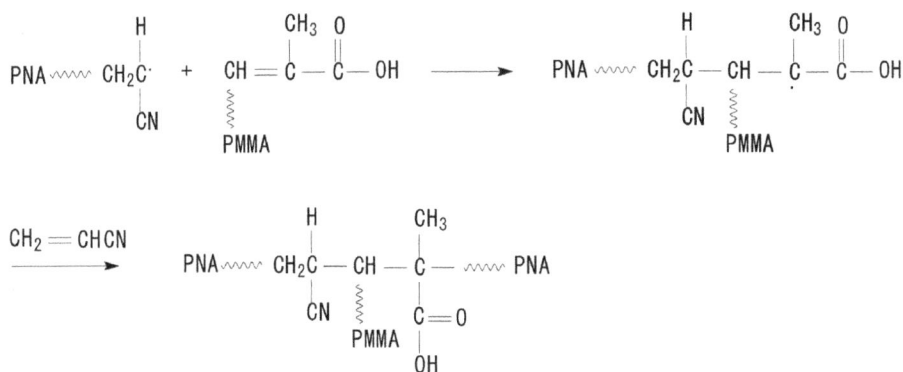

Fig. 8.3: Examples of graft copolymers prepared by addition polymerization.

The double bonds in the extra-enlarged molecules can be formed by the biradical disproportionation in the free radical polymerization process, or by the introduction of copolymers, or by chemical reaction of some genes in polymers. For example, when hydroxyl groups in cellulose are mixed with methacryloyl chloride during acetylation, unsaturated side groups can be introduced.

Natural rubber or synthetic polyisoprene can be dissolved in St, MMA or AN monomers, and added with azodiisobutyronitrile (AIBN) or other initiators, and then plasticized, a number of grafted polymers are obtained. The ratio of grafted polymer to homopolymer is related to the time of plasticization. If it is not molded, it is impossible to obtain grafted polymers, and only some homopolymer mixtures can be obtained even in nitrogen atmosphere.

Initiated by active hydrogen atoms linked to amines or amides in polyamide macromolecules, ethylene oxide or other epoxy compounds can be opened, thus grafting polyethylene oxide onto various polyamides. In addition, ring-opening graft copolymerization of propylene oxide or glycidyl acetate can also take place. Epoxides can be grafted onto vinyl monomers by appropriate chemical modification. For example, epoxides can be grafted by copolymerization of St, acrylate or methacrylate, which is difficult to hydrolyze, with more hydrolyzable monomers (such as vinyl acetate or propyl acrylate). Similar methods can also be used to graft ε-caprolactam onto polymers containing ester groups.

8.2.6 Ionic polymerization

Graft polymers can also be synthesized by ionic polymerization. For example, when ionic chain transfer occurs between PS-based carbon cations and poly(p-methoxystyrene), poly(p-methoxystyrene-g-PS) can be obtained. In addition, in carbon disulfide solution containing chloromethylated PS, aluminum bromide can initiate the carbocationic polymerization of isobutylene to form PS-g-polyisobutylene graft copolymer. When chlorinated polymers (such as PVC) exist together with ferric tetrachloride or aluminum trichloride, they form macromolecular carbon cations, which can also be used to initiate graft copolymerization of St monomers.

In addition, the grafted polymers can also be synthesized by anionic polymerization. For example, the aryl groups of ethylene polymers containing aromatic rings can react with alkali metals in solution to form a colored charge transfer complex, which can initiate the polymerization of epoxides to obtain grafted polymers. For example, when poly(4-vinyl biphenyl) or poly(5-vinyl naphthalene) is dispersed in tetrahydrofuran solution at −80 °C, blue and green solutions can be obtained by cesium mirror reaction. Adding ethylene oxide to these two solutions, the temperature rises to 0 °C. After 24 h of reaction, the activity of anion chain end was removed by methanol, and two types of graft polymers, poly(4-vinylbiphenyl-g-polyethylene oxide) and poly(5-vinylnaphthalene-g-polyethylene oxide), were obtained. They can be characterized by chromatographic analysis.

Another way to metallize aromatics is through the reaction of halogenated derivatives [e.g., poly(*p*-bromomethylstyrene)] with naphthalene–naphthalene complex in tetrahydrofuran solution at –78 °C. After metallization of aromatics and addition of monomers such as St or MMA, the corresponding grafted polymers can be obtained.

Anionic graft copolymerization can also be initiated by macromolecule nitrile and Green reagent. For example, copolymers of St and AN were dissolved in tetrahydrofuran to react with $(CH_3)_2$-SiH-CH_2MgCl (dimethyl-silylmethyl magnesium chloride). Then the corresponding monomers (such as 4-vinylpyridine, AN and MMA) were added to the above-mentioned cooling solution to prepare poly(styrene-acrylonitrile)-g-poly(4-vinylpyridine), poly(styrene-acrylonitrile)-g-poly(styrene-acrylonitrile)-g-poly(MMA) and other graft copolymers.

8.2.7 Macromer method

Macromer is an oligomer with polymerizable groups on the molecular chain. Its relative molecular mass is usually thousands to tens of thousands. This polymerizable intermediate is playing an increasingly important role in polymer design. For example, clear-structured graft copolymers can be obtained by the polymerization of macromonomers. They can synthesize completely opposite properties, such as soft/hard, crystalline/amorphous, hydrophilic/hydrophobic, polar/nonpolar and rigid/tough. Many macromonomers have been used in the preparation of functional polymers, so they have attracted increasing attention. At present, macromonomer technology has become an effective method for the preparation of graft copolymers.

Macromer is synthesized mainly by introducing polymerizable groups at the end of the oligomer chain. At present, anionic polymerization, cationic polymerization, radical polymerization, group transfer polymerization and atomic transfer radical polymerization (ATRP), which have been developed in recent years, are of practical application value. The following are some important examples of synthesis of macromolecular monomers (Figs. 8.4–8.9).

The copolymerization of macromers with other olefin monomers can produce graft copolymers with specified molecular structure, which is the main reason why macromers have attracted much attention. The branched chains of the copolymers prepared by other existing methods are usually uneven in length and distribution. The branched chain length of the graft copolymer prepared by macromers can be controlled when macromers are prepared. The graft distribution is uniform, the content of homopolymer is small and the quality of the product is high. The densities of branched chains in copolymers depend on the aggregation rates of macromers and small monomers.

$$n \ CH_2 = CH \text{—} \underset{\text{C}_6\text{H}_5}{} \xrightarrow{\text{BuLi}} Bu \text{—} (CH_2 - CH)_n^{} \ Li \quad \xrightarrow{\overset{CH_2 - CH_2}{\underset{O}{\diagdown \diagup}}}$$

$$Bu \text{—} (CH_2 - CH)_n^{} \text{—} CH_2 CH_2 OLi \quad \xrightarrow{CH_2 = CH_2 - COCl}$$

$$Bu \text{—} (CH_2 - CH)_n^{} \text{—} CH_2 CH_2 OCO \text{—} CH_2 = CH$$

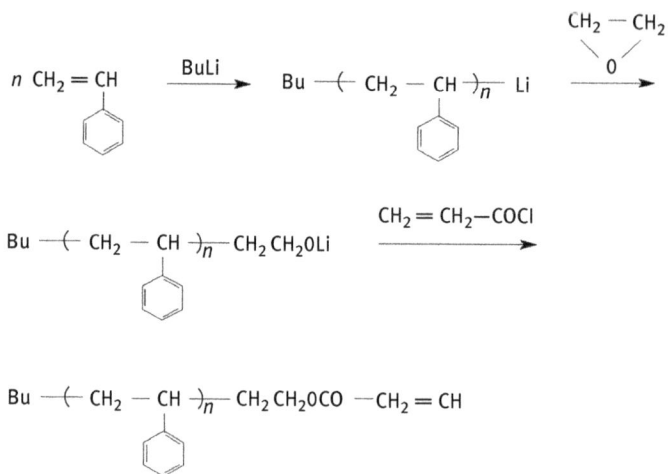

Fig. 8.4: Synthesis of acrylate polystyrene macromer.

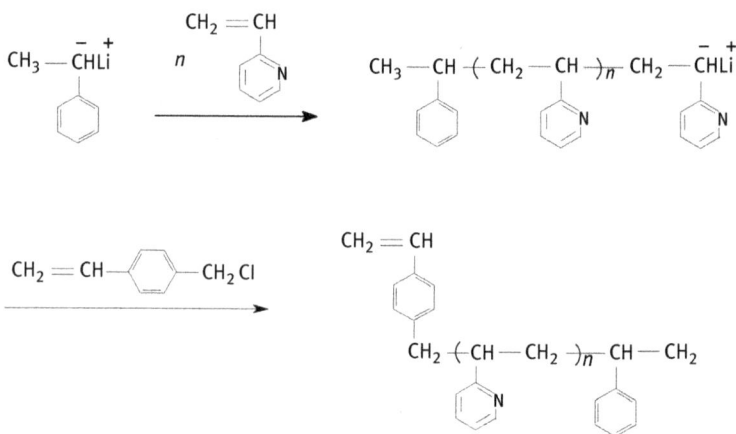

$$CH_3 \text{—} \overset{- \ +}{\underset{}{CHLi}} \quad n \quad \underset{}{CH_2 = CH} \longrightarrow CH_3 \text{—} CH (CH_2 - CH)_n^{} CH_2 \text{—} \overset{- \ +}{CHLi}$$

$$CH_2 = CH \text{—} \bigcirc \text{—} CH_2 Cl \longrightarrow$$

$$\underset{}{CH_2 = CH}$$

$$CH_2 (CH - CH_2)_n^{} CH - CH_2$$

Fig. 8.5: Synthesis of polyvinyl pyridine macromer of styrene type.

$$(CH_3)_2SiOLi + n \left[\begin{array}{c} CH_3 \\ | \\ (SiO)_3 \\ | \\ CH_3 \end{array} \right] \longrightarrow (CH_3)_2SiO \begin{array}{c} CH_3 \\ | \\ (SiO)_{3n-1} \\ | \\ CH_3 \end{array} \begin{array}{c} CH_3 \\ | \\ SiOLi \\ | \\ CH_3 \end{array}$$

$$\begin{array}{c} CH_3 \\ | \\ CH_2{=}C{-}C{-}O \; (CH_2)_3O(CH_2)_3SiCl \\ \| \quad\quad\quad\quad\quad\quad | \\ O \quad\quad\quad\quad\quad\quad CH_3 \end{array}$$

$$\longrightarrow \begin{array}{c} CH_3 \\ | \\ CH_2{=}CH{-}C{-}O \; (CH_2)_3O(CH_2)_3Si \\ \| \\ O \end{array} \begin{array}{cc} CH_3 \\ | \\ (SiO)_{\overline{3n-1}} \\ | \\ CH_3 \end{array} OSi(CH_3)_3$$

$$\begin{array}{c} CH_3 \\ | \\ CH_2{=}CH{-}\langle\bigcirc\rangle{-}SiCl \\ | \\ CH_3 \end{array}$$

or $\longrightarrow$
$$CH_2{=}CH{-}\langle\bigcirc\rangle{-} \begin{array}{c} CH_3 \; CH_3 \\ | \quad\;\; | \\ Si \; (SiO)_{\overline{3n-1}} \; OSi(CH_3)_3 \\ | \quad\;\; | \\ CH_3 \; CH_3 \end{array}$$

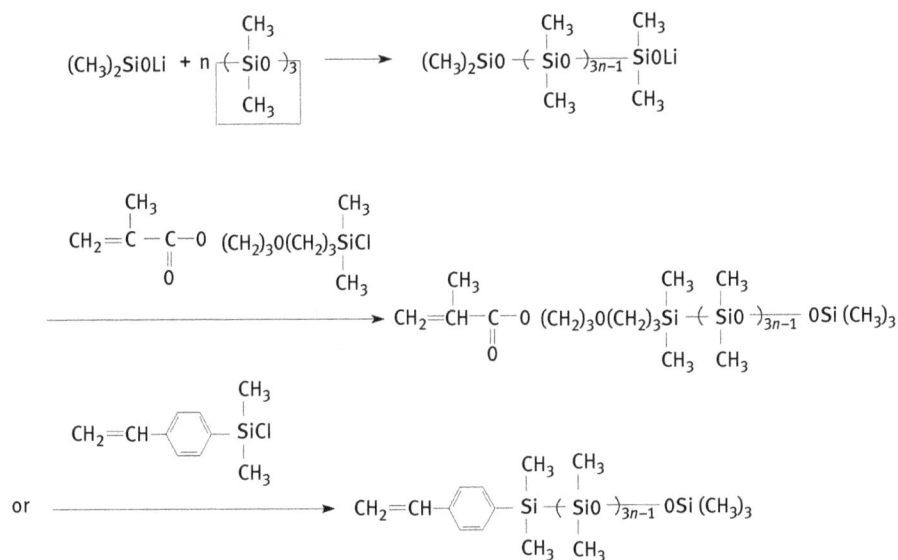

Fig. 8.6: Synthesis of polydimethylsiloxane macromer of styrene and methacrylate.

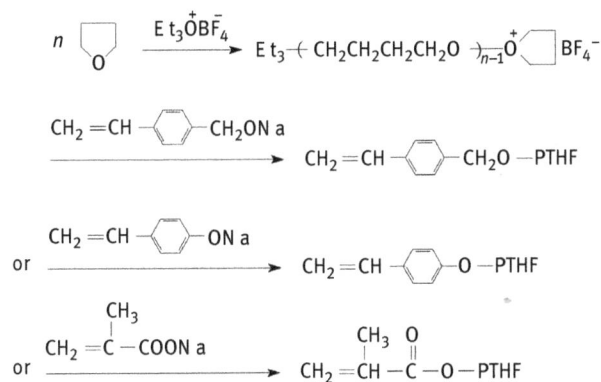

$$n \; \Big\langle\bigcirc_O\Big\rangle \xrightarrow{\;E t_3\overset{+}{O}B\overline{F}_4\;} E t_3 {+}CH_2CH_2CH_2CH_2O{+}_{\overline{n-1}} \overset{+}{O}\langle\bigcirc\rangle BF_4^-$$

$$CH_2{=}CH{-}\langle\bigcirc\rangle{-}CH_2ONa$$
$$\longrightarrow CH_2{=}CH{-}\langle\bigcirc\rangle{-}CH_2O{-}PTHF$$

or
$$CH_2{=}CH{-}\langle\bigcirc\rangle{-}ONa$$
$$\longrightarrow CH_2{=}CH{-}\langle\bigcirc\rangle{-}O{-}PTHF$$

or
$$\begin{array}{c} CH_3 \\ | \\ CH_2{=}C{-}COONa \end{array}$$
$$\longrightarrow \begin{array}{c} CH_3 \; O \\ | \quad \| \\ CH_2{=}CH{-}C{-}O{-}PTHF \end{array}$$

Fig. 8.7: Synthesis of methacrylate polytetrahydrofuran macromer.

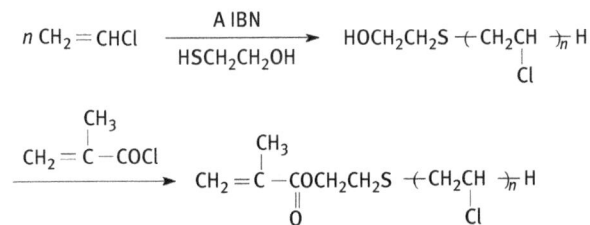

$$n \, CH_2{=}CHCl \xrightarrow[\text{HSCH}_2CH_2OH]{\text{AIBN}} HOCH_2CH_2S {+}CH_2CH{\tfrac{}{}}_n H$$
$$\begin{array}{c} \quad\quad\quad\quad\quad\quad\quad\quad | \\ \quad\quad\quad\quad\quad\quad\quad\quad Cl \end{array}$$

$$\begin{array}{c} CH_3 \\ | \\ CH_2{=}C{-}COCl \end{array}$$
$$\longrightarrow \begin{array}{c} CH_3 \\ | \\ CH_2{=}C{-}COCH_2CH_2S {+}CH_2CH{\tfrac{}{}}_n H \\ \| \quad\quad\quad\quad\quad\quad\quad | \\ O \quad\quad\quad\quad\quad\quad\quad Cl \end{array}$$

Fig. 8.8: Synthesis of methacrylate-type polyvinyl chloride macromer.

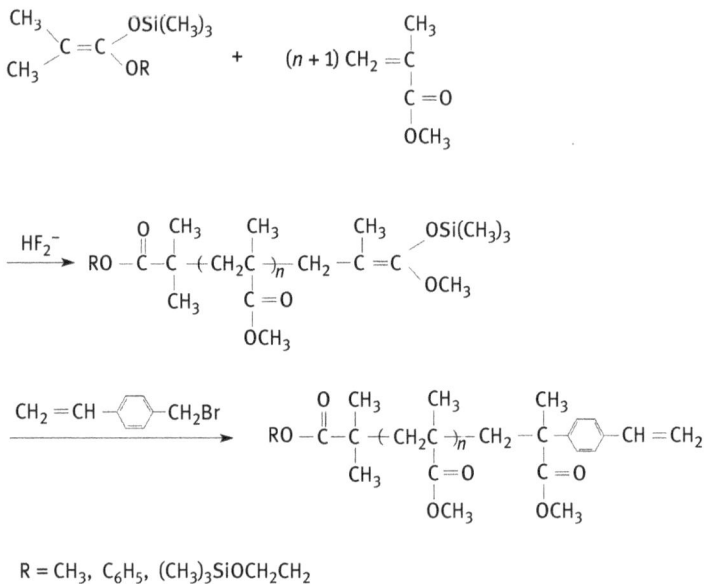

R = CH₃, C₆H₅, (CH₃)₃SiOCH₂CH₂

Fig. 8.9: Synthesis of styrene polymethyl methacrylate macromer.

The following is an example of the preparation of graft copolymers by macromers:

(1) Copolymerization of St-based polyethylene oxide macromers
As early as 1965, the copolymerization of St-based polyethylene oxide macromers with AN has been studied, although there was no concept of macromers at that time. The results show that styrene polyoxomane macromers can be stabilized with AN in water emulsion, and the product can improve the surface wettability of PAN film or fiber.

(2) Copolymerization of methacrylate polytetrahydrofuran macromers
The free radical graft copolymerization of methacrylate-based polytetrahydrofuran macromers with butyl methacrylate or St was studied using AIBN as initiator. The results showed that for St, the proportion of macromers in the graft copolymer was higher than that in the monomer feed. For butyl methacrylate, the opposite is true.

(3) Copolymerization of methacrylate PS macromers
Free radical copolymerization of methacrylate PS macromers with other monomers such as methacrylate, vinyl chloride and AN was carried out. Gas permeation chromatography measurements showed that when the conversion rate of small monomers was 80%, the macromers almost completely reacted and the grafted chains on the main chain were randomly distributed.

The graft copolymers obtained by copolymerization of methacrylate PS macromers with 2-hydroxyethyl methacrylate or perfluoroalkyl acrylate have strong hydrophilicity or lipophilicity.

(4) Copolymerization of methacrylate-type poly(alkyl methacrylate) macromers
When the conversion of methacrylate-2-hydroxyethyl methacrylate and perfluoroalkyl acrylate copolymerized with methacrylate-type PMMA macromers was 70%, the proportion of macromers in the graft copolymer was slightly lower than that in the monomer feed. The PMMA film containing the graft copolymer can improve the surface properties. Graft copolymer of 2-hydroxyethyl methacrylate as copolymer has good hydrophilicity. The graft copolymer of perfluoroalkyl acrylate as copolymer has good oil affinity. The free radical copolymerization of dodecyl methacrylate macromers with MMA can be used as dispersant and wetting agent.

(5) Copolymerization of methacrylate-type polydimethylsiloxane macromers
Using AIBN as initiator, methacrylate-type polydimethylsiloxane macromers were copolymerized with MMA. When the yield is 80%, the molar content of macromers in the graft copolymer is approximately the same as that in the monomer feed. The graft copolymer of polymethyl acrylate (PMA) and polydimethylsiloxane has good biocompatibility, and its surface energy is about 50% lower than that of PMMA, so it has potential medical application value.

8.3 Properties of graft copolymers

8.3.1 Chemical properties of graft copolymers

Graft copolymers are not only simple additions of two homopolymers in physical properties, but also new changes in chemical properties. For a branched chain, only the end of the branched chain participates in the formation of the grafting point. It can be inferred that the chemical properties of branched chain polymers are usually not greatly affected by grafting. However, there are always many sites on the main chain to participate in the formation of grafting sites, especially at high grafting density. The grafting sites occupy many reaction sites on the main chain, which makes the chemical properties of the main chain change significantly. Therefore, graft copolymerization mainly changes the properties of the main chain.

It has long been found that AN or methacrylic acid grafted wool fibers can improve the dyeing properties and they can resist mildew erosion. It was also reported that the alkali solubility of wool grafted with St could be greatly improved when the graft reached 10%. Cellulose grafted with AN can also greatly improve its microbial resistance. It is also reported that the tent fabric made from nylon grafted with St has high air aging resistance. These are examples of changes in the chemical properties of the main chain.

In recent years, in order to improve the thermal decomposition resistance of PVC, many methods have been reported, among which graft copolymerization is an effective way. For example, when acrylic acid and butyl acrylate are grafted with PVC under light initiation, the thermal decomposition temperature of PVC can be increased to exceed 200 °C.

Kennedy et al. have done a lot of research on PVC-g-PS graft copolymer prepared by cationic initiation. It was found that when St was grafted with PVC in the presence of Et$_2$AlCl, only one T_g appeared in the system, which indicated that the system has single phase structure. With the increase of St content, the thermal stability of the product has significantly improved. When the grafting amount of St reached 23%, there was almost no chlorine loss after 2.5 h at 165 °C. Even St containing only 4% delayed the release of chlorine, and the rate of chlorine loss was significantly slowed down.

Recent studies on the thermal stability of PVC indicate that the loss of a molecule of HCl in PVC means the formation of an allyl chloride, which is the most unstable form of chlorine. Therefore, the second and third chlorine are more likely to be lost one after another, and the free HCl further plays the role of catalytic dechlorination, resulting in many conjugated double bonds and cross-links, the color of the resin becomes darker, and the mechanical properties deteriorate. This indicates that cationic initiation grafting eliminates active perchlorine and replaces it with C–C bond with high thermal stability, which results in PVC becoming a heat-stable graft copolymer.

8.3.2 Solution properties of graft copolymers

8.3.2.1 Molecular morphology and properties of graft copolymer solution

When the graft copolymer exists in the solution with good solvent for both the main chain and the branch chain, the whole macromolecule is in the chain stretching state. When it is an undesirable solvent for one chain and a good solvent for another chain, the chain in the undesirable solvent will shrink and form a single or multimolecule micelle which is dispersed in the solvent, as described in Fig. 8.10.

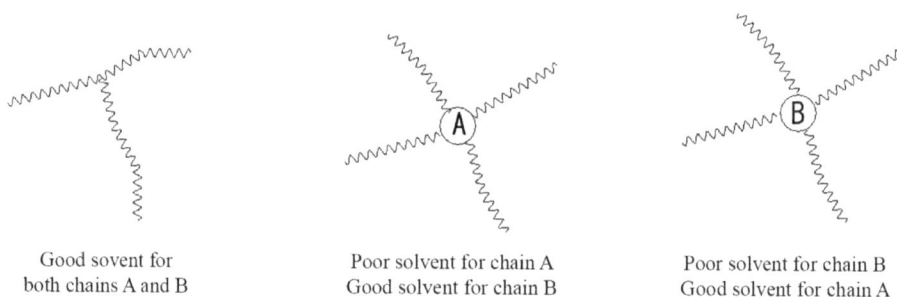

| Good sovent for both chains A and B | Poor solvent for chain A Good solvent for chain B | Poor solvent for chain B Good solvent for chain A |

Fig. 8.10: Morphology of graft copolymers in solution.

The morphology of the graft copolymers in different solvents showed that the properties of the graft copolymers depended on the solvents. If the solvent used is undesirable for the A segment and good for the B segment, the A segment is dispersed phase and the B segment is continuous phase. Conversely, the B segment is dispersed phase and the A segment is continuous phase.

There are many ways to understand the morphology of macromolecule in solution. The most effective methods are viscometry and light scattering. However, it must be pointed out that the molecular weight and radius of rotation measured by light scattering method are only an apparent value because the composition of the grafted copolymer is dispersive, and its physical significance is very complex. If the appropriate solvent is selected and the *dn/dc* value of one segment is zero, the light can be "transparent" to this segment, and the radius of rotation of another segment can be determined.

8.3.2.2 Emulsification of graft copolymers

Because the main chain and branched chain of the graft copolymer have different properties, the structure of the graft copolymer is similar to that of the surfactant. Facts have shown that many graft copolymers also have the function of emulsifier. For example, the addition of graft or block copolymer into incompatible low-molecular solvent mixture (castor oil–water system) can make both two phases form stable emulsion.

Incompatible polymer and low-molecular liquid phase can also be mixed by emulsification of graft copolymer. For example, PMMA can be dissolved in dimethyl sulfoxide (DMS), but a small amount of water can precipitate it. If a PMMA branched chain (DP = 1,810) is attached to the polyvinyl alcohol macromolecule with a degree of polymerization (DP) = 2,140, such a graft copolymer can effectively emulsify PMMA in the above solution and prevent precipitation. The amount of grafts needed for emulsification depends largely on the DP of PMMA homopolymer. When the DP of PMMA homopolymer is 1,040, 1,500, 1,700 and 3,500, the maximum number of emulsifying molecules per copolymer molecule is 264, 15, 11 and 3.

In polymer blends, graft copolymers can also be used as effective interfacial activators to disperse polymers. The two polymers can form a stable oil-in-oil dispersion system. For example, with vinyl acetate (DP = 2,000) as the main chain and a St branched chain (DP = 1,800), the graft copolymer was added to the PS–polyvinyl acetate–benzene system as emulsifier, and it was found that when the graft copolymer reached a certain quantity, it could form a stable emulsion. Electron microscopic observation showed that the graft copolymers aggregated at the particle interface, which stabilized the particles and prevented the agglomeration. The radius of dispersed phase particles is proportional to the concentration of dispersed phase polymer and inversely proportional to the concentration of graft copolymer. When the solvent evaporates, the morphological characteristics of one phase dispersed in another phase can be clearly seen.

8.3.3 Mechanical properties of graft copolymers

8.3.3.1 Relationship between morphology and mechanical properties of graft copolymers

The morphology of the graft copolymer in solution and the formation of micelles are described in the previous section. Obviously, if the natural rubber-g-PMMA graft copolymer is dissolved in common solvents such as benzene, a general polymer solution will be formed. At this time, if methanol is added to the rubber chain as a solvent, the rubber chain shrinks to form the rubber particles coated by the PMMA chain. Conversely, if petroleum ether is added without methanol, the PMMA chain shrinks to form rubber-coated particles. These two kinds of colloidal particles are called type A and type B, respectively. When the solvent is evaporated, in type A, rubber particle is dispersed phase and PMMA is continuous phase, while in type B, PMMA is dispersed in continuous phase. There are obvious differences in mechanical and physical properties between the two. Table 8.2 gives the measured results.

Tab. 8.2: Vulcanization properties of natural rubber–PMMA graft copolymer (weight ratio 70/30).

Performance	Polymer form		
	Emulsion copolymer	A	B
Tensile strength (MPa)	27.8	18.7	28.0
Modulus (100%) (MPa)	4.8	6.6	2.0
Modulus (300%) (MPa)	14.8	16.8	14.8
Elongation at break (%)	485	345	485
Hardness (BS°)	62	85	62
Tear strength (kN/m, 20 °C)	14	5.3	1.1

The results show that the overall properties of the material are close to those of the continuous phase.

For the multiphase structure mentioned earlier, the dispersed phase is customarily called island and the continuous phase is called sea. Therefore, the mechanical behavior of this island-structured graft copolymer system is mainly dominated by the marine phase. When the volume fraction of a chain in the system is between 0.3 and 0.7, the sea and island may interchange, and the performance will change greatly.

When components A and B form graft copolymers, the following five possible aggregation morphologies (similar to diblock copolymers) can be formed with the change of composition ratio:

(1) Component A is spherically dispersed in medium B
(2) Component A is rod shaped and dispersed in medium B
(3) The two components of A and B are arranged alternately in layers
(4) Component B is rod shaped and dispersed in medium A
(5) Component B is spherically dispersed in medium A

The graft copolymer PMA-g-PS composed of PMA and PS was blended into 3% benzene solution to prepare film. The morphological changes were observed under high power microscopy. When the volume fraction of PS in the graft copolymer increased from 10% to 80%, it was observed that PS was initially dispersed in PMA as spherical particles, and then dispersed in PMA as flat elliptical particles. Later, it was observed that PS and PMA were arranged in alternate lamellae. In samples containing 80% PS, it was clearly observed that PMA could be dispersed in PS medium as spherical particles. The rod structure was not observed.

8.3.3.2 Thermoplastic elastomers and toughened plastics

It is well known that when component A is dispersed in component B, with the increase in component A, the dispersed form of island A sea B type will change to island B sea A type. The maximum volume fraction of dispersed pellets can be deduced to be 0.74 according to the condition that the dispersed pellets reach the densest filling. When the volume fraction of a component is less than 0.26, it can only appear as a dispersed phase, but when it reaches 0.74 or above, it always appears as a continuous phase. Only when its volume fraction is between 0.26 and 0.74, it can occur in both island and sea phases with the change of sample preparation conditions. According to the introduction in the preceding section, the mechanical properties of the material mainly depend on the sea phase. When the sea phase is formed by a rubber-like main chain and the island phase is composed of plastic-like branched chains, the material will flow when the temperature is higher than the T_g temperature of the plastic phase or the melting point is T_m. When the material is cooled to below T_g temperature or melting point T_m, it loses its fluidity and presents the properties of vulcanizate. The material with this property is thermoplastic elastomer. Conversely, if the sea phase is plastic phase and the island phase is rubber particles, then the dispersed rubber particles can initiate and terminate cracks under stress, and the material presents high toughness, which is a toughening plastic.

So far, many toughened plastics have been prepared by grafting. However, at present, some important thermoplastic elastomers are mainly prepared by block copolymerization, such as ABA and $(AB)_n$ block copolymers. This is mainly due to the difficulties in controlling the degree of branched chain polymerization and the density of grafted copolymers. However, as the preparation of graft copolymers is much more convenient than block copolymers, efforts to use graft copolymers as thermoplastic elastomers still have attracted much attention.

Kennedy et al. reported the homogeneous reaction of St with chlorosulfonated PE rubber in the presence of Et$_2$AlCl. It was found that the graft chain was formed at the position of tertiary chlorine atom. The mechanical properties of the products show that the products can be used as both toughening plastics and thermal elastomers with the change of composition ratio. This system relies on the glass-like PS phase at room temperature to produce physical cross-linking. There are also examples of physical cross-linking by branched chain crystallization. The most typical example is graft copolymer of poly(terpentyl lactone) and rubber (as the main chain). Pariser used anion-initiated grafting to obtain crystalline branched chains of rubber (e.g., partially hydrolyzed polyethylene acrylate or carboxyl-containing ethylene–propylene–hexadiene terpolymer) with carboxyl groups in its main chain, which was synthesized by ring-opening polymerization of terpentyl lactone in the presence of quaternary amine salts. He found that when the DP of the rigid branched chain was more than 4, it could crystallize, and the melting point of crystallization increased with the increase in DP. The stress–strain curves showed that the properties of the obtained materials were basically similar to those of the corresponding chemically cross-linked or filled reinforced rubber. Generally, chemically cross-linked rubber has low strength without filling reinforcement, so the poly(terpentyl lactone) branched chain here also acts as an effective filler. In addition, it is noteworthy that the compression formability widely used in the rubber industry to estimate that the network stability is similar to that of chemical cross-linking and filling reinforced ordinary rubber. In this anionic grafting technology, the grafted chain is active chain, and there is no strongly chain transfer, so the degree distribution of the grafted chain is close to the Poisson distribution. Because of its clear structure, the graft copolymer is a good model polymer for the study of crystalline/amorphous thermoplastic elastomers.

8.3.3.3 Impact resistance of rubber-toughened plastics

One of the most valuable applications of graft copolymers is the development of toughened plastics. For example, modified PS (HIPS) with high impact resistance and ABS ternary graft copolymer are all multicomponent polymers that rubber particles disperse in matrix phase through grafting. Therefore, the degree of grafting, the molecular weight of branched chain, the number and size of rubber particles and the properties of the matrix will change the mechanical properties of the materials.

Merz et al. found that the impact properties greatly improved when the size of rubber particles was appropriately reduced. According to this fact, a model was proposed, which assumed that rubber particles can terminate cracks. The smaller the size of rubber particles, the greater the probability of terminating cracks and the higher the impact resistance.

In HIPS, when the size of rubber particles increases from small to 1 mm, for example, the impact performance decreases with the increase of particle size. However, there are also a lot of evidences indicate that for HIPS and ABS, the impact resistance increases with the increase of particle size within a certain range of particle size. Sudduth has proposed a new model for this purpose. Based on this model, the following three conclusions can be drawn:

1. The better the compatibility between the colloidal particles and the rigid matrix, the smaller the thickness of the graft layer and the smaller the average particle size D_{Smax}, which reaches the ideal maximum impact value. This shows when the two components are nearly compatible, the particles need not be grafted, and the smallest particles will give the best impact performance.
2. If good compatibility can be obtained with little grafting, the D_{Smax} depends mainly on the minimum thickness of the grafting layer achieved by the grafting reaction. When the particle size is smaller than that, if the degree of grafting increases, the impact value increases with the increase of particle size. If the particle size is larger than the D_{Smax} value and the degree of grafting remains unchanged, the impact properties will decrease with the increase of particle size.
3. If the compatibility is poor, then the degree of grafting and the thickness of the grafting layer will determine D_{Smax}. However, even with high degree of grafting, in the range of too large particle size, the impact properties will decrease with the increase of particle size.

The conclusion of this model can clearly explain the relationship between impact properties and particle size of toughened plastics. Unfortunately, the data provided in the literature are seldom correlated with D_{Smax}.

As mentioned earlier, in toughened plastics, the rubber phase is not necessarily a simple particle, which is often wrapped in dispersed plastic phase. For example, the size of the loss peak in the dynamic mechanical diagram of ABS is directly related to the size of the rubber particles encapsulated with St–AN copolymer. Therefore, only when the encapsulation morphology is fixed, the impact properties of toughened plastics can be well correlated with the area of the loss peak of the rubber phase.

As shown in Fig. 8.11, there are two kinds of interfaces between rubber phase and plastic phase, that is, external interface and internal interface.

It is assumed that there is a rubber particle in the impact-resistant PS material whose radius is r_1. If the concentration of grafting point in rubber phase is g, the concentration of grafting point on the interface is $r_1 g/3$.

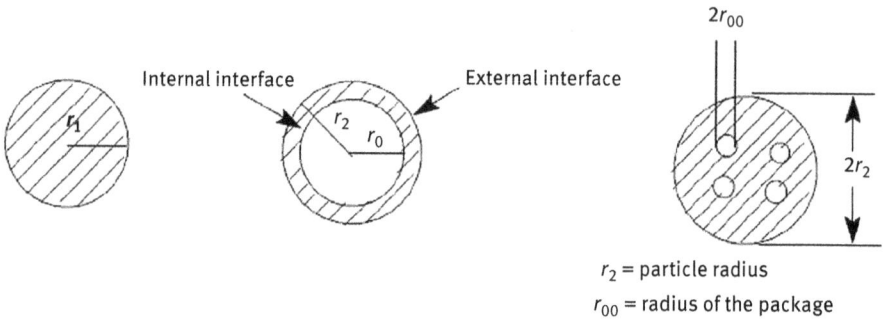

r_2 = particle radius
r_{00} = radius of the package

Fig. 8.11: Encapsulation phenomenon in HIPS dispersed phase.

If there is a small PS particle in the rubber phase, its radius is r_0, and the radius of the outer interface of the rubber particle is r_2, then the total volume of the rubber phase is still constant:

$$\frac{4}{3}\pi(r_2{}^3 - r_0{}^3) = \frac{4}{3}\pi r_1{}^3 \tag{8.1}$$

The total internal and external interface is $4r(r_0{}^2 + r_2{}^2)$, so the concentration of interface grafting point is

$$\frac{4}{3}\pi g r_1{}^3 / [4\pi(r_0{}^2 - r_2{}^2)] \tag{8.2}$$

Further consideration is given that a colloidal particle is occupied by n plastic particles with sizes of r_{00}. In this case, $4/3\pi r_0{}^3 = 4/3\pi n r_{00}{}^3$ and the sum of internal and external surfaces is $4\pi(r_2{}^2 + n r_{00}{}^2)$. Therefore, the concentration of interfacial grafting point is as follows:

$$g r_1{}^3 / [3(r_2{}^2 + n r_{00}{}^2)] = g r_1{}^3 / [3(r_2{}^2 + n^{-1/3} r^2)] \tag{8.3}$$

The above formula shows that when the total volume of the colloidal particles remains unchanged, the concentration of interfacial grafting point increases with the decrease of particle size. Therefore, for the interface bonding effect, it is better that there is no wrapping phenomenon. However, one of the most important functions of wrapping is to improve the volume effect of rubber dispersions. If the volume fraction of the dispersed rubber phase is f_R without encapsulation and f_D with encapsulation, then:

$$f_D/f_R = r_2{}^3/r_1{}^3 \tag{8.4}$$

For ordinary impact PS, the volume fraction of dispersed phase/rubber phase is about 4.5. If the relative density of the rubber phase is 1, the following formula can be derived:

$$r_0^3 = 0.78r_2^3 \tag{8.5}$$

This formula shows that if f_D and r_2 have been determined, the optimal particle size of the package can be obtained accordingly.

Although many literatures have discussed the influence of particle size and degree of grafting on impact properties, it is difficult to quantitatively predict the actual impact properties by theoretical calculation because of the complex morphology of the actual particles and the uneven distribution of the grafted particles on the surface. However, it can be concluded that larger rubber particles play a role in initiating microcracks, which are prevented by many dispersed rubber particles in the development of microcracks, making the formation of cracks impossible in a certain period. It is precisely because many microcracks consume a lot of energy during the formation process that the impact strength is improved.

The occurrence of ABS cracks and their impact resistance have been studied. For ABS with high degree of grafting of small particles (SPHG), low degree of grafting of large particles (LPLG) and their blends (SPHG/IPLG) at 1:1, the impact properties of the three systems were measured at room temperature. The results are shown in Fig. 8.12. As expected, the impact performance of SPHG is not good.

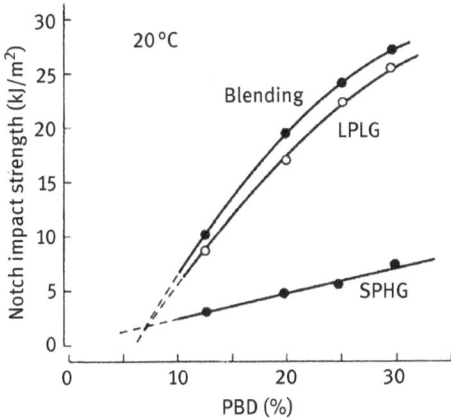

Fig. 8.12: The influence of rubber (PBD) content and particle size on notch impact strength.

The dependence of impact performance on temperature is shown in Fig. 8.13. Obviously, the impact performance of LPLG at low temperature is quite low, almost close to that of SPHG. However, the impact performance of LPLG and SPHG is obviously improved after blending.

The SPG/LPLG blend specimens with microcracks have been examined by scanning electron microscopy. It can be confirmed that the large particles play an important role in initiating microcracks. When the cracks develop in the matrix, the

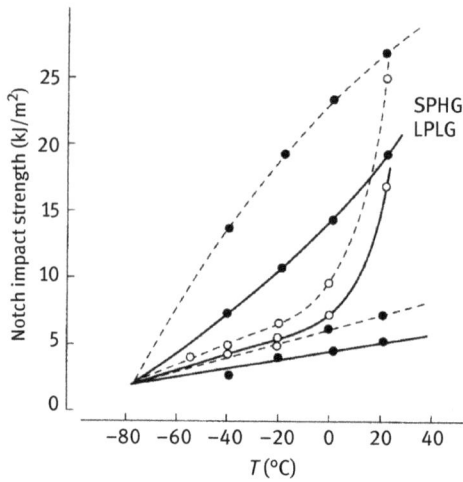

Fig. 8.13: Effect of temperature on notch impact strength.

matrix is filled with small particles of rubber, which forces the cracks to develop along the local area with smaller modulus. At the same time, branching occurs along the bending path, and is eventually terminated by small particles of rubber. It can be concluded that small rubber particles play a role of reinforcing unit, so their high degree of grafting is very important. The large particles mainly play the role of initiating cracks. For large particles, low degree of grafting can lead to uneven distribution of grafting, but it helps to initiate cracks at different interface positions at the same time. High graft makes crack initiation confined to the equatorial plane of particles. In the process of yielding, besides cracking, small particles can also make the shear deformation develop to a certain extent. These two kinds of deformation can improve the impact resistance of materials.

8.3.4 Gas permeability of graft copolymer

For polymer films, containers, shells and other products, the permeability of low molecular gas through polymer membranes or shells is an important property. Graft copolymerization can effectively change the permeability of polymers. Particularly important is that the grafted films have high permeability selectivity to certain gases in some low-molecular-weight gas mixtures, grafting monomers and appropriate grafting methods, which provides an effective new technology for gas separation.

The permeability is generally reflected by the permeability coefficient p, which is closely related to diffusion coefficient D and solubility coefficient S ($p = D.S$).

8.3.4.1 Solubility coefficient of graft copolymer

The solubility coefficient of the substance in the graft copolymer is closely related to the solubility of the substance in the matrix and the graft chain. It is the total reflection of their contribution to the solubility coefficient of the graft copolymer. However, the simple summation rule usually does not apply.

Grafting modification of PE film is the most studied system. PE film grafted with St was prepared by γ-ray irradiation. The relationship between the solubility coefficient of n-hexane and benzene in the graft and the graft percentage was determined, as shown in Fig. 8.14.

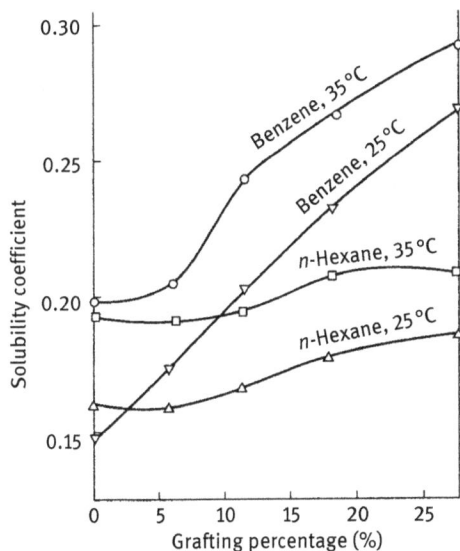

Fig. 8.14: Solubility coefficient of n-hexane and benzene in PE–PS graft copolymer.

Obviously, the solubility coefficient is a complex function of composition. Generally, the solubility coefficients of the above low-molecular-weight organic solvents in copolymers increase with the increase in the amount of grafted St. The increase of benzene is larger than that of n-hexane, which is due to the close solubility parameters of benzene and St. In order to improve the solubility coefficient of a substance in graft copolymers, monomers with similar solubility parameters are often selected intentionally.

In many cases, solubility coefficients have linear additivity of solubility coefficients of each component. Toi et al. measured the solubility coefficients of five simple gases, Hi, Ar, N_2, O_2 and CO_2, in the grafted films of high-density PE (HDPE) and St prepared by γ-ray irradiation. After removing PS homopolymer from the system by extraction, they found that the solubility coefficient of the grafted copolymer was almost equal to the volume fraction of PS branched chain and PE in amorphous region multiplied by the sum of their solubility coefficients. It is noteworthy that solubility coefficients cannot be calculated by additivity before extraction (in the presence of homopolymer), and the measured values are always lower than those calculated by additivity.

Kamel calculated that when St was grafted into low-density polyethylene (LDPE), a grafting point occurred only for about 10,000 CH_2 units, indicating that the grafting degree was very low. It is inferred that there is no significant difference in physical properties between the grafted PS and the blended PS in this system. However, the contribution of the two factors to the solubility coefficient is obviously different. Further differential scanning calorimetry measurements showed that the melting heat of the grafted copolymers varied in the same straight line with the composition, indicating that the extraction process did not change the crystallinity of the grafted copolymers. At present, it is difficult to give an exact explanation for the effect of these two PS. It may be that some morphologies of amorphous regions have changed after homopolymer extraction. Of course, further study on the real integral and other physical properties of the components in the amorphous region before and after PS homopolymer extraction will help to clarify this problem.

8.3.4.2 Diffusion coefficient of graft copolymer

Diffusion is a dynamic process, so the diffusion coefficient is closely related to the shape of amorphous region in polymer, more precisely to the free volume and chain motion of the region. Generally, the diffusion coefficients of grafted copolymers cannot be predicted simply by additivity of the volume fraction of each component in the amorphous region, but must consider the effects of aggregation and chain motion changes caused by grafting.

The diffusion coefficients of N_2 and CH_4 in PS-grafted PE (PE-g-PS) films were measured by Kanitz et al. The changes of permeability coefficients with the degree of grafting were investigated. The results are shown in Fig. 8.15. With the increase of grafting degree, the amorphous region of PE gradually decreases and the free volume decreases, which leads to the decrease of diffusion coefficient. However, with the further increase of grafting degree, the crystalline region is destroyed, the crystallinity of PE decreases, the grain size decreases sharply and the free volume increases again, so the diffusion coefficient increases significantly with the further increase of grafting degree. Therefore, a higher diffusion coefficient can be obtained by controlling the degree of grafting of copolymers. At the same time, the introduction of rigid branched PS into dispersed phase microzone is beneficial to improve the mechanical strength of thin film materials. Therefore, block copolymers and graft copolymers are widely used as membrane materials for gas separation.

Toi et al. measured the diffusion coefficients of St grafted onto PE membranes by radiation. The relationship between the diffusion coefficients and the volume fraction of St is shown in Fig. 8.16.

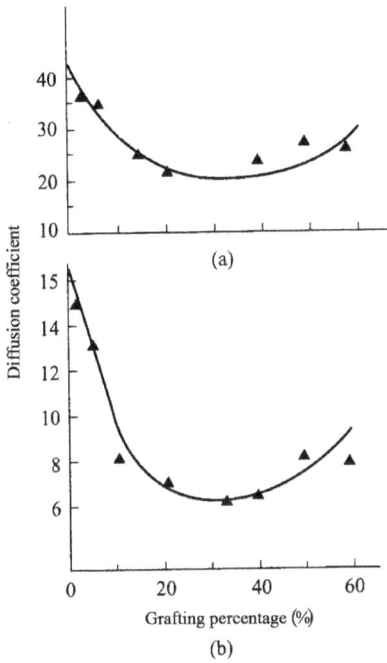

Fig. 8.15: Diffusion coefficients of PE-g-PS films vary with degree of grafting (30 °C): (a) CH_3 and (b) N_2.

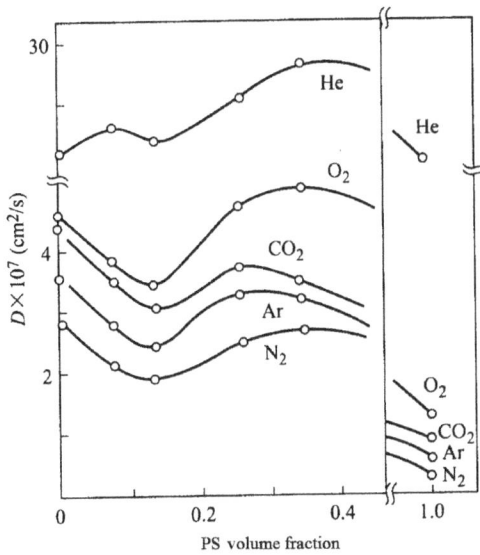

Fig. 8.16: The relationship between diffusion coefficient of PE-g-PS film and PS volume fraction (30 °C).

During the grafting process, St monomer entered the amorphous region of PE and grafted onto the macromolecular chain. In this way, PS branched chains occupy the original voids to reduce the free volume. At the same time, the activity of the grafted macromolecule chains is hindered, resulting in the increase of T_g. Therefore, with the increase of grafting degree, the free volume decreases and the diffusion coefficient decreases. However, Fig. 8.16 shows that the diffusion coefficient rises again after a minimum value. From the corresponding diffusion activation energy, there is also a phenomenon from low to high, and then decreased, as shown in Fig. 8.17. The results show that this is due to the excessive increase of the degree of grafting, which destroys the crystalline zone in the sample and produces new areas for diffusion.

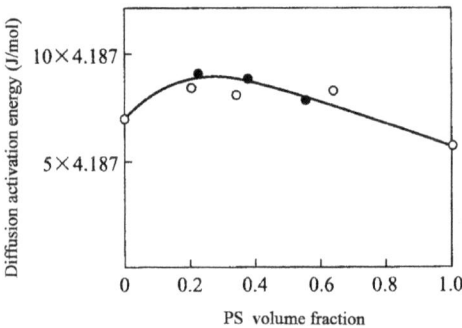

Fig. 8.17: The relationship between the diffusion activation energy of Ar in PE-g-PS films and the volume fraction of PS (15–40 °C).

However, the anomalous phenomenon that the change of diffusion coefficient of He deserves attention. From Fig. 8.16, when the degree of grafting is not high the diffusion coefficient of He increases with the increase of grafting amount. This phenomenon can be explained by the molecular scale of He and the characteristics of chain motion. As the degree of grafting increases, the chain motion is hindered, the vibration radiation decreases and the diffusion of larger molecules is obviously hindered. However, the relatively smaller He molecules can still pass smoothly through the interior. Therefore, with the increase of grafting degree, the amorphous region increases and the diffusion coefficient of He increases at the beginning. Only when the chain motion is very weak, the diffusion will be hindered and the diffusion coefficient will decrease.

Some results show that the free volume decreases with the increase in the amount of grafted St. Therefore, it can be concluded that the grafting generally occurs in the amorphous region, thus reducing the free volume for diffusion. At the same time, it is noteworthy that the free volume is affected by the type of diffuser. In theory, free volume is the property of polymer itself and should not be controlled by diffuser. A possibility is that the polymer is impregnated in the diffuser before determination. As a result, the grain size distribution changed, but the crystallinity remained constant. Obviously, this situation will affect the diffusion properties of polymers. As for completely amorphous polymers, the free volume is of course independent of the diffuser.

The free volume of amorphous polymers can be approximated as follows:

$$f_{am} = 0.025 + 4.8 \times 10^{-4}(T - T_g) \qquad (8.6)$$

where f_{am} is the free volume of amorphous polymer at temperature T, and T_g is the glass transition temperature of the polymer. The free volume of the crystalline polymer can be obtained by multiplying the above 100% amorphous free volume by the volume fraction of the amorphous region. The comparison of the measured free volume with the theoretical value is shown in Fig. 8.18. The longitudinal axis is the ratio of the measured value to the theoretical value (in percentage). As shown clearly in the figure, for ungrafted PE, the free volume decreased to 38% (benzene) and 55% (n-hexane) of the theoretical value. With the increase of grafting rate, the value continued to decrease. But when the degree of grafting is 5%, the free volume decreases greatly, and after this degree of grafting, the decline slowed down. A possible explanation is that the reaction is diffusion free at low degree of grafting. This means that enough soluble monomers can make full use of the grafting position, so the grafting rate is higher. At the same time, when the chain terminates, many short-branched chains are generated. On the contrary, many long-branched chains were formed at high degree of grafting. Moreover, the grafting process occurs more in the center of the membrane, which is easily controlled by diffusion, leading to a wide distribution of branched chain length. According to the results of molecular weight determination of the branched chain of cellulose–PS graft copolymer, the above inference is reasonable. Therefore, it can be concluded that high grafting rate and short-branched chains contribute greatly to the reduction of free volume. It can also be seen that the effect of grafting on free volume seems to be more

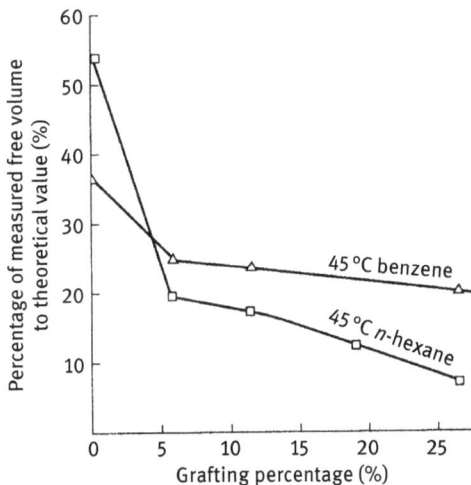

Fig. 8.18: The effect of grafting on the free volume of polymers.

significant than that of crystallization. Therefore, although the degree of crystallinity is reduced by grafting, the effect of the degree of crystallinity change is still secondary at least to low degree of grafting.

8.3.5 Liquid permeability of graft copolymers

The research of copolymers of PVA/AN-MMA used as permeation membranes for artificial kidney has achieved valuable results. The permeability and mechanical strength of the copolymers can be significantly changed by changing the mass ratio of PVA to AN. Table 8.3 shows that when PVA/AN is 1:1.5, the permeability is good, but the strength is not high enough (compared with the commonly used regenerated fibrous membranes of artificial kidney). When PVA/AN was 1:2, the tensile strength, water content and uric acid permeability of the grafted copolymer films were higher than those of regenerated cellulose, but the water permeability decreased.

Tab. 8.3: Effect of PVA/AN mass ratio on permeability of membrane.

PVA/AN (mass ratio)	Moisture content (%)	Uric acid permeation constant ($\times 10^7$/cm^2/s)	Water permeability constant ($\times 10^7$/cm^2/s)	Tensile strength (MPa)	Elongation rate (%)
1:2.5	39	3.3	–	39.6	198
1:2	50	7.3	6.0	25.4	235
1:1.5	69	5.7	14.8	8.7	212
Regenerated cellulose	50	5.9	8.0	Portrait 15.0	26
				Transverse 5.5	89

However, by changing the structure and morphology of PAN-MMA dispersed phase, the permeability of PAN-MMA dispersed phase can be greatly improved, but the strength of the film is not greatly affected. For example, the grafted copolymer film with PVA/AN of 1:3 was immersed in a mixed solvent of DMS and dimethylformamide (DMF) = 1:2 for 3 min (30 °C) and then washed to remove the solvent. Because of the swelling effect of solvent, the dispersed phase microarea becomes a porous structure, which greatly improves the permeability of liquid. The permeability of water and uric acid can be increased by 10 times. However, the strength decreased by only 15%. Figures 8.19 and 8.20 show the results of some studies. It can be seen from the graph that the strength and permeability of the membranes can be considered simultaneously by choosing the appropriate proportion of mixed solvents to make the dispersed phase porous.

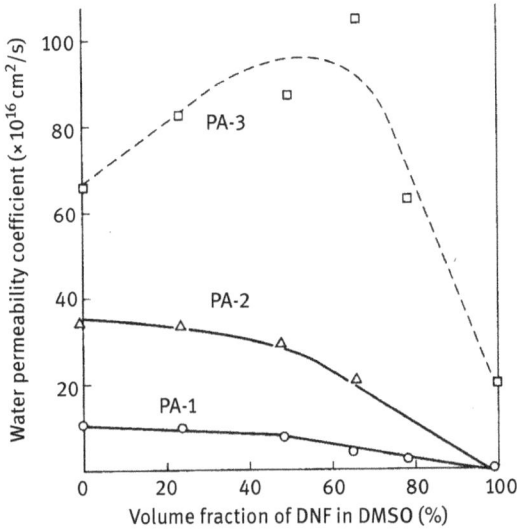

Fig. 8.19: The effect of solvent composition on water permeability of PVA/AN grafted membranes.
PA-1, PA-2 and PA-3 have PVA:AN of 1:1, 1:2 and 1:3, respectively.

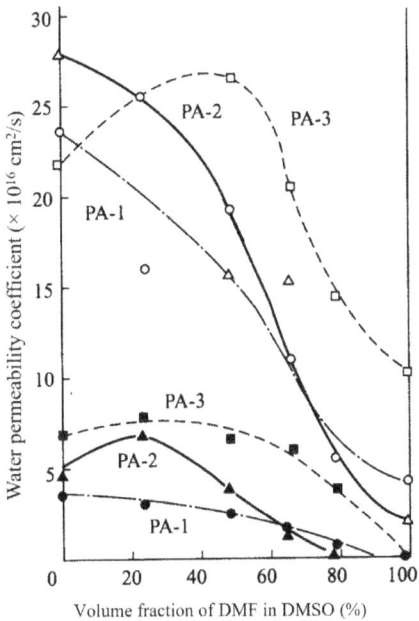

Fig. 8.20: Effect of solvent composition on the permeability of uric acid (O△□) and vitamin B$_{12}$
(●▲■) in PVA grafted membranes.
PA-1, PA-2 and PA-3 have PVA:AN of 1:1, 1:2 and 1:3, respectively.

8.3.6 Viscoelasticity of graft copolymers

Viscoelastic behavior is an important parameter to study the relationship between structure and properties of grafted or block copolymers. Usually, two glass transition points occur as a sign of incompatibility between components, that is, the presence of two-phase structure. In addition, in the dynamic mechanical measurements of grafted or block copolymers, besides the relaxation transition corresponding to the homopolymer with the same composition, new weak relaxation transition often occurs between these relaxation transition. Except for some special cases, such as partially miscible two-component systems or blends containing random copolymers as the third component, this new weak relaxation transition is a characteristic of graft or block copolymers, which is rarely seen in conventional mechanical blends. By studying this phenomenon, not only the mechanical properties of graft copolymers or block copolymers can be thoroughly understood, but also the important information of interfacial state of two phases can be obtained.

For example, the tensile relaxation behavior of PMA-g-PS graft copolymer was investigated. It was considered that there were two tensile relaxation processes: phase interface relaxation and structural flow relaxation. Tensile stress relaxation behavior of three graft copolymers containing 20%, 35% and 58% St was studied. It was found that with the increase of grafting degree, as the phase state of the system is dispersed from 20% PS spherical particles in the matrix of PMA, 35% alternating layer lamellar structure gradually develops to 58% continuous PS lamellar structure. As a result, the relaxation transition of PS phase becomes more and more obvious, while the relaxation transition of PMA phase decreases gradually.

If the temperature is plotted by the 100 s relaxation modulus of the sample, then Fig. 8.21 can be obtained. Data on the homopolymer of PMA and PS are also attached. The graph clearly reflects the relationship between composition and transformation. There are two phenomena related to weak relaxation in the graph. First, when the content of PS is 11% and 20%, the modulus of the plateau region of the sample decreases rapidly with the increase of temperature. Both phases are spherical PS dispersed in PMA. Second, when the content of PS is 35%, the modulus reduction zone corresponding to the glass transition of PS occurs at a lower temperature. This is similar to some biblock or triblock copolymers with alternating lamellar structure. It is noteworthy that the flow temperature of all graft copolymers is higher than that of both homopolymers. At the same time, the glass transition platform of PS becomes more and more obvious with the increase of grafting degree.

It is generally believed that there are three relaxation mechanisms in amorphous linear polymers: the local movement of molecular chains in the glass state, which is related to secondary glass transition; the intramolecular relaxation caused by short-range and long-range chain thermal movement, which is related to glass transition; and the thermal diffusion between molecules, which is related to the

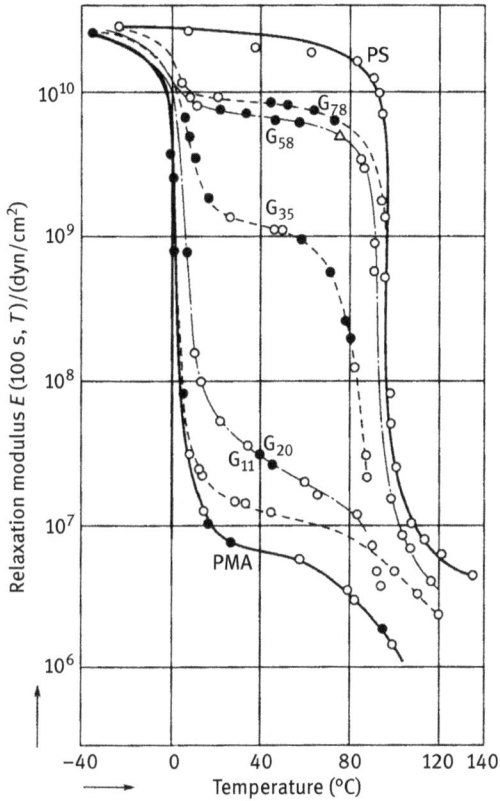

Fig. 8.21: A 100 s relaxation modulus of PS, PMA and their graft copolymers.
Samples with polystyrene content of 11%, 20%, 35%, 58% and 78% were indicated by G_{11}, G_{20}, G_{35}, G_{58} and G_{78}, respectively.

flow process of macromolecules. By decomposing the relaxation of the graft copolymer in the same way as earlier, two new relaxation processes can be found, which
are not found in the homopolymer. One relaxation reflects the weak relaxation between the vitrification transition (first-order transition) of two homopolymer chains
in the graft copolymer. It does not exist in the blend system, but is unique to the
graft copolymers or block copolymers with microscopic phase separation structure.
This is particularly evident in systems where hard segments are distributed in soft
chains. Therefore, it has been compared to the α-relaxation at the grain boundary
in semicrystalline polymers. Another new relaxation seems to be related to liquid-
ity. It was observed that the relative molecular weight of samples with 35% PS content was lower than that of samples with 58% PS content, while the flow
temperature of the former was higher than that of the latter. Therefore, it can be
considered that this flow is not intermolecular flow, but only a kind of "melted"
structural flow in the hard chain microregion. It seems that the graft copolymers

with 35% PS branched chain with alternating layered structure have higher elastic stability than those with 58% PS branched chain. From the high-temperature flow behavior of the graft copolymer containing 58% PS branched chain, the flow relaxation involving chain entanglement can only occur in the higher temperature region.

8.3.7 Melting and crystallization of graft copolymers

If the main or branch chains of the graft copolymers can crystallize, they can melt or crystallize during heating or cooling. If the crystallization of some chain in the main chain and the branch chain has not melted, the flow of materials becomes impossible. It has been reported in the literature that the softening temperature of HDPE or LDPE can be increased to about 40 °C when PAN is grafted with 25% of HDPE or 80% or LDPE, respectively. X-ray diffraction shows that besides the lattice diffraction of PE, the diffraction of PAN (a quasicrystalline structure) can also be found. Even when the grafted PAN is as high as 80%, it still can be proved that the two lattices coexist. The diffraction of PE crystals disappears when the temperature rises to 110 °C, but PAN crystals as branched chains still exist and the materials cannot flow. When cooling, the crystallinity of PE remains at the original level. This indicates that the branched chain of PAN can crystallize, which plays a physical cross-linking role and restrains the thermal movement of PE chain, resulting in corresponding changes in the properties of the main chain of PE.

The crystallization behavior of the main chain of crystalline polymer after grafting is greatly different from that of homopolymer. For example, polypropylene homopolymer can form perfect spherulite when it is cooled slowly from 180 to 80 °C for 4 h at high temperature, and the grain size reaches 300–500 μm. However, if 6.3% polyvinyl acetate is grafted, the spherulites could still reach the same size under the same crystallization growth conditions, but the small grain spots could be observed by polarizing microscope. When the amount of grafting reached 18.1%, the spherulites are obviously destroyed and many small grains are formed. When the amount of grafting reaches 41.5%, the large spherulites could not form, and there are cracks between the spherulites. The same trend also appears in the case of polypropylene-g-PS graft copolymer, and its effect is more significant than that of the former.

When random PS is grafted onto isotactic PS and grafting degree is in the range of 17–30%, the crystallization of isotactic PS could still form, but the crystallinity of isotactic PS become smaller and single crystal could not form. At this time, the orderliness of molecular chains only rests on the simple structure of fiber bundles, and there are 4–5 nm sized spheres between the bundles. Because the molecular weight of the branched chain is about 200,000, which is more than 80,000 of the main chain, the branched chain forces the main chain to form spherical particles.

The above facts show that the existence of branched chains is not conducive to the full expression of the original crystallization behavior of the main chain macromolecules. Therefore, it is difficult to grow the crystalline substructures (spherulites) and to develop higher structures (single crystals) even though they can still crystallize.

The crystallization behavior of branched chains is generally hindered by the binding of the host chain. An example is the crystallization of poly-ω-hydroxy undecanoate as a branch chain of polypropylene. When the degree of grafting is low (<10%), it is difficult to close and arrange regularly to crystallize due to the scarcity of branched chains. When the degree of grafting increases, crystallization begins to form and develop from small grains to spherulites. This indicates that the development of branched chains to normal crystalline substructures (spherulites) is limited due to the restraint of main chains.

However, it has been reported that the main chain promoted the crystallization of branched chains. An interesting example is that some difficult-to-crystallize polyesters can accelerate the crystallization of amorphous PMMA if they are grafted onto it. For example, at room temperature, the T_g of rubber-like poly(ethylene sebacate dithiovalerate) is −40 °C, which crystallize when it is placed for half a year. However, the branched chain obtained by grafting it onto PMMA has a crystal structure. Grafting of such flexible chain polyester improves its crystallinity on rigid polymer molecules. The flexible chain forms regular chain bundles with the support of rigid chain, thus accelerating its crystallization.

Usually, the main or branch chains of the grafted copolymers form cells according to their corresponding homopolymer lattices. But cell parameters are not constant. For example, when PE was grafted with amorphous PS or polyvinyl acetate, it was found that not only the crystallinity of PE changed with the properties, number and length of branched chains, but also the cell structure.

8.4 Application of graft copolymer

8.4.1 Preparation of polymers with high impact resistance

According to the toughening theory of plastics, proper introduction of rubber particles in brittle plastics can often restrain the growth of craze and improve the impact resistance of materials during the development of craze. Therefore, if the material in elastic state is evenly dispersed in the form of particulates into the glass-like polymer (e.g. PS), the impact resistance of the material should be increased. But in fact, sometimes when rubber particles are added to plastics, only one material with loose structure and greater brittleness is obtained, which is obviously due to the incompatibility between rubber phase and matrix phase, resulting in the formation of microphase separation.

If the grafted rubber is used as toughening particle, this problem can be completely solved, because the compatibility between dispersed particles and matrix materials can be greatly improved. For example, a toughening material with high impact resistance can be obtained by blending polybutadiene rubber as the main chain and PS as the branch chain of the graft copolymer into PS as the disperse phase. The results show that the impact resistance is the best when 9% rubber, 27% graft copolymer and 64% PS are contained.

High impact polymers such as HIPS and ABS are typical examples of the application of graft copolymers.

8.4.2 Reinforcement of natural rubber

Rigid branched chains in graft copolymers can be used as reinforcing units of rubber to improve the mechanical properties of rubber. This method is often better than direct addition of rigid polymers for blending. For example, the mechanical properties of natural rubber can be enhanced by grafting PMMA onto natural rubber. Its modulus of elasticity can reach more than 10^8 Pa at room temperature. This material can be used to make nonbreakable electrical plugs and sockets. Compared with products made of brittle thermoplastic materials or hard rubber, it has great advantages.

PS grafted onto natural rubber can also enhance the properties of natural rubber. In Tab. 8.4, after adding 30 phr copolymers, the hardness and modulus of elasticity of vulcanized natural rubber increase remarkably when the elongation at break and the strength of vulcanized natural rubber are not significantly reduced.

Tab. 8.4: Effect of poly-*cis*-1,4-isoprene-g-polystyrene on properties of natural rubber.

Rubber phr (quality)	Graft phr (quality)	Tensile strength (MPa)	Breaking elongation (%)	Elastic modulus at 300% elongation (MPa)	Shaw hardness (A)
100	0	23.79	650	34.1	47
80	20	21.65	600	83.7	55
70	30	20.34	600	97.7	65
60	40	13.31	530	103	80

8.4.3 Modification of fiber materials

Many synthetic fibers, such as PAN and polypropylene fibers, have the disadvantages of poor dyeability and low color retention. The dyeability of fibers can be improved by grafting some monomers with better dyeing properties. For example, acrylamide, *N*-vinyl pyrrolidone or methacrylonitrile can be grafted onto PAN or polypropylene, thus improving the dyeability of these fibers.

In addition, the physical properties of cotton fibers, including strength, wear resistance and wrinkle resistance, can be improved to some extent by grafting with vinyl monomers. For example, acrylate can be grafted onto the macromolecule of cotton fibers by radiation grafting to reduce the water absorption of cotton fibers.

8.4.4 Improvement of film-forming performance

The application of grafting technology can improve the film-forming properties of polymer materials. For example, the film-forming properties of PMMA homopolymer are very poor. MMA was grafted onto natural rubber by radiation grafting. When the mass fraction of natural rubber is about 5%, the PMMA has good film-forming properties.

8.4.5 Preparation of functional polymer materials

Many functional polymer materials can be prepared by grafting copolymers. For example, through high energy radiation, St can be grafted onto some polymers with better film-forming properties, such as PE, polypropylene and nylon. The grafted PS chains can be exchanged by some chemical treatments, such as sulfonation.

After grafting MMA or ethyl acrylate onto polyolefins, ester groups are hydrolyzed into carboxylic groups, which can be used as weak acid ion exchange materials. When 4-vinylpyridine and 2-methyl-5-vinylpyridine were grafted onto polyolefins and then quaternized with chloromethane, they could become strong base ion exchange membranes.

The examples of preparing functional polymers by graft copolymerization can be referred to the relevant functional polymer books, which are not discussed here.

8.4.6 As compatibilizer of polymer blends

Graft copolymers are the most important compatibilizers for polymer blends. As for this aspect, the second chapter of this book has been introduced in detail, which is omitted here.

8.5 Graft modification of PS plastics

PS is a kind of important plastic product with excellent electrical insulation and processing fluidity, excellent transparency, easy dyeing and high mechanical strength. Its biggest disadvantage is its brittle texture, easy stress cracking and poor impact resistance. The toughness and strength of HIPS modified by rubber grafting and ABS made from St are very good. It is the most successful model of polymer blends, and the yield ranks first among polymer blends.

8.5.1 High impact polystyrene

8.5.1.1 Preparation method of HIPS

HIPS can be prepared by mechanical blending and graft copolymerization. Mechanical blending method mainly uses SBR and PS blending. Toughening effect is shown when the amount of SBR in PS reaches 10–15%. When SBR exceeds 25%, the impact strength of the blends increases significantly, but the tensile strength decreases strongly.

Graft copolymerization is the main method to produce HIPS at present, mostly using *cis*-BR and St for graft copolymerization. The T_g of BR is lower and the modification effect is better than that of SBR. There are two main methods to produce HIPS: bulk suspension method and bulk continuous method.

1. Bulk suspension method

The bulk suspension polymerization process is carried out in two steps. Unvulcanized rubber was dissolved in St monomer and prepolymerized at about 100 °C under the action of initiator. Prepolymerization usually requires a good stirring effect. When the conversion rate reaches 30%, the prepolymer is poured into water containing dispersant and initiator, and dispersed into beads and suspended in water for suspension polymerization until the end of the reaction. The main characteristic of bulk suspension method is that it can control the size of rubber particles well, and the viscous liquid after bulk prepolymerization can suspend polymerization smoothly, which is easy to operate and control. However, the equipment utilization rate is low, and the process flow is long. Dispersant, centrifugal dehydration and drying need to be removed by washing. The energy consumption of the process is high.

2. Ontological continuity method

In recent years, most of the new HIPS devices adopt ontology continuity method, which has stable quality, low energy consumption, short process and relatively low cost.

The bulk continuous method can be divided into high polymerization degree (over 95% of the one-way yield of ST) and low polymerization degree (between 70% and 80% of the one-way yield of St). Controlling lower monomer conversion can make the reaction temperature stable and the molecular weight distribution uniform. Solvent recovery process takes away free monomers, so that the product is more pure and the quality is improved. At present, the production process of low conversion of polymerization in the presence of solvents is increasing in the proportion of bulk method, and there is a trend to replace solvent-free high conversion bulk method.

Regardless of which copolymerization method produces HIPS, most of the products of grafting reaction (HIPS) are homopolymer PS. The graft copolymer PB-g-PS is only a small amount, and others are some unreacted PB. So, HIPS is a blend system of homopolymer and copolymer.

8.5.1.2 Honeycomb structure of HIPS and its formation mechanism

In the third chapter, it is introduced that the morphological structure of HIPS is honeycomb structure. The dispersed phase of honeycomb structure is not a single component, but contains small particles composed of continuous phase components in the dispersed direction, so its shape is similar to honeycomb (sausage or cell) structure. Most of the graft copolymers have this phase structure.

Figure 8.22 is the phase-state electron microscopic photograph of grafted copolymer HIPS. The continuous phase is PS, the cell wall is composed of polybutadiene rubber and the cytoplasm (inclusion) is composed of PS. The graft copolymer PB-g-PS distributes at the interface of the two phases. The system contains only 6% rubber, but the volume fraction of rubber phase microzone is 22%. A small amount of rubber plays a role of larger volume fraction of rubber phase, which greatly strengthens the toughening effect of rubber. At the same time, the modulus of dispersed phase is obviously higher than that of pure rubber because PS is contained in rubber phase, so the modulus of HIPS cannot be reduced too much by the introduction of rubber. Therefore, the graft copolymer with the same composition and content has better modification effect than the mechanical blend.

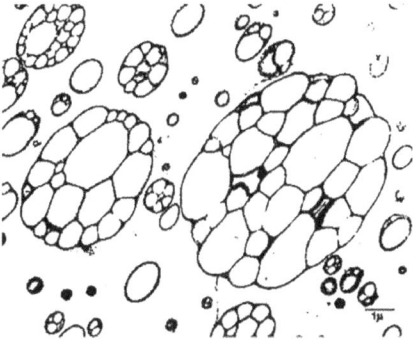

Fig. 8.22: Honeycomb structure of graft copolymerization of HIPS.

Why does HIPS form this special phase structure? The mechanism of the formation of this special structure is analyzed from the process of polymerization.

The phase diagram of St–PS–PB ternary system (see Fig. 8.23) was used to investigate the phase change during the polymerization process.

The initial composition of the system is that 6% of the PB is dissolved in the solution formed by the St monomer (*M* point in the phase diagram). The whole polymerization process is the conversion of St monomer to PS, and the content of PB remains unchanged. Therefore, the total composition should change along the *MS* line in the phase diagram (Fig. 8.23 is only a part of all phase diagrams of the system). When about 2% of St is aggregated into PS, the total composition reaches two-section line

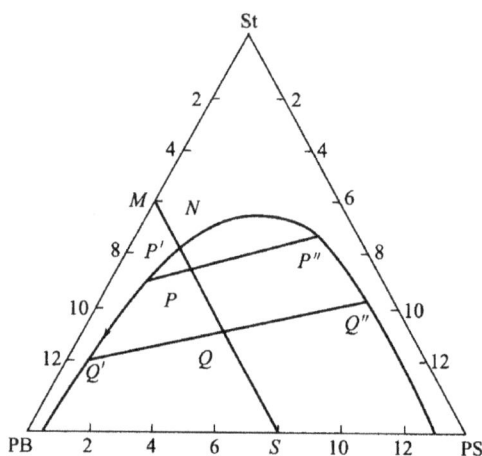

Fig. 8.23: Phase diagram of styrene–polybutadiene–polystyrene ternary system (figure shows volume percentage).

(N point), and the system begins phase separation. At point P, the two phases of the system are P' and P'', P' phase is a solution composed of PB and St, P'' phase is a solution composed of PS and St. According to the lever principle, the volume fraction of P' in rubber solution is larger than P'', so PB solution is still the continuous phase and PS solution is dispersed phase. As the polymerization proceeds, the two-phase composition changes along the two-segment line with the arrow pointing, the monomer concentration gradually decreases, the polymer concentration of each phase increases, and the volume of PS/St phase also increases. At Q point, the volume fraction of the two phases is close to the same (at this time, the conversion of St is about 9–12%). At this time, PS solution phase transfers from dispersed phase to continuous phase, while PB solution transfers from continuous phase to dispersed phase, that is, the system undergoes phase inversion. After that, the content of St in continuous phase and disperse phase is still quite high, and of course, the polymerization will continue. Therefore, PS will continue to be generated in the dispersed phase. Because the viscosity of the system is getting higher, and the graft copolymer formed during the polymerization process is concentrated on the surface of rubber particles, it is difficult for PS in the dispersed phase to migrate to the continuous phase. With the increase of PS content in the rubber phase, when the contained PS exceeds the homogeneous concentration limit, the PS in the dispersed phase will be separated, and finally the honeycomb structure of PS divided by the rubber network will be formed.

It should be emphasized that the polymerization process must have enough shear effect. If the reaction is carried out under static conditions without stirring, the above reversal phenomena will not occur until all St is converted to PS. Because of the high viscosity of the rubber phase and the cross-linking reaction between rubber molecules in the later stage of the reaction, the weight and volume of the rubber molecule in the continuous phase are lower than PS. There is not enough external force, the rubber will remain in the continuous phase, but at this time the rubber will become a continuous mesh, and the PS/St solution will be divided into many small areas. The rubber network in the continuous phase system is a sponge, which is difficult to process and has no practical value.

8.5.1.3 Factors influencing the morphological structure of HIPS

As pointed out earlier, it is necessary to exert a certain shear force on the polymer solution in order to realize the phase inversion. The shear force has a great influence on the phase structure. If the shear force is less than a critical value, the reversal cannot occur; if the shear force is too high, the dispersed phase particles will be too small to achieve the proper toughening effect. At the same time, the viscosity of the system should be controlled in a certain range when reversal occurs, and the viscosity is too small to establish enough shear force even if the shear rate is very high; the viscosity is too high and the stirring resistance is too large to provide enough shear effect, which also brings difficulties to phase inversion.

The rubber content in HIPS should not be too high, generally less than 12%. Excessive rubber content will delay the phase inversion to a higher monomer conversion. Because the volume fraction of PS in the system is more than 50%, it is possible for PS to transform into continuous phase. Therefore, with the increase of rubber content, the phase inversion will be delayed and will cause the phase inversion to occur in the case of high viscosity, which is not conducive to the phase inversion.

The PS content in rubber particles not only affects the volume fraction and particle size of rubber phase, but also affects the strength of rubber particles. When the number of inclusions is too large, the rubber particles will rupture during the processing. At the same time, it is possible to make the inclusion of PS craze, leading to craze through rubber particles, which is not conducive to the toughening effect. Therefore, it is also a very important issue to control the appropriate amount of inclusions.

The degree of grafting of copolymers is another important factor affecting the structure of final products. During the polymerization process, the formation of graft copolymers plays an important role in the emulsification and stabilization of the two phases, and plays a decisive role in the interfacial bonding force of the final product. At the same time, it has a direct impact on the size of colloidal particles. The higher the degree of grafting, the better the compatibility of the system and the smaller the size of rubber particles. This means that in order to control the size of rubber particles, it is necessary to control the degree of grafting.

The relative molecular mass of polybutadiene rubber used for graft copolymerization affects the viscosity of rubber phase. If the relative molecular weight is too high, the phase inversion will be delayed, and even the phase inversion cannot be achieved at the normal stirring rate. At the same time, with the increase of rubber viscosity, the size of rubber particles increases correspondingly and it is difficult to break into smaller rubber particles.

8.5.1.4 Effect of morphology and structure of HIPS on its properties

The mechanical properties of graft copolymerized HIPS are superior to those of mechanical blends, which are obviously determined by its special morphology and structure. To sum up, it can be attributed to the following reasons:

1. Because rubber particles contain a considerable amount of PS, compared with mechanical blends with the same amount of rubber, the volume fraction of rubber phase in honeycomb-like HIPS is 3–5 times larger, which greatly reduces the distance between rubber particles, makes a small amount of rubber play a role of larger volume fraction of rubber phase and strengthens the toughening effect.
2. In the HIPS system of graft copolymerization, PB-g-PS graft copolymer is dispersed at the interface of two phases, which improves the interfacial affinity, enhances the interfacial bonding force and is conducive to the improvement of mechanical strength and impact toughness.
3. The dispersion phase of grafted copolymerized HIPS contains PS, and its modulus is higher than that of mechanical copolymerized dispersion phase, which makes the modulus and strength of HIPS not reduce too much due to the existence of rubber phase.
4. HIPS with honeycomb structure will not produce harmful voids in the equatorial plane of rubber particles, which is beneficial to the formation and development of many crazes, to absorb more impact energy and greatly improve the impact strength.

From Tab. 8.5, the impact strength of the graft copolymerized HIPS is greatly increased (5–10 times higher than the general PS); at the same time, the tensile strength is not significantly reduced, and the processing fluidity and dyeing property are almost unchanged, but the transparency is obviously worse.

Tab. 8.5: Comparison of physical and mechanical properties of PS and HIPS.

Performance	Universal PS	Impact PS	High impact PS
Density (g/cm^3)	1.04–1.06	1.03–1.06	1.03–1.06
Tensile strength (MPa)	35–55	25–35	25–30
Bending strength (MPa)	70–100	40–65	25–35
Notch impact strength (kJ/m^2)	1–3	5–7	8–12
Rockwell hardness	65–80	80–85	60–75
Hot deformation temperature ($^\circ$C, 1.82 MPa)	85–95	80–90	80–95
Percentage elongation (%)	1.5–2	20–30	35–45
Transmittance (%)	87–92	–	–

HIPS can be used in injection and extrusion molding to produce various kinds of sheet metal, pipe, instrument shell, electrical equipment parts, household goods and so on, and its application field is much wider than PS.

8.5.2 ABS resin and its modification

ABS resin is the most widely used polymer blend product with the largest yield at present. It is a two-phase system consisting of random copolymer of St and AN (SAN) as continuous phase, polybutadiene or St–BR as disperse phase and copolymer formed by in situ graft copolymerization. It is a typical homopolymer and copolymer multicomponent blend.

ABS is widely used in many fields, such as mechanical and electrical, instrumentation, automobile, transportation and daily necessities. There are a lot of application examples, such as instrument shell, television recorder shell, automobile parts, decorative parts, various specifications and uses of sheet metal, pipelines and so on.

ABS has higher modulus, strength and toughness because of larger polarity of HIPS. As shown in Fig. 8.24, the yield strength and fracture strength of ABS are almost twice that of HIPS. Although from the stress–strain curve, the elongation at break of ABS is low, which is due to the local shear deformation of ABS during tension. The measured notch impact strength is usually nearly double that of HIPS.

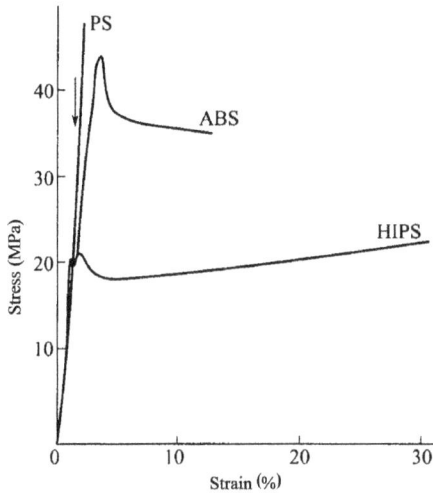

Fig. 8.24: Stress–strain curves of PS, HIPS and ABS.

8.5.2.1 Production technology of ABS resin

There are many methods of ABS production, and the main method of industrial production is graft copolymerization. There are many different methods in graft copolymerization, each with its own characteristics.

1. Emulsion graft copolymerization

The characteristics of emulsion graft copolymerization are simple operation and easy control. The amount of rubber can be changed in a large range, and the size of ABS particles is smaller. However, there are some shortcomings, such as the trouble of posttreatment and the difficulty of cleaning emulsifiers and coagulants, which affect the quality of products. In the actual production, the emulsion method is often changed into emulsion copolymerization-blending method, namely, polybutadiene-grafted SAN copolymer latex and SAN latex are first prepared, then two kinds of latex are mixed and then the ABS resin is prepared by posttreatment (cocondensation). This method not only has the characteristics of emulsion copolymerization, but also has the advantages of easy control of graft ratio of rubber, easy adjustment of ABS two-phase composition and so on. However, the trouble of posttreatment and the defect of impurity of the product still exist.

2. Ontological method and ontological suspension method

ABS production by bulk method is similar to HIPS bulk polymerization process. The product obtained is pure, with short process flow and high product quality. However, bulk ABS production has not been widely used up to now, mainly because of the difficulties in heat dissipation, rubber particle size, particle size distribution and morphology control during polymerization. At the same time, this method is not suitable for products with high rubber content. However, considering energy saving and eliminating the three wastes, the ontology method still has prospects for development.

Another method is bulk suspension, similar to the bulk suspension of HIPS introduced earlier. The ABS resin produced by this method is pure and has low production cost. But the amount of rubber should not be too high; otherwise, the system viscosity is too large to operate. Therefore, the ABS produced by this method has low gum content, large particle size and low impact toughness.

3. Emulsion suspension method

In recent years, the so-called emulsion suspension method has been developed, which combines emulsion method with suspension method. The first step is to graft copolymerization of latex. After a certain degree of reaction, the copolymer latex is then treated to obtain graft copolymer and then suspension polymerization is carried out under suspension polymerization conditions. The first stage of this method is easier to control than the bulk method, and the rubber content is not limited. The second-stage suspension polymerization can overcome the shortcoming of the emulsion method.

In addition, the method of mechanical blending of ABS produced by emulsion grafting and ABS produced by bulk suspension is used. The small size ABS produced by the emulsion method and the large particle size ABS produced by the suspension method will produce certain synergistic effect, which will further improve the workability and impact toughness of ABS. The blended ABS has better comprehensive properties than ABS produced by single method.

8.5.2.2 The structure of ABS

Compared with HIPS, the composition of ABS prepared by graft copolymerization is more changeable, the reaction is more complex and the influencing factors are more.

Because of the existence of styrene (St) and acrylonitrile (AN) monomers in the ABS system of graft copolymerization, besides the influence of St/AN ratio on the structure of the copolymer (SAN), there are also two kinds of monomers in the two phases. The affinity of St and AN to PB chains is different, and solvation between St and PB chains is better than that between AN and PB chains. Therefore, the content of St in the branched random copolymers of styrene and acrylonitrile (PSAN) formed around PB chains is higher than that in free PSAN (matrix). When the difference between the components of the grafted chain and free chain reaches a certain degree (about 40%), the two kinds of molecular chains will separate.

The difference between graft chain and free chain is also shown in their relative molecular weight. Generally, the relative molecular mass of the grafted chain is larger than that of the free chain. This is related to the environment in which the two kinds of molecular chains are formed. The grafted chains are formed near the rubber chains. Its local viscosity is high and the termination speed is low, so the relative molecular weight is high. In addition, the concentration of initiator is different. For example, the affinity of BPO to BP chain is higher than that of SAN, that is, the concentration of initiator in PB phase is much higher than that in SAN phase (selective dissolution). As a result, the primary free radicals of rubber phase initiator are mainly consumed by chain transfer reaction to form PB macromolecular chain free radicals. Because of the low activity of these free radicals, the rate of polymerization of PB macromolecular chain free radicals is obviously reduced. At the same time, the high concentration of PB radical increases the possibility of re-coupling, which also reduces the efficiency of initiator and the reaction speed, but improves the relative molecular weight of the branched chain.

ABS composition is not only affected by monomers and initiators, but also by the type and proportion of rubber, process conditions and production methods. Therefore, the actual composition of ABS varies greatly with the above-mentioned factors.

Figure 8.25 is an electron micrograph of emulsion ABS latex particle ultrathin section. It can be clearly seen that the shell formed by SAN on the surface of latex particles and the core containing SAN in the rubber phase. The ABS particles consisting of core–shell structure have a core diameter of about 0.5 μm and a shell

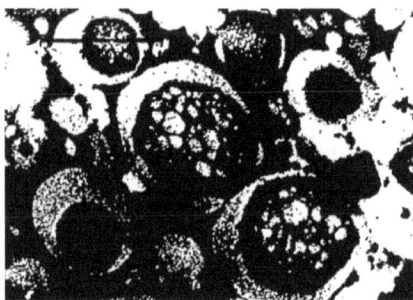

Fig. 8.25: Core–shell structure of ABS latex particles.

thickness of about 0.1 μm. The presence of homogeneous SAN particles without rubber core can also be seen in the electron microscopic photos, which is formed at higher surfactant concentration. SAN is formed on the surface of rubber particles or enclosed in the interior of rubber particles at low surfactant concentration. The variety of initiator has obvious influence on the morphology and structure of latex particles. Water-soluble initiator (such as persulfate ion) tends to make SAN forming a shell structure. Organic initiator can penetrate rubber particles swelled by monomers and initiate polymerization to form SAN inclusions.

The preparation methods, polymerization conditions and raw material composition of ABS differ greatly from those of other products. Therefore, there are many commercial ABS varieties, such as impact grade, high impact grade, super impact grade, flame retardant grade, heat resistance grade, low-temperature impact grade, transparent grade and plating grade. Table 8.6 lists the main physical and mechanical properties of several ABS varieties.

Tab. 8.6: Performance of ABS of different varieties.

Performance	High impact type	Impact type	Heat resistant type	Electro-plating type	Flame retardant type	Transparent type
Notched impact strength of cantilever beam (J/m, 23 °C)	367	300	230	367	300	131
Notched impact strength of cantilever beam (J/m, −40 °C)	105	63	58	68	52	21
Tensile strength (MPa)	42.1	42.1	51.0	37.2	40.2	42.1
Tensile modulus (MPa)	2,156	2,254	2,548	2,156	2,254	2,254
Bending strength (MPa)	725	775	882	676	676	706
Flexural modulus (MPa)	2,254	2,450	2,744	2,254	2,254	2,352

Tab. 8.6 (continued)

Performance	High impact type	Impact type	Heat resistant type	Electro-plating type	Flame retardant type	Transparent type
Rockwell hardness	102	108	111	103	99	103
Hot deformation temperature (°C)	89	84	94	89	88	77
Density (g/cm³)	1.04	1.04	1.05	1.06	1.22	1.07
Molding shrinkage (%)	0.7–0.9	0.7–0.9	0.6–0.8	0.5–0.7	0.5–0.8	0.5–0.7
Flame retardancy (UL-94)	HB	HB	HB	HB	V-0	HB

8.5.2.3 Factors influencing the structure and performance of ABS

ABS resin has a complex two-phase structure, and the factors affecting its properties are very complex, which can be mainly attributed to the following aspects:

1. Variety, composition and cross-linking degree of rubber phase

ABS mostly consists of polybutadiene as rubber phase. If nitrile–BR or St–BR is used, the T_g of rubber phase will increase and affect the cold resistance. Especially when AN is introduced into the copolymer, the influence of T_g on rubber is more serious, so it is seldom used as a raw material at present. ABS made of St–BR has good fluidity, so it is still used in some varieties, but the rubber content should not exceed 25%. Otherwise, cold resistance will be seriously affected.

The rubber used in ABS should have a moderate degree of cross-linking to ensure that the product has good impact toughness.

In addition, the content of rubber, the resin inclusion in the dispersed phase and the molecular weight of rubber all affect the properties of ABS resin. These factors are basically the same as those of rubber-toughened plastics.

2. Particle size of rubber

Many factors have opposite effects on impact strength and processing fluidity. For example, the impact strength of ABS resin increases with the increase of rubber content, but the fluidity and other properties decrease. The fluidity improves with the decrease in the relative molecular weight of SAN resin, but the impact strength decreases. The method of controlling ABS particle size to adjust the performance can not only improve the impact strength but also obtain better fluidity. The size of ABS resin produced by emulsion method is small, so the increasing latex particle size technology is often needed. However, the particle size should not be too large; otherwise, the emulsion system is unstable, and the performance is also unfavorable. In addition, blending ABS with different particle sizes prepared by different methods has been proved to be effective in improving impact toughness.

3. Grafting degree and grafting layer thickness of copolymers

Graft copolymers can improve the compatibility of the system and strengthen the toughening effect, and the degree of influence depends on the degree of grafting. Different particle sizes of rubber phase require appropriate degree of grafting to match them. Excessive or low grafting degree is not conducive to improving impact toughness (see Fig. 8.26).

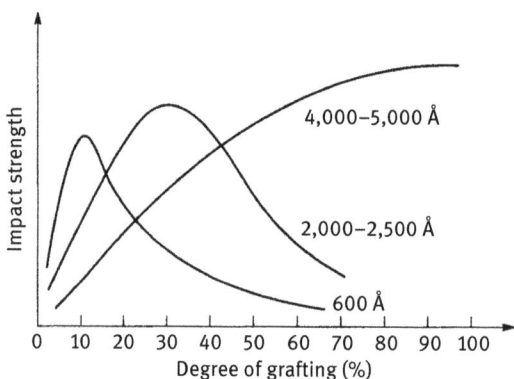

Fig. 8.26: Effect of grafting degree of rubber with different particle sizes on impact strength.

The thickness of the graft layer on the surface of rubber particles has an optimum range for the fluidity of ABS resin. Therefore, proper control of graft layer thickness is also an important key in ABS resin manufacturing technology.

8.5.2.4 Modification of ABS resin

In order to further improve some properties of ABS resin, physical and chemical methods can be used to modify ABS. Now a series of plastic alloys with ABS as the main body have been developed, and some new resins have been synthesized for some property defects of ABS resin.

1. Blend ABS with other plastics

ABS has poor flame retardancy, which limits its use as an electrical insulating material. In order to improve the flame retardancy of ABS, besides modifying with small molecular flame retardant, blending with high flame-retardant polymers such as CPE and PVC has obvious effect on improving the flame retardancy of ABS.

In order to improve the impact strength and abrasion resistance of ABS, polyurethane (PU) and ABS can be blended. The blend has good impact toughness, abrasion resistance, rigidity of ABS and good processability of PU.

The transparency can be improved by blending ABS with PMMA.

The heat resistance of ABS is not high enough. The thermal deformation temperature of general ABS is only about 90 °C. It is important to raise the thermal deformation temperature of ABS for improving the use grade of ABS and expanding its application field. Especially, the strength and heat resistance of the alloy blended with ABS and heat-resistant engineering plastics have been greatly improved, and can even be listed in the ranks of engineering plastic alloys. For example, the hot deformation temperature of ABS/polycarbonate (PC) alloy can reach 125 °C. The thermal deformation temperature of ABS/polysulfone is over 140 °C. At the same time, it has excellent impact toughness and processing fluidity.

ABS not only can be blended with many polymers, but also can be made into glass fiber-reinforced ABS alloy with glass fiber and other materials. This kind of composite material has high rigidity and dimensional stability, as well as excellent heat resistance and impact toughness. ABS is widely used in manufacturing the shell and frame of typewriters, photocopiers and other equipment, and the overall cost can be reduced to 20–30% compared with metal materials.

2. New resins of other St
In order to overcome some shortcomings of ABS, some polymer alloys similar to ABS in structure and properties have been developed. For example, AN- CPE–St (ACS) resin, AN-acrylate-St (AAS) resin, methyl methacrylate-butadiene-styrene(MBS) and methylmethacrylate/ co/ acrylonitrile/ co/ butadiene/ styrene copolymer (MABS).

AAS alloy obtained by graft copolymerization of polybutadiene with acrylate elastomer has better weatherability than ABS, and its impact strength and elongation hardly decrease after exposure for 9–15 months. AAS has a wide temperature range and can maintain good service performance in the range of –20 to 70 °C, and this is suitable for making outdoor appliances, such as automobile body, agricultural machinery parts, traffic signs, instrument housing and furniture.

ABS resin is easy to aging under the action of light and oxygen because of unsaturated double bonds in polybutadiene rubber. The weatherability of ACS resin prepared by replacing polybutadiene in ABS with CPE without double bonds can be greatly improved. Due to the introduction of chlorine atoms in ACS, the flame retardancy of ACS is also significantly improved than that of ABS, and the products have good dimensional stability and low shrinkage rate of forming. ACS is mainly used to replace wood and make outdoor products. MBS resin is characterized by good transparency, transmittance of up to 85% and other properties are similar to ABS.

8.6 Grafting modification of polyvinyl chloride plastics

8.6.1 Summary

Although there are many interesting graft polymers in the current literature, only a few systems have practical significance. This is mainly because it is not easy to control the performance of the final product and realize industrial-scale production when heterogeneous reactions are carried out. PVC graft copolymer is one of the most successful ones.

The graft copolymers of PVC can be divided into two categories: first is the product of vinyl chloride monomer grafting copolymerization on another polymer skeleton, and the second is the product of other monomer grafting copolymerization on the PVC skeleton. In most of the graft copolymers grafted with vinyl chloride monomer on another polymer skeleton, PVC is the main component, so it has the main characteristics of PVC. In most cases, the products of this kind of grafting reaction contain homopolymer PVC and grafted PVC, which provide the compatibility that homopolymer lacks in direct blending. Similar situations also occur when PVC is used as a skeleton for graft copolymerization with the second monomer. If none of the components of the graft copolymer has obvious advantages, the product will have completely new properties. In fact, except for a few grafted products with plasticized PVC properties, most of these products are molding materials made according to process requirements and have specific uses. In this section, we will also introduce the grafted polymers with low PVC content, which are generally seldom mentioned. The performance difference between it and the original polymers is not obvious. The degree of difference depends on the content of the grafted polymers.

8.6.2 Graft copolymer of vinyl chloride on soft polymer skeleton

As mentioned earlier, most of the reported grafted PVC are products with high vinyl chloride content, and their properties are similar to those of blended rigid PVC. In addition to the thermal deformation temperature, the surface quality and fluidity of the products are improved, especially the notched or nonnotched impact strength (especially the impact strength at low temperature). A two-phase system is usually formed after graft copolymerization of elastic polymers or copolymers with low glass transition temperature with vinyl chloride, which consists of a PVC matrix and a graft elastomer dispersed therein. Diolefin polymers or their copolymers and polyacrylates, which were previously used for high impact modification of PVC, have recently turned to use elastomers with aging resistance, such as polyolefin elastomers or olefin copolymers containing ethylene monomers.

8.6.2.1 Graft copolymers of vinyl chloride with polydiene and its copolymers
Polybutadiene, as the graft copolymerization object of vinyl chloride monomer, once attracted amount of attention, mainly in order to make use of their high elasticity. However, most of these grafted products still lack practical significance. The reason is that butadiene is an excellent inhibitor of vinyl chloride, and it is difficult for vinyl chloride to be polymerized in the presence of a small amount of butadiene

monomers. On the other hand, butadiene copolymers such as ABS or MBS graft co-polymers are particularly effective in improving the toughness of PVC and easy to blend, so it is not necessary to adopt grafting methods.

However, in recent years, due to the development of new synthetic technology, it is possible to graft vinyl chloride onto diene polymers. Therefore, it is reported that the toughness of PVC can be greatly improved by introducing a small number of elastomers (diene copolymers). Because the glass transition temperature of these elastomers is low (−90 to −50 °C), PVC products still exhibit high toughness and impact strength when the ambient temperature decreases. The impact properties of the obtained high impact PVC are better than that of the blends of the correspond-ing elastomer and PVC. For example, poly(butadiene/acrylonitrile)-g-PVC obtained by grafting has much higher tensile impact strength than the corresponding blends due to the presence of grafts, as shown in Tab. 8.7.

Tab. 8.7: Impact strength of graft copolymer and blend of PVC and NBR.

Sample name	Impact strength (J/cm^2)	Sample name	Impact strength (J/cm^2)
Graft compounds containing 2.5% rubber	5.63	Blends containing 5% rubber	6.15
Blends containing 2.5% rubber	4.97	Pure PVC	4.45
Graft compounds containing 5% rubber	6.43		

Kühne has proposed that the best structure of high impact PVC should be judged by T_g temperature, particle structure, particle size, particle size distribution of elastomer and interfacial compatibility between two phases. Figure 8.27 compares the toughness of the graft copolymers with polybutadiene as the main chain and PVC as the branch chain and the corresponding blends at different elastomer concentrations. The toughening ef-fect of the graft copolymer is much greater than that of the corresponding blends.

In recent years, some Japanese companies have introduced rigid PVC modified with butadiene or butadiene copolymer to the market. In addition to notable im-provements in toughness and tear resistance, some products have also been im-proved in terms of thermal stability and hardness. The presence of a small amount of polydiene is enough to reduce the melting viscosity of PVC and improve its process-ing behavior. PVC with high impact resistance can be obtained by using liquid poly-butadiene, degraded natural rubber and polypropylene chloride as graft skeleton.

High impact PVC can be used to manufacture sheets, profiles, bottles, pipes and films. Graft copolymers of vinyl chloride and acrolein on polybutadiene, poly-butadiene-g- (vinyl chloride-co-acrolein), are usually supplied in the form of water emulsion, and can be used as impregnating agents for paper, textiles and fur.

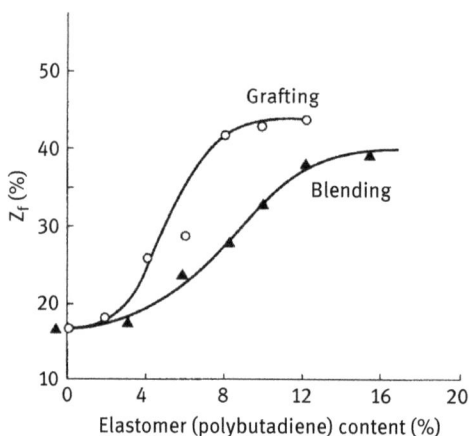

Fig. 8.27: Performance comparison of graft copolymers and blends. Z_f = elongation at break (%) × 100/impact yield strength × 9.81 (N/cm^3).

8.6.2.2 Graft copolymers of vinyl chloride with polyolefins and their copolymers

Olefin polymers have generally poor compatibility with PVC, but their T_g is low, so they are suitable for modification of PVC, especially in improving toughness. Due to the unique properties of this kind of graft copolymers, industrial products have been introduced into the market.

Graft copolymers of vinyl chloride and PE are rigid or semirigid products with high toughness, good tensile strength, low processing temperature and good fluidity, high thermal stability, good resistance to chemicals and solvents. In addition, they also have high ultraviolet radiation resistance. For example, a PE-g-PVC copolymer containing 11% (weight) of PE is reported to be much better in toughness, tensile strength, elongation and Young's modulus than mechanical blends of the same composition, as shown in Tab. 8.8. By improving the processing technology and adding monomers in batches, a series of products suitable for plastics, films, sheets and coatings with high toughness and good outdoor weather resistance can be produced. A small amount of vinyl chloride is grafted onto PE, and the product has some interesting properties like PE. For example, the printing performance of PE can be improved by forming PE-g-PVC graft copolymer. The coloring properties of PE can also be improved by grafting.

It has been reported that polypropylene is used to modify PVC to improve its impact strength. For example, modified polypropylene with improved rigidity and fracture properties can be obtained by grafting 20% vinyl chloride onto polypropylene, which is suitable for making films, fibers, filaments and molding parts. Table 8.9 lists the properties of this polypropylene-g-PVC product.

Ethylene-co-propylene copolymer is an effective grafting agent to improve toughness of rigid PVC. The mechanical properties of the graft copolymers containing 20%

Tab. 8.8: Performance comparison of mechanical blends and graft copolymers of polyethylene and polyvinyl chloride.

Project		Graft copolymer	Mechanical blends
Notched impact strength of cantilever beam (J/cm)		2.77	1.77
Tensile strength (MPa)	When yielding	49.2	35.1
	When cracking	42.1	34.5
Percentage elongation (%)	When yielding	4.2	4.0
	When cracking	140	27
Modulus of elasticity (MPa)		2.0×10^4	1.8×10^4

Tab. 8.9: Property changes induced by grafting a small amount of vinyl chloride onto polypropylene.

Project	Polypropylene	Graft copolymer	
		I	II
Vinyl chloride content (%)	0	0.4	0.5
Density (g/cm³)	0.9086	0.9125	0.9205
Brittle temperature (°C)	26	29	20
Intrinsic viscosities (tetrahydronaphthalene, concentration 0.25%)	1.70	1.15	1.14
Yield tensile strength (MPa)	31.3	35.2	35.9
Vicat softening point (°C)	144	161	–

ethylene–propylene copolymer are much better than those of the corresponding blends. The product is a two-phase system, in which the ethylene–propylene copolymer has greatly improved the compatibility with the PVC matrix by grafting, and plays the role of plasticizer in PVC to a certain extent, thus improving the elongation at break. The graft products, blends of the same composition and blends of different compositions of graft copolymers with PVC are compared. Among the three properties, the high impact strength of the graft copolymer is noticeable, as shown in Tab. 8.10.

Figure 8.28 depicts vividly the difference in impact strength between the graft copolymer (PE-co-polypropylene)-g-PVC and the corresponding mechanical blends. In Fig. 8.28, the impact strength of the graft copolymer is much greater than that of the

Tab. 8.10: Properties comparison of graft copolymers and blends of EP and PVC.

Sample	EP content (%)	Yield tensile (MPa)	Tensile fracture (MPa)	Bending strength (MPa)	Modulus of elasticity (MPa)	Notch impact (J/cm²)	Vicat softening point (°C)
Graft copolymer	16	22.5	22.5	37.5	1,500	3.50	70.5
	10	40.0	31.0	59.0	2,100	3.53	78.5
Graft copolymer + PVC	8	49.0	39.0	73.5	2,550	1.07	81.2
	5	46.0	38.0	66.0	2,550	0.88	79.0
	2.5	54.0	40.2	78.0	2,470	0.51	80.5
	0	56.0	42.0	72.0	2,950	0.43	81
Blend	25	14.2	14.0	21.0	850	0.32	56.2
	18	23.5	23.5	35.5	1,300	0.32	69.0
	12	36.0	30.5	49.5	–	0.70	76.5
	5	48.5	39.0	60.0	2,350	0.55	81.6

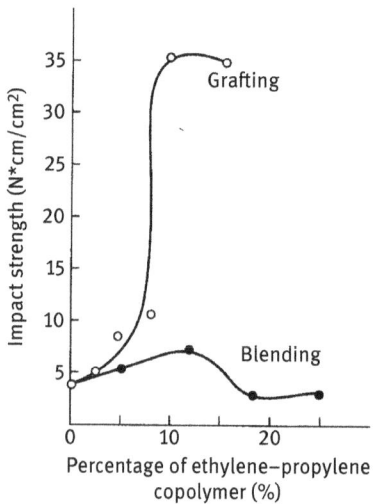

Fig. 8.28: Performance comparison of ethylene–propylene-g-PVC graft copolymer and mechanical blends.

corresponding mechanical blends. But for the two polymer alloys, the maximum toughness appears in the range of 8–10% of the content of ethylene–propylene copolymer.

8.6.2.3 Graft copolymer of vinyl chloride on halogenated polyolefins

In the graft copolymers of halogenated polyolefins and vinyl chloride, the graft products of CPE with low chlorine content are more important. The product has the characteristics of high impact PVC. The mechanical properties of CPE-g-PVC have been greatly improved compared with their corresponding mechanical blends. For CPE with chlorine content of 31.8%, the properties of the corresponding graft co-polymers and mechanical blends are listed in Tab. 8.11.

Tab. 8.11: Performance comparison of blends and graft copolymers of PVC and chlorinated polyethylene.

Chlorinated polyethylene content (%)		Mechanical mixture	Graft copolymer			
		10	4.8	9.2	13.5	17.0
Tensile strength (MPa)	When yielding	37.23	48.68	44.26	39.37	25.99
	When cracking	23.99	49.02	41.37	42.26	24.75
Percentage elongation (%)	When yielding	3.5	3.4	4.0	4.1	6.5
	When cracking	102	138	104	166	38
Tensile modulus (MPa)		1,931	2,689	2,344	1,999	1,586
Rockwell hardness		97	110	105	98	71
Notched impact strength of cantilever beam (J/cm)		8.90	8.90	5.34	6.67	8.45

Chlorosulfonated PE can also be used to replace CPE to graft onto PVC, and the prod-ucts are both high impact PVC. It can be used to prepare pipes, corrugated plates and various plastic products resistant to chemicals, solvents and ultraviolet radiation. Semirigid PVC materials can be prepared by increasing the amount of grafted skeleton.

CPE grafted with vinyl chloride was mixed with chlorosulfonated PE, and then combined with nitrile BR to produce a very soft resin with anticorrosive solvent, suitable for tank lining, insulating tape and so on.

Vinyl chloride can also be grafted on chloroprene rubber. The product is blended with PVC to form high impact resistance PVC.

8.6.2.4 Graft copolymer of vinyl chloride on polyacrylate

Polyacrylate with high elasticity has good compatibility with PVC, so PVC can obtain high impact resistance after grafting with it. In addition, PVC with internal plasticiza-tion can be obtained by graft copolymerization of polyacrylate and vinyl chloride. The grafted skeleton is prepared by copolymerization of butyl acrylate and diethylene glycol dimethacrylate. The final product contains 40–50% (weight) PVC and has the characteristics of plasticizing PVC. At the same time, it has excellent migration resistance and extraction resistance, low temperature performance and light stability.

By grafting copolymerization of vinyl chloride onto polyacrylic acid and its esters, some water-soluble polymer dispersions with complex composition can be prepared into protective and decorative coatings, which have good gloss and high wear resistance and are suitable for wood, metal, paper and other substrates.

8.6.2.5 Graft copolymer of vinyl chloride on polyester, polyether and polyamide

After the grafting of polyether and polyester with vinyl chloride, they can be used as final products and intermediate products in the production of PU.

Because polyether and polyester lack the necessary elasticity, they cannot be used in the production of high impact PVC. However, the impact strength of PVC can be improved successfully by grafting vinyl chloride onto amorphous polyether soluble in vinyl chloride monomer.

Linear block copolymers can be synthesized by reaction of polyether glycol with diisohydrogenate and condensation with diols containing azo bonds. In the presence of vinyl chloride monomers, these copolymers generate free radicals from the splitting of their azo groups, resulting in graft copolymerization and formation of block copolymers consisting of linear PVC and polyether sequences, that is, polyethylene oxide-b-PVC. The dynamic properties of the above-mentioned block copolymers containing 22% (weight) PVC were tested. It was found that the peak value of the dynamic loss factor (tan δ) appeared at -35 °C, indicating that the copolymers had polyether sequence.

The vinyl chloride is first grafted onto the polymer containing hydroxyl, and then the reaction product and water are combined with isocyanate to form foam material. Table 8.12 shows that although the density of grafted products is low, the tensile strength, elongation, impact resilience, compressive stress and compressive deformation of the grafted products are improved.

Tab. 8.12: Properties of polyurethane foam with graft PVC.

Sample	Density (kg/cm³)	Tensile strength /MPa	Elongation at break (%)	Impact resilience (%)	Compressive stress (kg/cm²) Compression ratio (%)			Compression residue (%)
					20	40	60	
Graft sample*	31	0.11	415	32	21	24	38	23
Control sample	32	0.02	60	28	13	14	21	84

*Every 100 phr of polypropylene glycol (molecular weight 2,000, hydroxyl value 61) was grafted with 3.5 phr of PVC, and then reacted with 3 phr of water and 38.5 phr of toluene diisocyanate.

8.6.2.6 Graft copolymer of vinyl chloride on polysiloxane

Polysiloxane with high molecular weight can also be used as the skeleton for vinyl chloride grafting. In most cases, vinyl chloride is grafted with linear polysiloxane before cross-linking. For example, vinyl chloride (25–75%) was first grafted onto hydroxyl-terminated polysiloxane with viscosity ranging from 0. 1 to 30 Pa s, then tetraethyl orthosilicate was added as cross-linking agent, and the product was cross-linked under the influence of atmospheric humidity. The product can be used for casting and molding to manufacture electrical insulators. Vinyl chloride was grafted onto hydroxyl-terminated liquid polysiloxane to produce thixotropic products in solvents, which could be used to make sealing putty after compounding. For some special paper, antisticking paper can be prepared by treating it with polydimethylsiloxane solution, which has been grafted with hydroxyl-terminated vinyl chloride or with polysiloxane graft products containing alkoxy or other hydrolyzable groups.

8.6.2.7 Graft copolymers of vinyl chloride on cellulose and other natural products

Cellulose can be grafted with vinyl monomer after pretreatment. For example, when 20% vinyl chloride is grafted onto paper, it can absorb water, but when the degree of grafting reaches 100%, the paper has excellent water resistance. The degree of grafting depends entirely on the swelling pretreatment of the fibers. X-ray diffraction analysis shows that the structure of PVC grafted on cellulose is still equivalent to that of bulk or suspended polymers, and cellulose matrix had no orientation effect.

The tensile strength and wet strength of vinyl chloride-grafted rayon are improved, but the flexibility is reduced. Grafting of vinyl chloride onto cotton, wood pulp and paper can also improve the wet strength of these materials. Cellulose grafted with vinyl chloride and its comonomers and partially hydrolyzed is suitable for paper coating and soil additives.

Wood, silk, gelatin and casein can be used as skeleton polymers for grafting vinyl chloride. When vinyl chloride is grafted with water-soluble gelatin and casein, hydrophilic comonomers are often chosen. There are two purposes: one is to produce materials that can be used to prepare adhesives and thickeners and the other is to improve the dyeability of fibers.

8.6.3 Graft copolymer with vinyl chloride as framework

The object of this section is the graft copolymer of free radicals of PVC formed by free radicals attacking PVC and the polymerizable components in the system. This kind of graft product is usually prepared by high-energy radiation treatment of PVC. The reaction of the grafted PVC formed is almost nonhomogeneous during the grafting process, and the PVC is always dispersed in the suspended graft phase in a suspended or swollen state.

In addition to high impact PVC products (containing a high proportion of PVC, which are equivalent to products grafted through vinyl chloride monomer in meaning and performance), the reported PVC graft reaction and reaction products basically add powdered PVC to the polymerization system, and the mechanical properties of the obtained materials can be controlled from soft plasticized PVC (thermoplastic) to thermosetting and PVC foam materials.

8.6.3.1 General performance

With regard to the grafting reaction of PVC as the skeleton, the main purpose is to develop graft copolymers with special properties and applications. Therefore, it is necessary to understand the basic properties of these materials, such as the degree of grafting, the phase distribution of skeleton polymers in the grafted monomers, their thermal stability, degradation and dynamic mechanical properties. The practical synthesis reaction is radiation-induced graft copolymerization.

In the grafting reaction of PVC-g-polybutadiene, the particle size of the grafted polybutadiene component decreases with the increase of grafting temperature by means of electron microscopy. For *cis*-1,4-polybutadiene grafted PVC, the thermal stability of the grafted product is significantly improved compared with that of pure PVC homopolymer. The reason may be that *cis*-1,4-polybutadiene acts as an antioxidant, thus preventing the oxidative degradation of PVC. The thermal stability of PVC-g-*cis*-1,4-polybutadiene and *cis*-1,4-polybutadiene-g-PVC containing homopolymer PVC is compared in Fig. 8.29. The results show that the tendency of dehydrochlorination of these two graft copolymers is not as obvious as that of homopolymer. In addition, differential thermal analysis (DTA) test showed that compared with nongrafted PVC, the T_g of products with grafting degree of more than 50% decreased significantly. Heterogeneous graft copolymers can be obtained by mixing PVC with different monomers. Of course, block copolymerization may also occur. The thermal stability of grafted or block copolymers is generally better than that of homopolymers. However, alkaline monomers such as 5-vinylpyridine can reduce their stability, even lower than matrix polymers. The degradation behavior of PVC graft copolymer obtained from plastics was studied. It was found that the dehydrochlorination reaction of PVC was accelerated greatly under alkaline conditions.

According to this characteristic, in order to increase the number of active points and grafting points on the molecular chain of PVC, HCl can be removed from PVC in alkaline solution to form dehydrochlorinated PVC. Then solution polymerization is carried out to complete the grafting process, and a high grafting rate of PVC grafting copolymer (the grafting rate can be more than 50%) can be obtained.

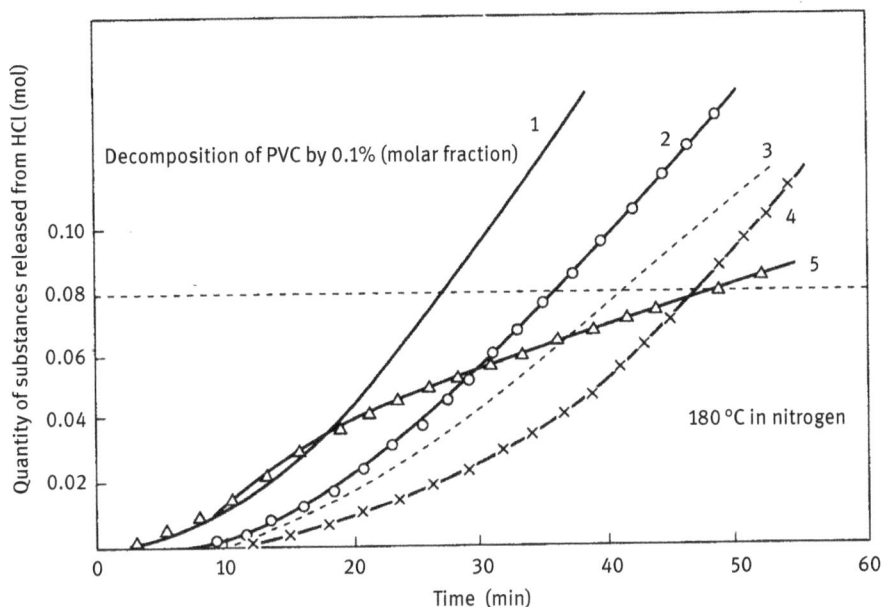

Fig. 8.29: Comparison of thermal stability between PVC-g-polybutadiene graft copolymer and PVC. 1. General suspension PVC; 2. PVC with heat stabilizer; 3. PVC grafted with polybutadiene; 4. PVC grafted with polybutadiene with heat stabilizer; 5. *cis*-1,4-polybutadiene-g-PVC containing homopolymer of PVC.

8.6.3.2 Impact-resistant rigid PVC

By grafting vinyl chloride onto the elastomer skeleton, the high impact resistance of PVC can be achieved. On the contrary, when PVC is dissolved or swelled in conjugated diene, AN, acrylate, methacrylate and other monomers (which can produce elastomer products when they are polymerized alone or copolymerized), graft copolymers or blends containing graft copolymers with good impact resistance can also be obtained by graft copolymerization. The graft copolymers prepared contain a high elastic component necessary for high impact modification and can be processed directly. If the content of high elastic component is high, then the graft copolymer is no longer rigid PVC, and can be compounded with PVC for further processing to high impact materials. Since the grafting reaction is almost always carried out in the aqueous dispersion of PVC, that is, suspension, in most cases, the graft copolymer products are provided to users in the form of fine particles, which is convenient for processing. Therefore, this kind of graft copolymer has reached the level of large-scale industrial production.

In the PVC-g-(butadiene-co-acrylonitrile) graft copolymer, the content of free homopolymer PVC is very low, so the graft rate is very high. The particle size of the main components in the product is less than 0.1 μm, and its refractive index is similar to that of PVC. The main mechanical properties of these grafted products are listed in Tab. 8.13. Similar products can be obtained by grafting butadiene and acrylate monomers onto PVC.

Tab. 8.13: Properties of butadiene–acrylonitrile copolymer-g-PVC graft copolymer.

Composition of graft copolymer (Wt)			Tensile strength (MPa)	Elongation (%)	The temperature at which the following modulus (MPa) is obtained (°C)		
Vinyl chloride	Butadiene	Acrylonitrile			945.0	70.0	12.5
5	3	2	13.0	100	−27	22	47
5	4	1	13.0	170	−46	18	43
5	2	3	25.0	150	8	28	46

The grafted product of PVC-g-(butadiene-co-acrylonitrile) is usually not used alone, and it is usually used in combination with homopolymer PVC. Compared with unmodified PVC, notched impact strength of these composites increases a lot, but their tensile strength and elastic modulus decrease slightly. As shown in Tab. 8.14, the toughness of the composite increases with the content of the graft copolymer.

Tab. 8.14: Properties of PVC/PVC-g-(butadiene-acrylonitrile) blends.

Performance	Sample				PVC
	1	2	3	4	
Dosage of graft copolymer (phr)	5	10	15	20	0
Notched impact strength (simply supported beam) (J/cm^2)	0.31	0.64	0.92	1.54	0.32
Tensile strength (MPa)	60.50	54.50	52.2	62.1	62.1

Because of the butadiene component, the graft copolymer can be cross-linked by free radical reaction, resulting in higher thermal deformation temperature. Similar grafting products can be obtained by adding divinyl compounds as comonomers, but their processability has been improved. PVC-g-(butadiene-co-acrylonitrile) can be used to improve the flexibility of chlorinated PVC, to produce transparent films, plates and pipes and to produce injection plastics by compounding with PVC.

A mixture of St or its derivatives with monomers such as dienes and acrylates, once grafted onto PVC, can also produce a high impact-grade PVC. The characteristic of this kind of product is that the Vicat softening temperature also increases while the impact performance is improved, as shown in Tab. 8.15. In addition, they have good transparency and aging resistance. If the inner plasticized monomers such as vinyl fatty acid ester or maleate are grafted onto PVC, it will have synergistic effect with the outer plasticizer.

Tab. 8.15: Properties of the products grafted by different monomers on PVC.

Sample	Composition (phr)						Properties of resins	
	PVC	MMA	AN	MSt	Butadiene	Isoprene	Tensile impact strength (J/cm^2)	Notch impact strength (J/cm^2)
1	100	6	–	2.5	5	–	54.94	2.45
2	100	–	6	2.5	5	–	51.99	2.26
3	100	6	–	2.5	–	5	53.14	1.67
4	100	4	–	4	–	4	44.15	1.18
5	100	3	–	2	–	3	34.34	0.78
Contrast	100	–	–	–	–	–	9.81	0.25

8.6.3.3 Soft, cross-linking and thermosetting PVC

The same series of different monomers and vinyl compounds with multifunctional groups, or unsaturated polyester–monomer mixture containing free radical initiator, can be used for the grafting reaction of PVC. These products are difficult to classify because of the complex nature of the copolymer components.

Some of these materials are modified PVC, while others are unable to make such a simplified definition because of the lack of dominant specific structural units.

When PVC is grafted with a monomer mixed with polymer plasticizer, the grafted product is soft PVC. Compared with ordinary soft PVC, it can improve the tensile strength and reduce the extractability of plasticizer. In the presence of comonomers, the graft copolymers of PVC and diene can be transformed into cross-linked rigid or soft PVC by high-energy radiation, which can help to inhibit the migration of plasticizers.

PVC is dispersed in the form of powder in the polymerizable monomer. If it can be swelled by the monomer, the mixtures can be polymerized under the conditions of free radical initiation, high-energy radiation initiation or thermal initiation. For example, when PVC and methylstyrene, AN, unsaturated polyester or ethylene glycol dimethacrylate are mixed together and irradiated, rigid impact plastics can be obtained. Table 8.16 summarizes the typical performance of the system.

By grafting methylstyrene and acrylate onto PVC with peroxide initiator, coatings for glass fiber fabrics or transparent plastic films can be prepared. They have

Tab. 8.16: Properties of PVC grafted with different monomers.

Sample	Component (wt%)					Performance			
	PVC	MSt	Ethylene glycol dimethacrylate	Unsaturated polyester	Acrylonitrile	Bending strength (MPa)	Flexural modulus (MPa)	Notch impact strength (J/cm²)	Hot deformation temperature (°C)
1	50	33.4	3.0	13.6	–	251.1	26,610	0.11	63
2	50	30.1	2.7	12.2	5	440.2	27,900	0.24	58
3	50	26.7	2.4	10.9	10	597.99	29,295	0.31	64
4	50	23.4	2.1	9.5	15	613.80	29,925	0.45	84
5	45	22.0	2.0	9.0	22	682.22	33,480	0.38	72
6	40	24.0	2.2	9.8	24	750.82	33,480	0.41	73

high hardness, tensile strength, flexibility and thermal deformation temperature. Table 8.17 lists the properties and compositions of some of these materials.

In the presence of PVC and polymer plasticizers (polyadipate, polysebacate, etc.), a kind of polyester resin can be prepared, which will neither shrink nor crack during the curing process.

Tab. 8.17: Properties of PVC grafted with methylstyrene and acrylate.

Sample	Component (wt%)				Performance			
	PVC	MSt	Cyclohexyl acrylate	Stannous octanoate	Bending strength (MPa)	Notch impact strength (J/cm²)	Hot deformation temperature (°C)	Transmittance (%)
1	60	35	5	0	62.05	1.13	72	33.2
2	60	35	5	2	107.56	3.05	74	83.1
3	60	30	10	0	70.55	1.24	73	47.5
4	60	30	10	2	107.55	3.38	76	84.6
5	60	25	15	0	65.50	1.35	74	66.1
6	60	25	15	2	96.53	2.60	72	88.2
7	60	20	20	0	77.22	1.58	73	82.6
8	60	20	20	2	86.87	2.48	73	88.3

Epoxy resin can be prepared from PVC which has been grafted with glycerol methacrylate or vinyl glycerol ether to form a kind of imitative enamel paint. Flame retardant resin can be obtained by graft copolymerization of monofunctional and bifunctional unsaturated monomers with vinyl chloride copolymers and phosphate esters.

EVA copolymer can be grafted with PVC by free radical initiator in the process of forming at 50–200 °C, thus forming a resin suitable for coating on textiles, paper, wood and fur.

8.6.3.4 PVC foaming material

Through special grafting reaction on PVC, cross-linking system products with excellent properties can be prepared. For example, when maleic anhydride is copolymerized with another monomer and heated with powdered PVC in the presence of foaming agent and diisocyanate, the grafting can be initiated. At the same time, gelation of PVC occurred, and the foaming agent also decomposed. Then the cooled foam is immersed in water, and the reaction is achieved by reaction between the two kinds of amine formed by hydrolysis and the anhydride group in graft copolymerization. The foam is closed cell with good dimensional stability and solvent resistance. There is also no need for foaming agent, so that the reaction is carried out in a warm, pressurized closed die. The product is expanded into honeycomb structure by carbon dioxide released from hot water treatment at the end of the grafting reaction, and the cross-linking reaction of anhydride and amine occur at the same time. The resulting foam is completely insoluble. Compared with traditional PU foaming materials, it has better thermal conductivity and higher compressive strength at low temperature, as shown in Fig. 8.30.

8.6.3.5 Polyurethane modified by grafting PVC

Using free radical initiator or high-energy radiation, the graft copolymers containing hydroxyl polyether and polyester (synthetic PU block) and PVC can be prepared. Then the graft product is introduced into PU synthesis, and the elastomers and foams with improved properties can be obtained through reaction with diisocyanate. For example, PU elastomer is synthesized first, then treated with benzoyl peroxide or γ-ray in the presence of PVC to form a graft. In this process, PVC and polyols generally form free radicals, which combine to form grafted macromolecules. The hardness and tensile strength of the product can be improved obviously, which is about twice as much as that of the unmodified material (Tab. 8.18).

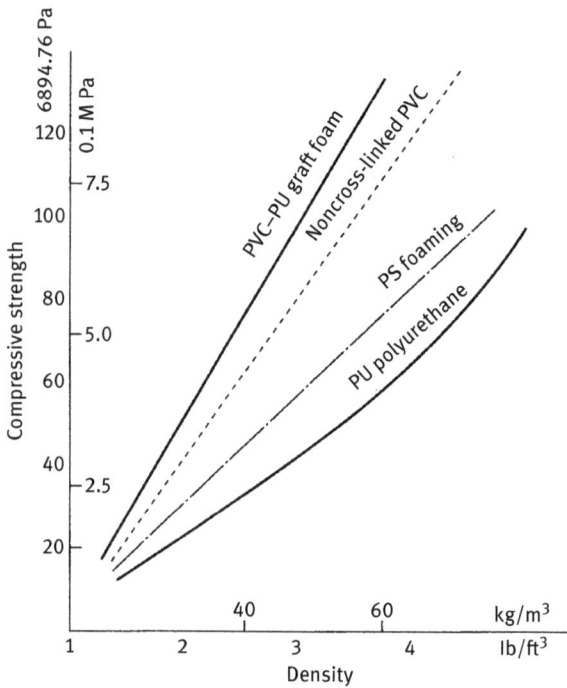

Fig. 8.30: Performance comparison of PVC-g-polyurethane graft copolymer foaming materials with other foaming materials.

Tab. 8.18: Properties of PVC grafted polyurethane.

Sample	Volume of PVC in mixture (%)	The amount of polyols in the mixture (%)	Reaction conditions		Catalyzer	Properties of polyurethane		
			Mixing time (h)	Mixing temperature (℃)		Tensile strength (MPa)	Tensile modulus (MPa)	Elongation rate (%)
1	0	100	–	–	–	0.83	2.07	60
2	0	100	–	–	γ-Ray	0.34	1.79	47
3	20	80	5	45	–	1.10	0.62	375
4	20	80	5	45	2% BPO	2.0	4.62	100
5	20	80	5	45	γ-Ray	2.39	4.69	100
6	20	80	1.5	105	1% AIBN	2.28	3.03	155

8.6.3.6 Dispersion system for coatings and adhesives

Waterborne dispersions capable of drying film can be prepared by grafting vinyl chloride/AN random copolymer with diene and AN monomers. The film formed has high water resistance and hydrocarbon solvent resistance. The water dispersion system of vinyl chloride/ethylene random copolymer is used as the grafting skeleton copolymer, and the mixed monomers of St, vinylidene chloride and acrylate are grafted to form a dispersant, which is suitable for latex paint as a binder with high filler loading capacity.

Epoxy monomers, such as glycerol dimethacrylate or epoxidized butadiene, can be grafted and copolymerized in the toluene or acetone dispersions of PVC. The grafted products are suitable for baking paint.

Grafting vinylpyridine onto PVC fiber can improve its dye affinity. In the presence of a gas capable of forming free radicals, the adhesion, dyeing, coloring and water resistance of PVC fibers and films can be effectively improved by electrospark discharge treatment and grafting with St, isoprene, AN and other monomers.

8.7 Graft copolymers of inorganic materials

8.7.1 Summary

Silicate materials and other inorganic materials can form a layer of chemically bound organic polymers on the surface of their particles through graft polymerization. Similar grafting can be used not only to improve the interfacial compatibility between inorganic materials and organic polymer matrix, and the mechanical properties of composites, but also to prepare highly dispersive pigments, inorganic fillers with coupling function and high-performance solid lubricants.

At present, the grafting of inorganic materials has become a new field of graft copolymerization technology. However, the binding mode and principle of inorganic matrix and organic polymer are still unclear in many cases. Some of them belong to real chemical bonding, accompanied by the formation of new chemical bonds, while others are only some kind of adsorption. Of course, these two ways of combination will also exist at the same time.

The methods of graft copolymerization of inorganic materials can be roughly divided into four types: coupling grafting, chemical initiation grafting, radiation initiation grafting and mechanochemical initiation grafting. These methods are described further.

8.7.2 Preparation of inorganic material graft copolymer

8.7.2.1 Coupling grafting

Some of the prepolymer molecules and the inorganic material can react with functional groups to form coupling bonds, thereby obtaining a graft. In order to achieve this reaction, the functional groups that can react with the matrix of inorganic materials are required necessarily. Otherwise, they must be chemically treated to make them have the necessary functional groups.

In the process of polymer synthesis, the available end groups are often left at the end of the polymer. These end groups can be further converted into other forms to interact with the groups of inorganic materials as required.

There are some active functional groups on the surface of many inorganic materials, such as Si–OH group on the surface of glass, silica gel and Al–OH group on the surface of mica. Si–OH group can be converted into Si–Cl group by using thionyl chloride, thionyl chloride or chlorine gas as a chlorinating agent. The Si–Cl group is very active. It can further react with the end group of the polymer to realize surface grafting.

It is a practical method to treat inorganic materials with organosilicon compounds. When chlorosilane is used, Si–O–Si bonds can be formed by the condensation of Si–Cl bonds with Si–OH groups on the surface of inorganic materials.

For surface-inert materials, silicone compounds containing Si–Cl bonds can hydrolyze and condensate on the surface by the action of trace water, forming a layer of firmly adsorbed organic film, and at the same time can lead available groups to the surface. Other forms of organosilicon, such as silanol and silicic acid, are also widely used in surface treatment of inorganic materials.

For example, thionyl chloride is used to introduce Si–Cl group onto the surface of SiO_2, and then sodium alcohol with different length carbon chains is used to form graft copolymers with alkane chains on the surface. If SiO_2 containing Si–Cl bond reacts with PS-based lithium-active macromolecule, SiO_2–PS graft copolymer can be obtained. Compared with amorphous SiO_2 with a surface of 200 m_2/g, stable dispersion can be formed in aromatic solvents when the graft weight gain is 50%. Further studies on this kind of graft copolymerization showed that besides the chemical bond formation, there was also irreversible adsorption in the product. But the amount of chemical grafting is 10–100 times of the irreversible adsorption. Grafting weight gain is affected by reaction time, temperature and molecular weight of prepolymer. The size of the prepolymer macromolecule can be controlled by changing the solvent or counterion, and the effect of dissociation can be adjusted. The reaction rate of coupling grafting is related to the diffusion rate of prepolymer molecules through the polymer surface.

The cationically active prepolymer can react with the hydroxyl groups on the surface of the inorganic material to form a graft. For example, positive carbon ions are formed by butadiene–PS copolymers and $AlCl_3$, and then grafted with asbestos.

When the grafting process is carried out in dilute solution and the asbestos/polymer ratio is small, it can be proved that the reaction belongs to the first-order reaction. However, in concentrated solution, grafting limited by other factors such as cations may be inactivated and macromolecular aggregates with poor reactivity appear.

8.7.2.2 Chemical initiation grafting

The copolymerization of monomers initiated by free radical initiator with inorganic materials that have unsaturated groups has been used to modify cadmium pigments. First, the pigments are treated with vinyl-containing organosilicon monomers, such as $CH_2=CH-Si(OCH_3)_3$ isopropanol solution, and then the pigments are grafted with AN, vinylidene chloride and other monomers initiated by potassium persulfate. The products are easily dispersed in organic media (including polymers). Similar reactions such as the treatment of kaolin with γ-methyl allyl propoxy trimethoxysilane and the copolymerization of kaolin with St have been reported. The swelling of the grafted product in benzene has been reported; thus, the viewpoint of three-dimensional cross-linking structure of clay has been put forward. The former Soviet Union scholars modified the silica gel with vinyl trichlorosilane $CH_2=CH-SiCl_3$ and then grafted with heterocyclic monomers containing vinyl groups. It was found that the rate constant of grafting increased with the increase of temperature, and the activation energy was obtained.

Another way is to involve the matrix in the formation of free radicals, so that branched chains are formed in the matrix. For example, carbon black was dispersed in benzene, lauroyl peroxide was added and then reacted with benzene solution of butyl rubber. Finally, 12.8% of the grafting amount was obtained. The above method is also suitable for the grafting of carbon black with St–BR. The results showed that carbon black peroxide was first formed in the reaction, and then the free radicals were decomposed and acted on rubber to produce rubber free radicals, resulting in the formation of grafts between carbon black and rubber.

Grafting of AN or MMA onto mica surface was initiated by free radicals formed by hydroxyl groups under the action of Ce^{4+} salts. The homopolymer was removed by solvent extraction at room temperature for 72 h. Infrared spectrum characterization confirmed that there were 2,250 cm^{-1} absorption peaks corresponding to AN or 3,050, 1,730, 1,450 and 1,240 cm^{-1} absorption peaks corresponding to MMA, indicating the success of grafting.

Graft copolymerization of methacrylonitrile, St, MMA and vinyl acetate with silica, asbestos, mica, kaolin and bentonite can be realized by using ceric ammonium nitrate as initiator. It is reported that the monomer conversion rate can reach 88%, and the weight of the grafted product after extraction with methyl ethyl ketone is more than 30%.

Many reactions in polymer chemistry and organic chemistry, such as chain transfer reaction and ionic polymerization, have been used to prepare inorganic graft copolymers.

8.7.2.3 Radiation-induced grafting

SiO_2 was treated with tert-butyl hydrogen peroxide, and then vapor-phase grafting of AN, MMA and St was initiated by ultraviolet light. After irradiation for 1 h, the weight gain of grafting was about 25%. This initiation process may be achieved by $SiOOC-(CH_3)_3$ on the surface of inorganic substrates, which is generated by SiO free radicals produced by photochemical fracture.

Litov et al. used SiO_2 adsorbed quinone to initiate AN polymerization under ultraviolet light to obtain SiO_2 grafted with PAN. When quinone was present, the grafting amount was 0.

Silica treated with dimethyldichlorosilane (Me_2SiCl_2) for 1 h at 350 °C can be grafted with AN in gas phase by ultraviolet light initiation. The reaction of Me_2SiCl_2 with -OH group on SiO_2 surface has been confirmed by spectral analysis. The adsorption of monomers on treated silica gel surface can be proved by weight gain. The amount of graft formation (not adsorption) can be judged by the absorption change of -CN group in the infrared spectrum of the product after 3 h extraction by boiling DMF. The initiation process may be due to the formation of Si and Cl free radicals by Si–Cl bond under ultraviolet light.

It has also been reported that the grafting of macromolecules onto inorganic substrates by radiation. For example, by irradiating SiO_2 with liquid polydiethylsiloxane by X-ray or ultraviolet light, it is found that the adsorption amount of SiO_2 increases. It has been proved that free radicals are formed on the surface of the irradiated surface, so it is believed that there may be a process of surface grafting.

Dispersion of SiO_2 in cyclohexane solution containing PS or St–butadiene copolymer shows that grafting can be achieved under ultraviolet irradiation. At the initial stage of irradiation, the molecular weight of PS decreased from 200,000 to 145,000. Then the molecular weight is basically constant. After a long time of radiation, the molecular weight increased.

High-energy radiation is also used for preparation of graft copolymers of inorganic materials. For example, lamellar aluminum silicate glass treated with potassium permanganate or potassium dichromate and MMA vapor were irradiated simultaneously by γ-ray. The results showed that most of the products were homopolymers, but a small amount of them were grafted onto the glass. Electron spin resonance studies showed that the grafting site may be at the structural defects of inorganic substances in the matrix.

Other scholars have reported a series of examples of grafting of SiO_2, zeolite and activated alumina onto adsorbed St by gamma rays. Figure 8.31 shows the relationship between monomer conversion and total irradiation dose. All the experimental

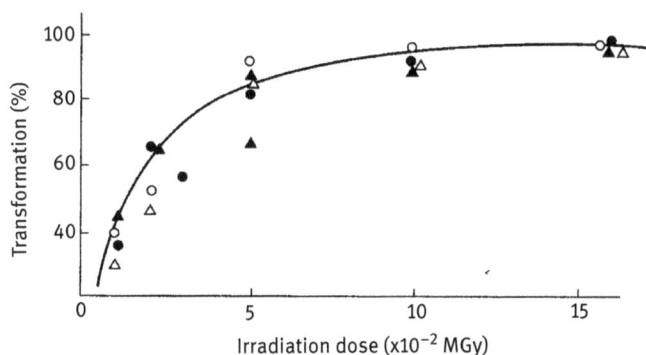

Fig. 8.31: The relationship between monomer conversion and irradiation dose on inorganic materials. Inorganic materials: SiO_2, zeolite, activated alumina.

points are basically on the same curve regardless of the type of matrix, which has not been satisfactorily explained at present. However, at least it can be considered that the reaction rate of monomers is not affected by the type of matrix because of the influence of the adsorption force field of inorganic matrix, which accelerates the polymerization rate.

It is also an effective method to initiate grafting of inorganic materials with polymers under irradiation. For example, the graft copolymer with 14% weight gain can be obtained by irradiation of PS adsorbed on SiO_2, the molecular weight of the branched chain of the copolymer is similar to that of the original PS. For example, PS was deposited or grafted onto SiO_2 to study the effect of irradiation on it. It was found that when PS was deposited on SiO_2, the grafting of PS and PS could occur during irradiation, but the degradation of the grafted PS could occur after irradiation. Degradation can be inhibited by the presence of homopolymer. This conclusion supports the following argument: at the later stage of radiation-induced grafting, the reason for the decrease of molecular weight with the increase of grafting amount is due to the secondary effect of radiation on polymers.

8.7.2.4 Mechanochemical initiation grafting

When the polymer is degraded under the action of force, macromolecular free radicals (molecular chain fragments) can be formed. When polymer and inorganic filler are molded together, the generated macromolecular free radicals can interact with inorganic materials to form grafting or cross-linking networks. The interaction between rubber and carbon black is an example. PE and carbon black have similar effects. Styrene BR, nitrile BR or silicone rubber are ground with asbestos in ball mill and vibration mill. Rubber can be grafted onto asbestos. The affinity of these grafted products to organic media is significantly improved, and they can disperse well in the resin matrix.

When external forces act on metals, salts, oxides and other materials, fracture surfaces can be formed. There are different active centers on these new fracture surfaces that can initiate polymerization, such as defects, dislocations, electron transfer or electron cloud defects. Therefore, it can be used as a force initiator to initiate monomer grafting on the surface of inorganic materials. At present, the study of this mechanochemical process is still insufficient. According to the fact that there is no obvious postactivation effect in this process, it can be concluded that the initiation of grafting is generally ionic free radical mechanism.

It is found that KCl, NaCl, CaF_2, LiF and other inorganic materials can release electrons during the crushing process, and at the same time form defects with charges on the surface. These electrical defects are often the active sites that initiate the graft polymerization of monomers. Grafting of St monomers on solid crushed surfaces such as KCl, LiF and CaF_2 has been reported. As far as the degree of grafting is concerned, LiF is one order of magnitude larger than KCl and CaF_2 is one time larger than LiF. This order is related to the difference of bond energy in the three salt crystals. After studying the mechanism of the above reaction, other scholars believe that the graft copolymerization has an ionic initiation mechanism, and the active center is the potential-carrying position on the surface of inorganic materials.

In addition, it was found that the humidity of the medium had a great influence on the grafting. NaCl crystals were ground in a closed vibration mill for 60 min, then ground in a certain humidity of argon atmosphere for 30 min and then contacted with the monomer to form the grafting product. It was found that the maximum grafting yield could be obtained when the humidity was 70–80%. Activation of water molecule is the result of dissociation of water molecule adsorbed on the charged surface. On the contrary, if it is ground in vacuum, it cannot initiate grafting, and even heating after grinding is useless.

Oxides such as SiO_2, Al_2O_3, Cr_2O_3, TiO_2, Fe_2O_3 and MgO can also be used as force initiators. Force activation is due to the emergence of free radicals and ions on the newly formed surface. When the metal is crushed, a layer of oxide film is formed on the surface, so the grafting mainly depends on the oxide film.

In the inorganic material–monomer system, the breakage of inorganic materials results in electron emission, which can be captured by the monomer to generate anionic radicals and initiate polymerization. This has been confirmed by adding benzoquinone inhibitor. For example, the polymerization of acrylamide under mechanical crushing of NaCl and $BaSO_4$ has free radical mechanism, while the polymerization of AN under iron crushing seems to have ionic mechanism.

It has been reported that standard sand (containing 92.4% SiO_2, 34.1% Al_2O_3 and 48–115 meshes in size) can be grafted with methyl methacrylate (MMA) or methacrylic acid (MA) during grinding. The reaction mechanism is free radical polymerization. Experiments show that the polymer of 1 mg/g sand is formed, and the hydrophobicity of the obtained sand is obviously enhanced. However, after 11 days immersion in acetone, the hydrophobicity was greatly weakened. This indicates that most of the polymers generated in this process have been dissolved, and they may not really achieve chemical grafting, but only form a polymer adsorbed on the surface of sand particles.

The research in this area is still very preliminary, and no conclusive law has yet been drawn.

8.7.3 Structure of inorganic material graft copolymer

After graft copolymerization of inorganic materials, the grafted organic polymers are usually covered on the surface of inorganic materials. It was found that the grafted inorganic particles retained the shape of the original inorganic particles after irradiation-initiated polymerization, indicating that the grafted polymers were uniformly deposited on the surface of the particles. If the inorganic material inside the particle is removed, the organic polymer can be separated from the particle as a composite shell of the particle. The detached polymer shell can be dissolved in benzene without gel. This indicates that no cross-linking reaction occurs under irradiation. The results of the coupling grafting of SiO_2 with anionic active PS were investigated by electron microscopy. It was also considered that the surface of the particles was evenly covered with the polymer.

Inorganic particle surface grafting can sometimes produce uneven coverage. It has been reported that the grafting of St or vinylpyridine with SiO_2 in the presence of mercaptonaphthylazophenol. It is found that the grafted PS and polyvinylpyridine have partially contracted, partially penetrated and deformed ellipsoidal structures, and their cross-sectional area is smaller than that of the entangled ellipsoidal clusters in solution.

Similar structures were also found in the graft products of St or vinylpyridine with silica gel chloride initiated by lithium butyl. Similarly, the products of LiF, NaCl or glass grafted with St after crushing were investigated by electron microscopy. It was also found that PS hemispherical droplets appeared on the surface of particles.

In addition, inorganic particles, as multifunctional materials, are enough to form cross-linking structures during the grafting process. The rubber filled with carbon black has this structure after being molded.

Some Japanese scholars have studied the grafting of St onto kaolin treated with γ-methacryloxypropyl trimethoxysilane. It is found that the grafted product only swells but does not dissolve in benzene, which is attributed to the three-dimensional structure of clay cross-linking.

Bridger et al. have studied the products of the coupling grafting of SiO_2 with anionic active PS, analyzed the PS on the unit area of SiO_2, calculated the corresponding coverage area of each molecule and compared the projection area of the mean square radius of rotation of random PS clusters with the same molecular weight from benzene solution. It was found that the coverage area of the grafted PS macromolecule was smaller. Accordingly, the macromolecule is more extensible in the grafting state than its corresponding random coils, and its long axis is perpendicular to the surface of inorganic particles. The possibility of such molecular conformation in the product of SiO_2 grafted with polyethylene oxide has also been proved.

8.7.4 Properties and applications of inorganic material graft copolymers

Grafting of inorganic materials usually occurs on the surface of particles, so grafting plays an important role in the change of surface properties. When AN was grafted onto macroporous silica gel with a contrast surface of 50 m^2/g and a pore diameter of 50 nm, it was found that the specific surface area decreased with the increase of graft weight, thus the retention time of polar substance gas in the chromatographic column was greatly reduced. Such a filler can give not only chromatographic peaks for mixtures of organic acids, alcohols, esters and aldehydes, but also retention peaks for water. Japanese scholars also reported that the efficiency of organic matter separation could be improved by grafting St with silica gel by γ-radiation.

The physical adsorption and graft copolymerization on the surface of inorganic materials are completely different. The adsorption of polyacrylic acid or polymethacrylic acid on kaolin surface and radiation-grafted kaolin were compared and analyzed by potentiometric titration. The results showed that the adsorbed macromolecules underwent desorption and rearrangement during neutralization. Grafted polymers are different. Although the activity of macromolecules increases with the degree of grafting, it is impossible to rearrange on the surface of kaolin particles due to chemical grafting.

Inorganic particles can change their affinity to surrounding media by surface grafting. SiO_2 can be stably dispersed in water for a long time after it is grafted with hydrophilic polymer polyethylene oxide. On the contrary, if the hydrophobic polymer PS is grafted, the organic phase (including alkanes, aromatic hydrocarbons and other solvents) can keep stable dispersion for a long time. Many inorganic materials are used as plastic fillers. Grafting of fillers can improve their compatibility with plastics.

Styrene was grafted onto mica by plasma surface treatment. The modified mica can be used as a filler of PS to improve the tensile and impact properties. If mica is grafted successively through St and ethylene vapor, then such mica filler can improve the properties of incompatible PE–PS blends. Obviously, the modified mica acts as coupling agent and compatibilizer.

Grafting carbon black with St, butyl methacrylate, dodecyl methacrylate and vinyl stearate can greatly improve the compatibility of carbon black and PVC, and the melt fluidity is also improved. The conductivity of PVC filled with carbon black was improved due to the improvement of dispersion.

Solid lubricants can also be modified by grafting. For example, after the grafting of fine crushed molybdenum disulfide, graphite with St, MMA, isobutyl vinyl ether and N-vinylpyridine, the film-forming property, adhesion and friction coefficient of the product in dry state were improved, and the modification of molybdenum disulfide was the most obvious.

A new method of coal liquefaction has been developed by graft copolymerization of ethylene monomers with coal. After grafting monomers with coal, the molecular structure of coal has changed, which can be easily converted into liquefaction state. It is said that this soluble coal will become a real solution when mixed with other liquids.

It is possible to obtain some functional inorganic materials by grafting technology. Former Soviet Union scholars have prepared 5-methyl-5-vinylpyridine-grafted silica gel, and found that it has a linear dependence on iodine exchange capacity and degree of grafting. The material with 15% weight gain can achieve maximum exchange capacity within 10 h, while the common anion exchange resin needs 50 h, and the former can desorb 70% iodine within 15 min, while the latter can desorb 61% within 6 h.

8.8 Graft copolymer of carbon nanomaterials

8.8.1 Types, structures and characteristics of carbon nanomaterials

In the mid-1980s, following graphite and diamond, American scientists R. E. Smalley, R. F. Curl and British scientist H. Kroto first discovered the third crystal form of carbon element, whose molecular formula is C_n, in which n value is even. These carbons are known as carbon cage clusters or fullerenes. Among the various fullerenes, C_{60} has been studied most deeply because it is one of the most abundant and stable fullerenes in the fullerene family. Because of its special structure and peculiar optical, electrical and magnetic properties, fullerenes have become the focus of scientists all over the world. The discovery of C_{60} has brought the understanding of carbon chemistry to a new level. Until then, chemists had seldom used elemental carbon as starting material for synthetic research. In less than 30 years since the discovery of C_{60}, its impact on physics, chemistry, materials science, life science and medical science has shown great application value and potential prospects.

C_{60} is a football body consisting of 12 regular pentagons and 20 regular hexagons. Each vertex (or apex) represents a carbon atom, and each carbon atom is located at a junction of a pentagon and a hexagon, as shown in Fig. 8.32.

With the development of research work, many C_n with high n value have been found. Furthermore, it is found that fullerenes are no longer soccer-like circles, but gradually develop into ellipses with the increase of n value. When the value of n is large enough, the shape of C_n turns into a tubular carbon tube in the middle and a hemispherical carbon tube at both ends, as shown in Fig. 8.33. Because the diameter

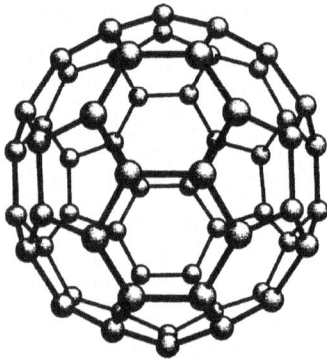

Fig. 8.32: Sketch of C_{60} structure.

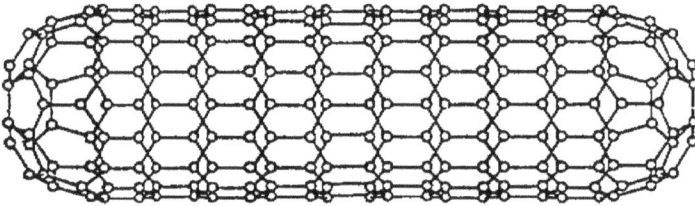

Fig. 8.33: Chart of shape and structure of carbon nanotubes.

of these tubular C_n is nanometer-sized, they are called CNTs. CNTs were first discovered in 1991 by Iijima of NEC Company in Japan. Because of their unique structure, good electrical, thermal and mechanical properties, CNTs have become a research hotspot in recent years.

CNTs have nanometer diameter and micrometer length, and can be considered as tubular products from the convolution of single-layer or multilayer two-dimensional graphene sheets. The ideal structure is a seamless hollow tube surrounded by a hexagonal carbon atom grid, with two ends covered by large fullerenes. According to the number of carbon atom layers in the tube wall, it can be divided into multiwalled CNTs (MWNT), which consists of two to dozens of carbon atom layers, and single-walled CNTs (SWNT). MWNTs are composed of several coaxial cylinders, and the spacing between layers is approximately equal to that of graphite layers (0.34 nm). The diameter and length of MWNTs are 2–30 nm and 0.1–50 μm, respectively. SWNTs have only one layer of cylinder with a diameter of 1–2 nm and a length of 1–50 μm, respectively. Both CNTs have very high aspect ratio, generally 100–1,000, up to 10,000. Figure 8.33 shows the shape and structure of CNTs.

From the structure of fullerenes and CNTs, they are composed of single-layer carbon atoms. If they are cut apart and laid flat, they are single-layer graphite. Since spherical fullerenes and tubular CNTs can be synthesized artificially, planar single-layer graphite should also be manually prepared.

In 2004, Andre K. Geim, a physicist at the University of Manchester, UK, and his colleague Konstantin Novoselov used a special plastic tape to bind the sides of the graphite sheet, rip it apart and split the sheet into two. Repetition of this process lead to thinner and thinner graphite sheets, which eventually consisted of a single layer of carbon atoms, named Graphene. Gaim and Novosholov used this primitive method to prove that graphene can exist independently and stably. They also won the 2010 Nobel Prize in Physics for their pioneering experiments in two-dimensional graphene materials.

Graphene is the thinnest but hardest nanomaterial in the world. It transmits electrons faster than any conductor known at room temperature. Perfect graphene is two-dimensional, it only includes hexagons (equiangular hexagons). If pentagons and heptagons exist, graphene defects will be formed. The connection between carbon atoms in graphene is very flexible. When applied to graphene, the surface of carbon atoms will bend and deform, so that the carbon atoms do not need to be rearranged to adapt to the external forces, thus maintaining a very stable structure.

Fullerenes, CNTs, graphene and other carbon nanomaterials have perfect structure, so they have high mechanical strength and ideal elastic modulus. They are a kind of excellent nanomaterials. Their properties are superior to any current granular materials, fibrous materials and planar materials. Therefore, carbon nanomaterials can be used as reinforcing materials for advanced composites. Because carbon nanomaterials combine the semimetallic properties of graphite with the quantum laws of energy levels and electron waves and have nanoscale, they have a very broad application prospect in the field of electronics. Carbon nanomaterials have a large specific surface area and pore structure that can absorb a large amount of hydrogen, so carbon nanoparticles can be used as reinforcing materials. As the best hydrogen storage material, materials have also become the focus of research. Because of the unique pore structure and adsorption properties of carbon nanomaterials, they can be used as catalyst carriers to maximize the catalytic effect of catalysts, and have shown good application prospects in catalysis.

However, there are some difficulties in using carbon nanomaterials as practical materials. Carbon nanomaterials are easy to agglomerate because of their huge specific surface area and surface energy and have poor dispersion in most polymer materials. Modification of carbon nanomaterials by chemical methods can change the surface state and properties of carbon nanomaterials, to change or improve the dispersion of carbon nanomaterials in polymer materials. Chemical modification of carbon nanomaterials by graft copolymerization is the most commonly used method at present.

CNTs are the most popular and mature graft modification of various carbon nanomaterials and have typical significance. Therefore, the following is an example of CNT grafting modification.

8.8.2 Graft modification of carbon nanotubes

The wall of CNTs is composed of six-membered rings of carbon. The carbon atoms in each six-membered ring are dominated by sp^2 hybridization. Each carbon atom overlaps with the sp^2 hybridization orbits of the carbon atoms in the adjacent six-membered ring to form a carbon–carbon σ bond. Due to the formation of space structure, the hybrid orbits of sp^2 are deformed, forming a hybrid structure of $sp^{2.38}$ between sp^2 and sp^3. In addition, each carbon atom has one remaining orbital, and the electrons in these π orbitals form large, highly delocalized π bonds. The aromaticity of the nonplanar pion electronic structure is small, and there are a lot of topological defects in the wall of CNTs, hence, a certain reaction activity.

The systematic development of fullerene chemistry indicates that their additive reactivity depends greatly on the curvature of fullerenes. The larger the curvature of the carbon structure is, the more the sp^2 hybrid structure is in the system, and the easier it reacts with other groups. CNTs are open at both ends and have graphite carbon layers on their walls, which are usually defective. Therefore, there is no region with large curvature on CNTs as the active center for direct addition. Therefore, only the groups with high reactivity can functionalize CNTs. After a lot of exploration and research, it has been proved that small organic molecules or polymers with some characteristics can be covalently bonded onto the surface of CNTs. So far, many methods have been developed to modify the surface of CNTs, such as acidification, carbene addition, free radical reaction, electrochemical reaction or thermochemical reaction, 1,3-dipole moment cycloaddition, azide reaction, electrophilic addition reaction and mechanochemical reaction.

8.8.2.1 Grafting CNTs via acidification and acidification-derived reaction
Carboxyl groups can be produced on the surface of CNTs by oxidizing CNTs with strong acid, which is the simplest and most mature method for surface modification of CNTs.

CNTs were mixed with concentrated sulfuric acid and concentrated nitric acid, treated by ultrasonic wave for 24 h in a constant temperature water tank, then diluted, filtered and washed with water. The CNTs were mixed with concentrated sulfuric acid and 30% hydrogen peroxide with a volume ratio of 4:1, stirred for 0.5 h, then filtered and washed to collect the products and the surface-acidified CNTs were obtained. The absorption peak of carboxyl group was found at the wave number of 1,719 cm^{-1} by infrared spectroscopy. The X-ray absorption fine structure spectrum analysis showed that the oxygen atoms on the surface of CNTs accounted for about 6–9% of the total CNTs.

Surface carboxylic-acidified CNTs can react with other compounds to form abundant grafted CNTs. For example, surface-acidified SWNTs were chlorinated by sulfoxide chloride and reacted with 11-mercaptoundecylamine to obtain CNTs containing mercaptan groups. CNTs with octadecylamine were synthesized by reaction of SWNTs with octadecylamine at 90–100 °C for 96 h. The CNTs can be dissolved in chloroform, dichloromethane, aromatic solvents and carbon disulfide, and the solubility in carbon disulfide is more than 1 mg/mL. CNTs with hydroxyl groups on the surface can be prepared by reacting acyl chloride CNTs with ethylene glycol at 120 °C and controlling the appropriate feed ratio. CNTs grafted with NBR can be prepared by further reacting previous and pedal nanotubes with low-molecular-weight carboxyl-terminated NBR. The CNTs can be used for strengthening and toughening modification of epoxy resin.

The modified CNTs can be dissolved in strong polar solvents and water by esterification of polyvinyl alcohol with acidified CNTs. In addition, hyperbranched polymers were prepared by the reaction of long alkyl chains with amino or hydroxyl end groups and polyvinyl alcohol oligomers, and then reacted with CNTs with acyl chloride to graft hyperbranched polymers onto the surface of CNTs. CNTs modified by this method are soluble in water and general organic solvents such as cyclohexane. The copolymer of propionyl vinylimide and vinylimide can be grafted onto the surface of CNTs by a similar method to improve the solubility of CNTs. CNTs were chlorinated with norbornene, and then the ruthenium-based olefin exchange catalyst was added to initiate the ring-opening exchange polymerization of norbornene. The CNTs grafted with polynorbornene showed good solubility in organic solvents.

8.8.2.2 Grafting CNTs by atomic transfer radical polymerization

ATRP reaction is one of the most attractive controllable living polymers. The length of polymer chains on the surface of CNTs can be controlled by using this method to adjust the solubility of CNTs and their dispersion in other polymer matrices.

MWNT-Br, which can initiate ATRP reaction, was synthesized by acidizing CNTs, re-acylating them, hydroxylating them with ethylene glycol at 120 °C and introducing bromide of 2-bromo-2-methylpropionate as ATRP initiator. Polymerization of MMA or St was initiated by using the CNTs. CNTs grafted with PMMA or PS were obtained. This method is called "grafting method." The length of polymer chains on the surface of CNTs can be adjusted according to the ratio of the number of monomers introduced to MWNT-Br. Figure 8.34 is a schematic diagram for the preparation of PS-grafted CNTs by ATRP.

Brominated PS with bromine atom at the end can also be synthesized first, then azide group can be introduced through azide reaction and then graft PS onto CNTs through cycloaddition reaction. This method is called "insertion method." Figure 8.35 is the diagram of PS-grafted CNTs prepared by insertion method.

Raman spectroscopy, near-infrared spectroscopy and thermogravimetric analysis (TGA) were used to characterize CNTs. It was found that the grafting efficiency of CNTs by grafting method was higher than that by the insertion method. For example, in these two examples, a polymer chain can be grafted onto every 448 carbon atoms

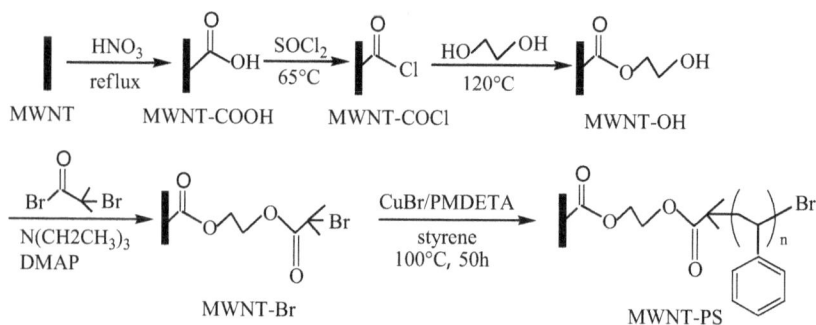

Fig. 8.34: Grafting carbon nanotubes onto polystyrene by ATRP.

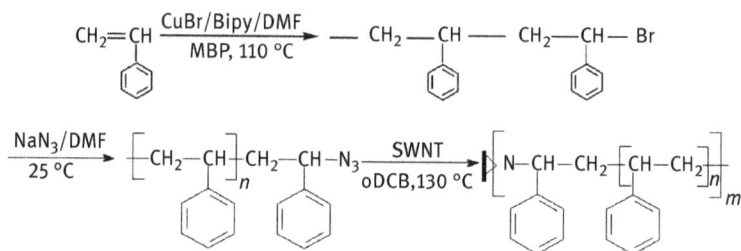

Fig. 8.35: Schematic diagram of polystyrene functionalized carbon nanotubes prepared by insertion method. (Bipy is bipyridine, MBP is methyl bromopropionate, oDCB is o-dichlorobenzene.)

using the grafting method. When the insertion method is used, a polymer chain is grafted onto every 1,240 carbon atoms. This may be related to the larger steric hindrance effect of polymer chains synthesized in advance by the insertion method.

PS-grafted CNTs prepared by the earlier two methods can be well dispersed in xylene, and can be well dispersed in ABS and PC.

8.8.2.3 Grafting CNTs by ordinary free radical polymerization

CNTs and sodium 4-styryl sulfonate were placed in water and potassium persulfate was used as initiator. The remaining catalyst, amorphous carbon and unreacted sodium styryl sulfonate were removed by centrifugation after 48 h reaction at 65 °C. Finally, 68 mg sodium PS sulfonate-grafted CNTs were obtained in 100 mL aqueous solution. The polymerization process was shown in Fig. 8.36. It was found that the CNTs grafted with sodium PS sulfonate prepared by in situ radical polymerization could be deposited layer by layer to prepare electrode films with high conductivity.

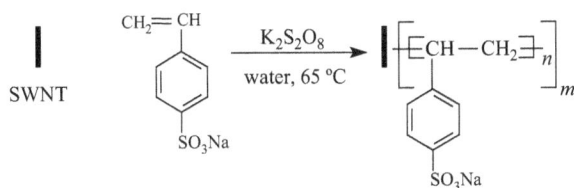

Fig. 8.36: Grafting carbon nanotubes onto sodium polystyrene sulfonate by in situ radical polymerization.

Hydrophilic polyvinyl pyrrolidone-grafted CNTs can also be prepared by similar methods.

8.8.2.4 Grafting CNTs via fluorination and fluorination-derived reaction

Under certain pressure, CNTs can be fluorinated by adding hydrogen fluoride as catalyst to the mixture of F_2 and He. Fluorinated CNTs can be further derived. For example, CNTs without fluorine can be obtained by defluorination reaction, and methoxy CNTs can also be obtained by reaction with sodium methoxide. The fluorinated nano-CNTs were added into hexane or ether solution of alkyl lithium, stirred in ice acetone bath at −40 °C for 1 h, stirred at room temperature for 12 h, then cooled in ice bath to remove unreacted lithium reagents and then filtered with tetrafluoroethylene membrane. Alkylated CNTs were obtained after vacuum drying. Ammoniated CNTs can be obtained by the reaction of fluorinated CNTs with two-terminal amino compounds (see Fig. 8.37). The aminated CNTs can further condensate with acyl chloride compounds to produce nylon CNTs. This reaction can also be used for other coupling reactions, such as connecting amino acids and DNA to CNTs.

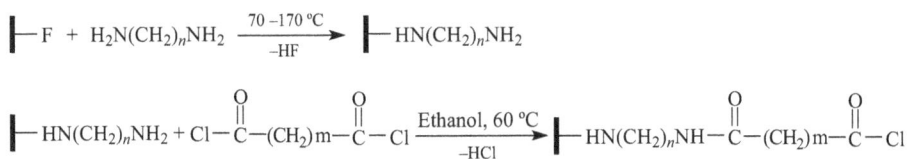

Fig. 8.37: Schematic diagram of double-ended amino compound grafted carbon nanotube.

8.8.2.5 Grafting CNTs by cycloaddition reaction

The cycloaddition reaction of CNTs was developed based on the cycloaddition reaction of C_{60}. Cycloaddition reaction refers to the addition of two molecular olefins or conjugated polyolefins into cyclic compounds. Cycloaddition reactions can be divided into [2 + 1] cycloaddition, [2 + 2] cycloaddition and [4 + 4] cycloaddition according to the number of p electrons of the reactants.

Cycloaddition is a promising method for surface modification of CNTs due to their mild conditions, high yield, good stability and easy to obtain monoaddition products. Among them, azide cycloaddition is the most commonly used method to modify CNTs.

The azide ring addition reaction is a reaction between sodium azide and polymer containing halogen atom end groups. The azide group is introduced into the end group position of the polymer. Then the CNTs and azide group are cycloaddition to form a five-membered ring intermediate. The intermediate loses N_2 under the action of heat, thus the polymer is connected to the surface of CNTs (see Fig. 8.38). Long alkyl chains, aromatic groups, hyperbranched polymers, crown ethers and ethylene glycol oligomers have been covalently bonded to the side walls of CNTs by azide cycloaddition. The carbon tubes modified by this method can dissolve in tetrachloroethane, DMS and 1,2-dichlorobenzene. CNTs with crown ether or ethylene glycol oligomer units have the highest solubility in tetrachloroethane and DMS, reaching 1.2 mg/mL.

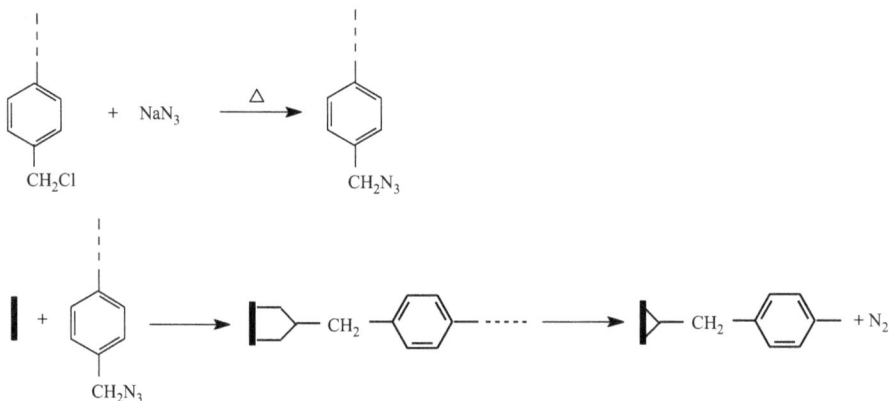

Fig. 8.38: Schematic diagram of preparing grafted carbon nanotubes by azide cycloaddition reaction.

8.8.3 Application of polymer grafted carbon nanotubes

CNTs are ideal toughening and reinforcing materials for composites. Their super mechanical properties can greatly improve the strength and toughness of polymer matrix composites. However, due to the high specific surface area of CNTs, it is very easy to agglomerate and difficult to disperse in polymer matrix. Therefore, many methods have been used to modify the surface of CNTs to improve their dispersion in polymer matrix. The research in this field has achieved very gratifying results. However, the problem is that the conjugate structure of CNTs is often destroyed by surface chemical modification, which reduces the mechanical properties of CNTs.

In order to solve this new problem, the author's research group used hyper-branched polymer to modify CNTs. By using the huge coverage area of hyper-branched polymer, the damage to CNTs structure could be minimized under the same modification effect. For example, amino-terminated hyperbranched PU was used to modify MWNTs. TGA showed that when the modification amount was 28%, there was a modification point on about 1,550 carbon atoms, while triethylene-tetramine was used to modify CNTs. There was a modification point on about 540 carbon atoms when the modification amount was 28%. Obviously, the structural damage of CNTs modified with hyperbranched polymers is much lower. Unmodified CNTs, amino-terminated hyperbranched PU-modified MWNT-(PNH$_2$) and triethylene-tetramine-modified MWNT-(4NH$_2$) were added to epoxy resins with the amount of 1% resin, respectively. The experimental results are listed in Tab. 8.19.

Tab. 8.19: Effect of different modified carbon nanotubes on mechanical properties of epoxy resin.

Types of carbon nanotubes	Tensile strength (MPa)	Bending strength (MPa)	Impact strength (kJ/m^2)	Elongation at break (%)
Noncarbon nanotubes	26	78	8	2.8
Unmodified MWNT	65	92	22	4.5
MWNT–4NH$_2$	82	108	40	11.2
MWNT–PNH$_2$	98	118	56	13.5

It can be seen from the table that when the content of CNTs is 1%, the tensile strength, impact strength and elongation at break of epoxy resin prepared by MWNT-PNH$_2$ are 276%, 600% and 382% higher than those of pure resin. The tensile strength, impact strength and elongation at break of epoxy resin prepared by MWNT-4NH$_2$ are 215%, 400% and 300% higher than those of pure resin, while the tensile strength, impact strength and elongation at break of epoxy resin prepared by unmodified CNTs are only 150%, 175% and 60% higher than those of pure resin.

Figure 8.39 is a scanning electron microscope photograph of epoxy resin modified by unmodified CNTs and MWNT-PNH$_2$. It can be seen from the figure that the dispersion of CNTs in Fig. 8.38(b) is better than that in Fig. 8.38(a). The obvious wire drawing phenomenon may be due to the chemical bonding between the amino group on CNTs and epoxy resin, which enhances the interfacial bonding force between the resin and CNTs. Therefore, when the external force acts on the composite material, CNTs and the resin encapsulated outside the tube are pulled out together. In addition, CNTs in Fig. 8.38(b) have a certain orientation arrangement, which may be due to the uniform distribution of CNTs stretched along the stress direction when the cross section is subjected to external forces, resulting in the orientation of CNTs. It is evident that this is also related to the interfacial bonding between amino groups on CNTs and epoxy resin matrix. The results show that the interfacial bonding force between CNTs and epoxy resin matrix is increased, so that the load-bearing effect is strengthened.

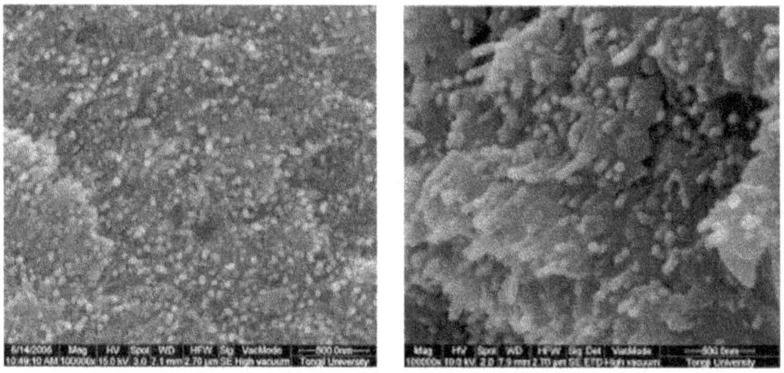

(a)   (b)

Fig. 8.39: Scanning electron microscope photographs of epoxy resin modified by carbon nanotubes. Modification of (a) unmodified carbon nanotubes and (b) MWNT-PNH$_2$.

The author's group modified CNTs with Boltorn H20 and H40, two kinds of commercialized hyperbranched polyesters produced by Perstorp, Sweden. The products can be used for strengthening and toughening modification of PVC. Table 8.20 shows mechanical properties of PVC materials without CNTs. Compared with pure PVC, the tensile strength, flexural strength, impact strength and elongation at break of 1% CNTs modified PVC can be increased by 236%, 381%, 175% and 358%, respectively, and both the toughening and reinforcing effects can be achieved.

Tab. 8.20: Effect of different modified carbon nanotubes on the mechanical properties of PVC.

Types of carbon nanotubes	Tensile strength (MPa)	Bending strength (MPa)	Impact strength (kJ.m^{-2})	Elongation at break (%)
Noncarbon nanotube	25	16	24	3.6
Unmodified MWNT	45	23	26	3.8
MWNT-H20	72	56	48	12.3
MWNT-H40	84	77	66	16.5

The author's group also used hyperbranched PU to modify CNTs and then to modify epoxy resin. When the content of CNTs was 1%, the tensile strength, impact strength and elongation at break of epoxy resin increased by 276%, 600% and 382%, respectively, compared with pure resin material, and the purpose of strengthening and toughening was also achieved. The tensile strength, impact strength and elongation at break of epoxy resin prepared by unmodified CNTs are only 35%, 22% and 60% higher than those of pure resin materials, respectively.

CNTs were acidified and reacted with ethylenediamine to obtain aminated CNTs (see Fig. 8.40). The CNTs can replace some polyamines to participate in the curing cross-linking reaction of epoxy resin, so that the epoxy resin can be joined by chemical bonds, and the toughening effect is very significant. For example, when aminated CNTs were mixed with epoxy resin (0.5% of the mass of epoxy resin), the tensile strength, impact strength and Vicat temperature were measured by cross-linking reaction at 80 °C (addition of 2% m-phenylenediamine) for 16 h. The results are shown in Tab. 8.21.

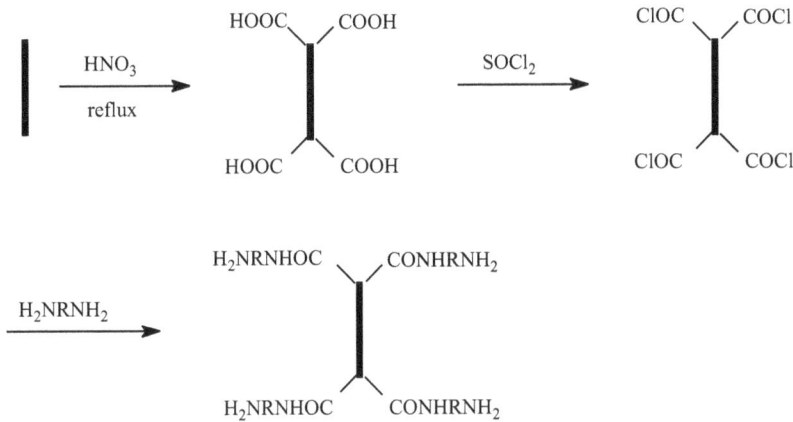

Fig. 8.40: Schematic diagram of preparation process of aminated carbon nanotubes.

Tab. 8.21: Effect of carbon nanotubes on mechanical properties of epoxy resin castings.

Types of carbon nanotubes	Tensile strength (MPa)	Impact strength (kJ/cm²)	Vicat softening temperature (°C)
Noncarbon nanotube	26	8	81
Unmodified MWNT	34	16	82
Amino carbon nanotubes	76	82	126

From the table, the toughening effect of using aminated CNTs as epoxy resin cross-linking agent is remarkable, and the impact strength increases by 925%, which is far greater than that of 600% when using hyperbranched PU-modified CNTs. This is evidently due to the chemical bonding between aminated CNTs and epoxy resin.

After modifying CNTs with polymers, the dispersion of CNTs in the resin matrix is improved, and the conductivity of the resin is also greatly improved. For example, polycaprolactone-modified CNTs were used for antistatic modification of PVC materials. When the CNT content is 0.5%, the volume resistance of PVC decreases from 10^{14} Ω to 10^8 Ω, and when the same amount of unmodified CNTs are used, the volume resistance of PVC decreases to only 10^{12} Ω.

Exercises

1. Please outline the common methods of ABS copolymer modified by chain transfer.

Answer:
The two most commonly used methods are suspension polymerization of butadiene-eye rubber in St monomer solution, swelling of polybutadiene latex by solution of AN and St mixed monomer, and then grafting polymerization. The initiator used may be organic peroxide or redox initiator system.

2. Please give examples to illustrate the synthesis methods and respective mechanisms of the graft copolymers.

Answer:
(1) Chain transfer method: Chain transfer method is to prepare grafted polymers by chain transfer reaction in free radical polymerization. Through chain transfer reaction, the activity of free radicals can be transferred to monomers, solvent molecules, transfer agent molecules (such as mercaptan) and polymer molecules. For example, MMA grafted onto the PS.
(2) Active group method: Some reactive monomers can be introduced into the polymer chain as units through chemical treatment of the polymer in the skeleton main chain. For example, a perester can be formed by treating acrylic polymers or copolymers with phosphorus pentachloride and reacting with hydrogen peroxide.
(3) Radiation method: Ultraviolet radiation can activate polymers and form active sites for grafting. For example, chlorine atoms at the side groups can be photolyzed to obtain active sites for grafting.
(4) Additive polymerization and ring-opening polymerization: Macromolecular free radicals can undergo additive reactions when they encounter additional

macromolecules containing unsaturated bonds. Then react with the monomer, and eventually form a graft polymer. For example, PAN-g-PMMA is obtained by adding PAN macromolecular free radicals to the double bonds at
the end of the molecule.

(5) Ionic polymerization: Graft polymers can also be synthesized by ionic polymerization. For example, when ionic chain transfer occurs between PS-based carbon
cations and poly(p-methoxystyrene), poly(p-methoxystyrene-g-polystyrene) can
be obtained.

(6) Macromer method: Macromer is synthesized mainly by introducing polymerizable groups at the end of the oligomer chain. For example, copolymerization of
St-based polyethylene oxide macromers.

**3. What is the emulsification of graft copolymers? How to make the two poly-
mers form a stable oil-in-oil dispersion system?**

Answer:
Because the main chain and branched chain of the graft copolymer have different
properties, the structure of the graft copolymer is similar to that of the surfactant.
Therefore, many graft copolymers also have the function of an emulsifier. In polymer blends, graft copolymers can also be used as effective interfacial activators to
disperse polymers, the graft copolymers aggregated at the particle interface, which
stabilized the particles and prevented the agglomeration, and then the two polymers can form a stable oil-in-oil dispersion system.

4. Please outline three relaxation mechanisms in amorphous linear polymers.

Answer:
The local movement of molecular chains in the glass state, which is related to secondary glass transition; the intramolecular relaxation caused by short-range and long-
range chain thermal movement, which is related to glass transition; and the thermal
diffusion between molecules, which is related to the flow process of macromolecules.

5. Please list the applications of graft copolymers.

Answer:
(1) Preparation of polymers with high impact resistance
(2) Reinforcement of natural rubber
(3) Modification of fiber materials
(4) Improvement of film forming performance
(5) Preparation of functional polymer materials
(6) As compatibilizer of polymer blends

6. Please outline the morphological structure and formation mechanism of HIPS.

Answer:
The morphological structure of HIPS is a honeycomb structure. The dispersed phase of honeycomb structure is not a single component, but contains small particles composed of continuous phase components in the dispersed direction, so its shape is similar to honeycomb (sausage or cell) structure.
  The formation mechanism is detailed in Section 8.5.1.2.

7. What is the relationship between morphology and properties of HIPS?

Answer:
(1) Because rubber particles contain a considerable amount of PS, compared with mechanical blends with the same amount of rubber, the volume fraction of rubber phase in honeycomb-like HIPS is 3–5 times larger, which greatly reduces the distance between rubber particles, makes a small amount of rubber to play a role of larger volume fraction of rubber phase and strengthens the toughening effect.
(2) In the HIPS system of graft copolymerization, PB-g-PS graft copolymer is dispersed at the interface of two phases, which improves the interfacial affinity, enhances the interfacial bonding force and is conducive to the improvement of mechanical strength and impact toughness.
(3) The dispersion phase of grafted copolymerized HIPS contains PS, and its modulus is higher than that of mechanical copolymerized dispersion phase, which makes the modulus and strength of HIPS not reduce too much due to the existence of rubber phase.
(4) HIPS with honeycomb structure will not produce harmful voids in the equatorial plane of rubber particles, which is beneficial to the formation and development of many crazes, to absorb more impact energy and greatly improve the impact strength.

8. What are the methods of graft copolymerization of inorganic materials? Please list and describe them separately.

Answer:
(1) Coupling grafting: Some of the prepolymer molecules and the inorganic material can react with functional groups to form coupling bonds, thereby obtaining a graft.
(2) Chemical initiation grafting: The copolymerization of monomers can be initiated by free radical initiator with inorganic materials that have unsaturated groups.

(3) Radiation-induced grafting: High-energy radiation is also used for preparation of graft copolymers of inorganic materials.

(4) Mechanochemical initiation grafting: When the polymer is degraded under the action of force, macromolecular free radicals (molecular chain fragments) can be formed. When polymer and inorganic filler are molded together, the generated macromolecular free radicals can interact with inorganic materials to form grafting or cross-linking networks.

Chapter 9
Block copolymers

9.1 Basic concepts of block copolymers

Block copolymers are copolymers formed by head-to-tail connections of two or more macromolecules with different chemical structures and properties. Each macromolecule chain has at least dozens of repetitive units earlier. Usually, A, B, C are used to represent the chain segments composed of different units. Therefore, block copolymers can be expressed as

$$\text{AB} \quad \text{ABC} \quad \text{ABA} \quad (\text{AB})_n \quad (\text{AB})_n \text{X}$$

Among them, AB represents diblock copolymers; ABC represents triblock copolymers, and each segment is polymerized by different monomer units; ABA also represents triblock copolymers, but the middle segment is B block with A block at both ends; $(\text{AB})_n$ represents alternating block copolymers; $(\text{AB})_n \text{X}$ represents star block copolymers with n arms and X cores, and each arm is AB block copolymers.

In the early stage of the development of polymer science, the concept of block copolymers has been proposed and some preparation techniques have been developed. However, the concept and technology of active anionic polymerization were put forward by Szwarc in 1956. Szwarc reported in 1956 that under the condition of strict removal of active impurities, when styrene was polymerized in tetrahydrofuran with sodium naphthalene as initiator, the red color of styrene anion would not disappear by itself. After the monomer was consumed, the monomer was added and the polymerization could continue. So Szwarc called this kind of polymerization as living anionic polymerization. Since the anionic activity will not disappear by itself, the block polymer will be obtained if the second monomer is added after the monomer is consumed. After Szwarc invented this active anionic polymerization method, his students Milkovich et al. quickly developed the technology of preparing triblock copolymer SBS, and realized the industrialization of this thermoplastic elastomer in the mid-1960s.

The main characteristic of active anionic polymerization is that, because of the stability of anions, no chain termination reaction occurs when impurities in the reaction system are removed, that is, chain activity can be maintained for any long time. Generally, anion initiation reaction is very fast, so once the initiator reaches the appropriate conditions, all of them will be converted into active species and initiate monomers. Thus, at the end of the growth reaction, the monomer is evenly distributed to the initiator molecule. Therefore, the molecular weight of the product is uniform and can be synthesized according to the expected value. By changing the feeding mode, selecting different coupling agents and so on, a series of copolymerization hooks with specified macromolecule structure and well-defined structure can be

https://doi.org/10.1515/9783110596335-009

produced. Active anionic polymerization has not only great application value in practical production, but also an ideal research object for polymer physicists because of its regular structure. Since the late 1960s, there has been an upsurge in the study of the synthesis, structure and properties of such copolymers, with tens of thousands of relevant literatures.

In 1961, researchers at Shell Chemical's Synthetic Rubber Laboratory began to study the cold flow and low raw rubber strength of high *cis*-1,4-structured polyisoprene and high *cis*-1,4-structured polybutadiene, both of which were produced using lithium alkyl as initiator. As part of this research project, they prepared low-molecular-weight polyisoprene and polybutadiene with short polystyrene blocks, that is, SI and SB diblock copolymers. It was found that there was no cold flow in these two block copolymers. Thereafter, they prepared SIS and SBS triblock copolymers, and hot-pressed the two block copolymers into plates to determine the mechanical properties of their rubber. It is found that the tensile strength, elongation at break and resilience of SIS and SBS samples are very high. Although they are not vulcanized, their properties are like those of conventional vulcanized rubber. In addition, the sample can also dissolve in toluene, which indicates that no chemical cross-linking has been formed.

Further studies soon identified these triblock copolymers as a class of thermoplastic elastomers with great potential for development, and explained their properties with the microzone theory: in solid state, due to the lack of enough compatibility between polystyrene and polybutadiene or polyisoprene, rigid polystyrene terminal blocks aggregated into hard dispersed phase (i.e., microzone). The softer segments of polybutadiene or polyisoprene form a continuous phase. In this system, polystyrene microarea acts as physical cross-linking point, thus forming a cross-linking network similar to vulcanized rubber. SIS and SBS not only have excellent mechanical properties, but also can be processed by thermoplastics, so they have great industrial prospects.

Although the discovery of SIS and SBS triblock copolymer thermoplastic elastomers is contingent, it is based on many basic research backgrounds. In October 1965, styrene–diene triblock copolymer thermoplastic elastomers were commercialized. In 1967, the microblock theory was extended to other types of block copolymers.

From 1950s to 1960s, scientists from DuPont and Goodrich began to study thermoplastic block polyurethane (TPPU) elastomers and high modulus elastic fibers, and gradually introduced a number of polyurethane thermoplastic elastomers. At present, it has been recognized that the formation of TPPU into thermoplastic elastomer is also due to its phase separation and formation of microstructure.

TPPU consists of two kinds of blocks, one is hard block, which is formed by adding chain extender such as butanediol (BDO) to diisocyanate such as 4,4′-diphenylmethane diisocyanate (MDI), and the other is soft block, which is composed of soft long-chain polyether or polyester embedded between two hard segments. At room temperature, the soft segment with low melting point is incompatible with the hard segment with high melting point and polarity, which leads to microphase separation. Another part of the driving force for phase separation is the crystallization of hard segments. When heated above the melting point of the hard segment, the TPPU forms a homogeneous melt, which can be processed by thermoplastic processing technology, such as injection molding, extrusion molding and blow molding. After cooling, the hard and soft segments are reseparated to restore elasticity.

Soft segments form continuous phases and give TPPU elasticity, while hard segments act as physical cross-linking points and reinforcing fillers. When heated or in solvents, the physical cross-linking point disappears, cools or evaporates, and the cross-linking network is reconstructed.

In the past 20 years, the research of active polymerization has been very active, which provides more methods and ways for the synthesis of block copolymers. The study of the relationship between the synthesis, structure and properties of block copolymers has also provided strong support for the development of reactive polymerization, and has become a hot topic in the field of polymer science.

9.2 Synthesis method of block copolymer

From the current situation, block copolymers can be basically divided into two categories, one is olefin block copolymers prepared by living controlled polymerization, and the other is polyurethane block copolymers. The preparation methods of these two block copolymers are quite different. The following are introduced separately.

9.2.1 Preparation of block copolymers by living anion polymerization

Among various block copolymer materials, styrene block copolymer thermoplastic elastomer is one of the earliest thermoplastic elastomers studied, and its output ranks first at present. Early studies on block copolymers focused on polystyrene/polydiene triblock copolymers. In recent years, interesting progress has been made in the study of triblock copolymers with polyisobutylene as elastic segments. As polystyrene/polydiene triblock copolymers have clear molecular structure and good reproducibility in preparation, they are used as model polymers to predict the properties of other block copolymers. The domain theory of polystyrene–polydiene–polystyrene block copolymers has been extended to all ABA block copolymers, where A represents a polymer block that is glass or crystalline at service temperature and can flow at high temperature, and B represents an elastic polymer block at service temperature. The study of polystyrene/polydiene triblock copolymers opens a new field for polymer science.

SBS, SIS and similar block copolymers are prepared by anionic polymerization. Anionic polymerization of styrene (including substituted styrene), butadiene and isoprene is usually carried out in inert hydrocarbon solvents such as cyclohexane and toluene. Oxygen, water and other impurities that can react with active growth centers must be removed completely. Under these conditions, the relative molecular weight of the polymer can be accurately controlled. The relative molecular weight distribution of block copolymers obtained by other methods is generally wider. The best initiator for anionic polymerization is organic lithium compounds.

The preparation methods of triblock copolymers can usually be divided into three types:

(1) Sequential polymerization: The first monomer is initiated by a single active initiator, and then the other monomers are added to the first monomer for sequential polymerization.

(2) Coupling method: The first monomer and the second monomer are polymerized by a single active initiator, and then the active chain is linked by a coupling agent.

(3) Multifunctional initiator method: the first monomer is initiated by initiator with two or more active centers, and then the second monomer is added to continue the polymerization.

Take the preparation of SBS as an example. For the first polymerization method, butyl lithium is a good initiator because it can easily initiate styrene polymerization. Compared with the growth reaction, the initiation rate is very fast and the relative molecular mass distribution of the obtained polymer is very narrow. The product of chain growth is called polystyrene-based lithium, which is expressed by $S^- Li^+$. If diene monomers were added further, $S^- Li^+$ could initiate further polymerization, and the active center would be polybutadiene lithium ($S^- Li^+$).

In the earlier example, the polymerization of diene monomers is carried out in the form of 1,4-addition. At least 90% of monomer units are 1,4-structure in inert and nonpolar solvents, and the rest are l,2-structure (for butadiene) or 3,4-structure (for isoprene). $SB^- Li^+$ can also be regarded as an initiator. If styrene is added, the end of the active chain will continue to polymerize.

Because the reaction rate of styrene initiated by $SB^- Li^+$ is lower than the growth rate of styrene, the relative molecular mass distribution of the resulting polystyrene block is widened. In extreme cases, even after the consumption of styrene, some $SB^- Li^+$ has not yet reacted. This problem can be solved by adding solvent such as ether with strong solvation ability before adding styrene. When solvent with strong solvation ability was added, the initiation rate increased and the relative molecular mass distribution of the resulting polystyrene block narrowed. However, solvents with strong solvation ability cannot be added in advance, because this will change the microstructure of polydiene and reduce the content of l,4-structure.

After completing the last step mentioned above, proton compounds such as alcohols can be added to deactivate SBS⁻ Li⁺.

The whole preparation process of SBS can be represented by Fig. 9.1.

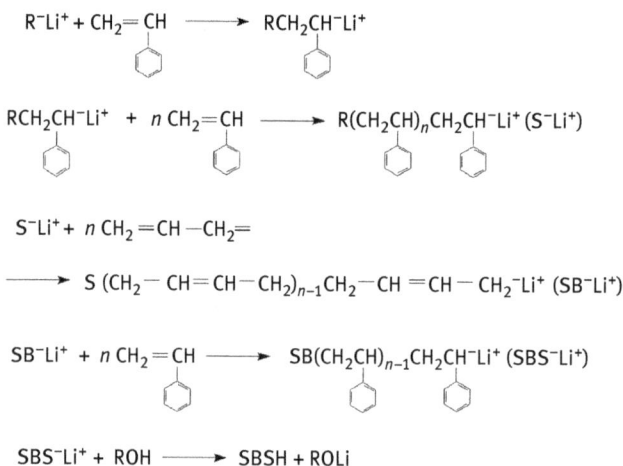

$$R^-Li^+ + CH_2{=}CH \longrightarrow RCH_2CH^-Li^+$$

$$RCH_2CH^-Li^+ + n\,CH_2{=}CH \longrightarrow R(CH_2CH)_nCH_2CH^-Li^+\,(S^-Li^+)$$

$$S^-Li^+ + n\,CH_2{=}CH{-}CH_2{=}$$
$$\longrightarrow S\,(CH_2{-}CH{=}CH{-}CH_2)_{n-1}CH_2{-}CH{=}CH{-}CH_2^-Li^+\,(SB^-Li^+)$$

$$SB^-Li^+ + n\,CH_2{=}CH \longrightarrow SB(CH_2CH)_{n-1}CH_2CH^-Li^+\,(SBS^-Li^+)$$

$$SBS^-Li^+ + ROH \longrightarrow SBSH + ROLi$$

Fig. 9.1: Schematic diagram for preparation of SBS triblock copolymer.

If triblock copolymers are prepared by coupling method, the first three reactions will remain unchanged, but SB⁻Li⁺ is no longer used to initiate styrene polymerization, but reacts with coupling agents:

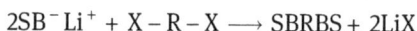

$$2SB^-Li^+ + X{-}R{-}X \longrightarrow SBRBS + 2LiX$$

There is functional group R between the middle two B segments, which accounts for a small proportion, so the performance of the whole chain is generally not affected. The coupling agents used include esters, organic halides and silicon halide. The coupling agent in the above formula is a bifunctional coupling agent, or a multifunctional coupling agent (such as $SiCl_4$) and can be used to obtain branched or star-shaped polymer $(SB)_nX$. If SB⁻Li⁺ reacts with divinylbenzene (DVB), highly branched polymers can be obtained.

The third method for preparing triblock copolymers is the multifunctional initiator method, for example, the bifunctional initiator Li⁺ ⁻R⁻ Li⁺ initiates the polymerization of diene first (take butadiene as an example):

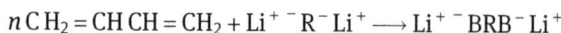

$$n\,CH_2{=}CH\,CH{=}CH_2 + Li^+\,{}^-R^-\,Li^+ \longrightarrow Li^+\,{}^-BRB^-\,Li^+$$

The latter two steps are similar to the sequential polymerization method described earlier, that is, Li⁺ ⁻BRB⁻ Li⁺, which is obtained by adding styrene to Li⁺ ⁻SBRBS⁻ Li⁺, and finally adding proton compounds to deactivate the active chain, resulting in block copolymer SBRBS. In principle, besides bifunctional initiator, multifunctional initiator can also be used, so that star block copolymer can be obtained, but it has not been reported in the literature.

Compared with the first two methods, the multifunctional initiator method did not attract attention at first, although early researchers used sodium naphthalene as a dual-functional initiator. Because there is a serious defect in the multifunctional initiator method, that is, when the polymerization ends in the hydrocarbon solvent, the end of the active chain is associated with each other, and it is easy to produce gel at the initial stage of polymerization. Although the association can be inhibited by adding solvators such as ethers, as mentioned earlier, this in turn changes the microstructure of polydienes. Recently, there has been renewed interest in multifunctional initiator methods, such as the synthesis of triblock copolymers with styrene/α-methylstyrene copolymers as terminal blocks.

Block copolymers with saturated elastic blocks can be obtained by hydrogenation of SIS and SBS. For SBS, it is necessary to add a microstructure regulator in the process of polymerization, so that the middle block of polybutadiene contains a considerable amount of 1,2 structure, so that the copolymer SEBS containing random copolymer block of vinyl butadiene can be obtained after hydrogenation, as shown in Fig. 9.2.

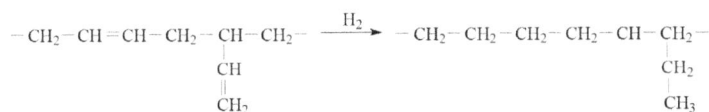

Fig. 9.2: Schematic diagram of SEBS block copolymer prepared by hydrogenation.

Similarly, the copolymer SEPS containing ethylene–propylene alternating copolymer block can be obtained by hydrogenation of polyisoprene block in SIS.

SEBS and SEPS triblock copolymers have excellent stability and are two very important thermoplastic elastomers of polyolefins.

Almost all the end blocks of the above-mentioned block copolymers are polystyrene. Substituted polystyrene can also be used as terminal blocks, such as poly(α-methylstyrene), copolymer of α-methylstyrene/styrene and poly(p-tert-butylstyrene). The advantages of these three polymers are higher glass transition temperature, which can increase the upper limit temperature of block copolymers. However, the disadvantage of α-methylstyrene is that the polymerization rate is slow and the upper limit temperature is low. The disadvantage of poly(p-tert-butylstyrene) is that it has some compatibility with polydiene, and the phase separation structure can only be formed when the relative molecular weight of poly(p-tert-butylstyrene) is high.

The ratio of different microstructures (*cis*-1,4, *trans*-1,4 and 1,2 addition) of poly-butadiene varies greatly with the change of synthesis conditions. Because the T_g of these three microstructures are quite different, the change of microstructures will affect the properties of SBS products. Some of the results are summarized in Tab. 9.1.

Tab. 9.1: Effect of solvents and corresponding ions on microstructure of butadiene polymerization (0 °C).

Catalysts (corresponding ions)		Microstructure		
		Cis-1,4-structure (%)	Anti-1,4 structure (%)	1,2-Structure (%)
Polymerization solvent: pentane	Alkyl-Li	35	52	13
	Alkyl-Na	10	25	65
	Alkyl-K	15	40	45
	Alkyl-Rb	7	31	62
	Alkyl-Cs	6	35	59
Polymerization solvent: tetrahydrofuran	Naphthalene-Li	0	4	96
	Naphthalene-Na	0	9	91
	Naphthalene-K	0	18	82
	Naphthalene-Rb	0	25	75
	Naphthalene-Cs	0	25	75

From a practical point of view, the polymerization of lithium alkyl in nonpolar solvents is the most advantageous, because the 1,4-polymerization microstructures of PB block in the product are close to 90%, and T_g is the lowest at this time. The change of the microstructures with the properties of the corresponding ions and solvents may be related to the association of anions. Anions in nonpolar solvents always associate with dimers, which limits the configuration of newly grown butadiene monomers. When a small amount of polar solvent (such as THF) is added, the association dissociates and the probability of 1,2-polymerization is greatly increased.

In the anionic copolymerization of styrene with dienes, an interesting phenomenon was also found, that is, the anionic copolymerization of styrene and butadiene (or isoprene) mixed monomers in nonpolar solvents with BuLi as initiator would not result in random copolymers, but in block copolymers. Kinetic studies show that when the two monomers coexist, butadiene always polymerizes first until it is nearly depleted, and styrene begins to polymerize. Generally speaking, the rate of anionic polymerization of butadiene is much lower than that of styrene, so the above results are not easy to understand at first. In fact, there are four growth responses in mixed monomers, as shown in Fig. 9.3.

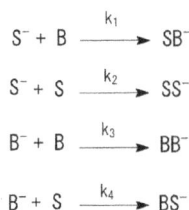

$$S^- + B \xrightarrow{k_1} SB^-$$

$$S^- + S \xrightarrow{k_2} SS^-$$

$$B^- + B \xrightarrow{k_3} BB^-$$

$$B^- + S \xrightarrow{k_4} BS^-$$

Fig. 9.3: Diagram of growth reaction of anionic polymerization of styrene and butadiene.

The velocity relationship between them is $k_1 \geq k_2 > k_3 > k_4$. Therefore, S^- ions always make monomer B react with k_1 to form B^- ions. The reaction rate of B^- with B monomer is faster than that of S monomer. In this way, butadiene in the system naturally forms blocks first, and then styrene blocks. However, when monomer B is almost exhausted, a transition chain is generated, that is, the proportion of unit B decreases and the proportion of unit S increases, which is called "tapered block." The existence of this part has a great influence on the size of the interface area of S and B when they are separated. Because people are interested in the study of the interfacial zone, it is necessary to prepare block copolymers which can control the length of the gradual block chain. Because the changes of the above four reaction rate constants are closely related to the reaction conditions (solvents, catalysts, etc.), the gradient blocks with specified structure can be prepared by changing these factors. For example, when mixed monomers of styrene and isoprene are polymerized in benzene with BuLi, isoprene is first polymerized, so block copolymers with transition segments are obtained. However, if the same catalyst is used to polymerize in THF at −78 °C, the opposite is true. The block copolymers with transition segments are also obtained by prepolymerization of styrene with isoprene, but the two transition segments are obviously different. Tsukahara et al. used this characteristic to copolymerize styrene and isoprene with benzene containing 1% THF, so that the rate of entry of the two monomers into the polymer was relatively close. Gradual block copolymers were prepared, which consisted of pure S block through a long transition chain and then converted into pure I block. Annighoefer and Gronski proposed a method to prepare triblock copolymers. The two ends of the triblock copolymers are PI and PS, and the central segment is composed of random copolymers of S and I. Thus, the PI and PS blocks are pure homopolymer chains, and the length of the central block can be controlled, that is, the size of the interface zone can be controlled. The block copolymers were prepared by using BuLi as initiator, 80/20 benzene/triethylamine as solvent, adding pure isoprene, 50/50 I/S mixed monomer and pure styrene sequentially at 25 °C.

9.2.2 Preparation of star block copolymers by living anion polymerization

The characteristics of active anionic polymerization are that the copolymers prepared have uniform-molecular weight and that different blocks can be linked together in a specified way to obtain linear diblock copolymers and triblock copolymers. In addition to linear block copolymers, the synthesis of star-shaped dobby copolymers has attracted much attention in the past 20 years.

There are two ways to prepare star-shaped copolymers by active anionic polymerization: one is to couple block anions with multifunctional electrophilic reagents, and the other is to add dimeric functional monomers, such as DVB.

In the first method, there are many coupling agents available. Silicon tetrachloride is the most commonly used coupling agent for the preparation of four-arm star copolymers. Taking styrene–butadiene anion as an example, the reaction is as follows:

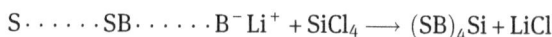

$$\mathrm{S\cdots\cdots SB\cdots\cdots B^- Li^+ + SiCl_4 \longrightarrow (SB)_4 Si + LiCl}$$

Many halogenated alkanes or halogenated silanes can be used as coupling agents, but the coupling efficiency or the number of arms forming products are often less than the number of halogenated groups in coupling agents.

The second method is to quantitatively add DBV to the block anion system. Since two vinyl benzene can be added to block anions on one hand and can react with itself on the other hand, it can form a microgel core with multiple block copolymer chains. Controlling the molar ratio of DBV to block anion can control the number of arms of block. Star block copolymers of styrene–isoprene with an average arm number of 5–29 have been synthesized by this method. Of course, the molecular weight of block copolymers prepared by this method is no longer monodisperse.

The most interesting practical property of multiarm block copolymers is their low melt viscosity. Even when their molecular weight is much higher than that of linear systems, the melt viscosity does not increase much. Therefore, such copolymers are no longer only of theoretical significance. K-resin is an industrial product of styrene–butadiene block copolymer, which is a transparent and tough packaging material. In theory, dobby block copolymers can be used to systematically study the structure of macromolecules and their effects on physical properties.

The biggest limitation of active anionic polymerization is the limited variety of monomers available. Sometimes, due to the structure of monomers, active anions cannot be formed, or too many side reactions cannot get a homogeneous molecular weight of the active chain. The most typical one in this respect is methyl methacrylate (MMA). The ester group of MMA monomer can react directly with initiator or with macromolecule anion, which reduces the efficiency of initiator or terminates the active chain, because it is not easy to obtain the product of average molecular weight. In order to avoid the influence of side reactions, it is suggested to polymerize at low temperature in order to increase the stability of its anions, and in some cases to achieve a predetermined molecular weight and narrow molecular weight distribution. In the preparation of block copolymers, the active chain of the first monomer is usually required to initiate the second monomer smoothly, which is not always satisfied. In the preparation of copolymers of MMA and styrene, since MMA anion cannot initiate hydrocarbon monomers, it is necessary to initiate the polymerization of styrene before it initiates the polymerization of MMA. Because of these limitations, although active anionic polymerization can theoretically produce copolymers with uniform molecular weight of specified structure, only a few monomers, such as α-methylstyrene, vinylpyridine, ethylene oxide and caprolactone, have been studied carefully in practice, except styrene and dienes.

In recent years, with the development of active controllable polymerization, the preparation of block copolymers has gone far beyond the scope of active anionic polymerization, such as styrene–isobutylene–styrene (SIBS) triblock copolymer prepared by cationic polymerization. Cationic polymerization enlarges the choice of soft and hard blocks. The middle block can be polyisobutylene, and the two end segments can choose various polyaromatic hydrocarbons. Saturated polyisobutylene rubber blocks can be obtained directly by cationic polymerization of isobutylene, which omits the hydrogenation step when diene is used. The cationic polymerization of styrene derivatives such as halogenated styrene can also be carried out. The glass transition temperature of the obtained polymer is higher than that of polystyrene, so the upper limit temperature of block copolymer is increased. Other methods such as atom transfer radical polymerization, group transfer polymerization and living ring-opening polymerization have been successfully used to prepare block copolymers.

9.2.3 Preparation of polyurethane block copolymer

9.2.3.1 Introduction of polyurethane block copolymer

Block polyurethane elastomer is the first elastomer that can be processed by thermoplastics. At present, it is still playing an important role in the rapid development of thermoplastic elastomer industry.

Otto Bayer of I. G. Farbenindustrie (now Bayer A. G.) and his colleagues (1937) were the first to carry out polyurethane research. Their initial goal was to improve the properties of polyamide synthetic fibers. Later, DuPont and ICI discovered the elasticity of polyurethane. In the 1940s, polyurethane was put into industrial production, then known as "I-rubber," but its performance was very poor, which was thought to be caused by the irregular network of elastomers. To this end, polyurethane elastomer composed of linear polyester and 2-nitro-4,4-diisocyanate biphenyl was synthesized. After chain extension with water, the product has good elasticity. Similar results were obtained when 2-nitro-4,4-diisocyanate biphenyl was replaced by 1,5-naphthalene-diisocyanate. These results can be explained by the following reaction processes: first, a strictly linear hydroxyl-terminated polyester reacts with excessive diisocyanate to form α, ω-diisocyanate prepolymer. Then, the prepolymer is extended by water to form urea bond, which reacts with diisocyanate to form an elastic network. Later, further use of diols as chain extenders could be a breakthrough in the synthesis of polyurethane elastomers, which Bayer registered as Vulkollan. In the USA, the early Vulkollan polyester polyurethane was developed by Seeger et al. of Goodyear Tire and Rubber Company. Its commercial name is Chemigum SL. DuPont developed polyether polyurethane branch, which is called Adiprene.

Early polyurethane block copolymers consisted mainly of the following three basic components:
(1) Polyester or polyether high-molecular-weight diols
(2) Chain extenders such as water, short-chain diols or diamines
(3) Diisocyanates such as 1,5-naphthalene-diisocyanate (NDI)

However, these polyurethane block copolymers are not thermoplastic elastomers because their melting point is higher than the decomposition temperature of urethane bonds. Subsequently, significant breakthroughs were made when MDI was used instead of NDI in the above three-component system. In 1958, Schollenberger of B. F. Goodrich Company, USA, produced TPPU. Prior to this, DuPont announced the production of a polyurethane elastic fiber Lycra-based on MDI. In the early 1960s, Goodrich developed Estane, Mobay Texin and Upjohn Pellethane. In Europe, Desmopan and Elastollan were also produced by Bayer and Elastogran. Thereafter, the relationship between the structure and properties of polyurethane thermoplastic elastomers has become a hot topic in polymer science.

TPPUs are generally prepared from long-chain polyols with average molecular weight of 600–4,000, chain extenders with molecular weight of 60–400 and polyisocyanates. Because the ratio of hard segment to soft segment can be adjusted in a wide range, the obtained TPPU can be either a soft elastomer or a hard and high modulus plastic.

9.2.3.2 Soft segment in TPPU

The long soft segments of TPPU mainly control the low-temperature performance, solvent resistance and weather resistance of polyurethane.

There are two kinds of important soft segments: hydroxyl-terminated polyester and hydroxyl-terminated polyether.

1. Polyester

Typical hydroxyl-terminated polyesters are synthesized by the reaction of adipic acid with excessive diols (e.g., ethylene glycol, 1,4-butanediol, 1,6-hexanediol, neopentyl glycol or a mixture of these diols).

The synthesis of polyester from other dibasic acids such as azelaic acid or $O(p)$-phthalic acid has also been reported. These dicarboxylic acids can be used alone or in combination with adipic acid. Generally, the glass transition temperature of polyester can be increased when aromatic or aliphatic rings are contained in dibasic acids or alcohols.

In addition, there are two special types of polyesters with commercial value, namely polycaprolactone and aliphatic polycarbonate.

Polycaprolactone is obtained by ring-opening polymerization of ω-caprolactone initiated by difunctional compound 1,6-hexanediol. The properties of this polyester are similar to those of polyadipate.

Aliphatic polycarbonates have good hydrolytic stability. They can be prepared by the reaction of dibasic alcohols such as 1,6-hexanediol and phosgene. They can also be prepared by transesterification of dibasic alcohols with low-molecular-weight carbonates such as diethyl carbonate or diphenyl carbonate.

2. Polyether

There are two main types of polyethers used in industrial TPPU elastomers: polypropylene oxide polyether diols and polytetrahydrofuran polyether diols.

Polypropylene oxide diol was synthesized by ring-opening polymerization of propylene oxide catalyzed by alkali using bifunctional compounds such as propylene glycol as initiator. Because propylene oxide is used, the end of the polyether chain is mainly secondary carbon hydroxyl group. If a certain amount of ethylene oxide is added, especially ethylene oxide is added at the later stage of the reaction, the primary carbon hydroxyl group can be introduced. Because of side reactions, the functionality of polypropylene oxide is always lower than that of initiator, and with the increase of the relative molecular weight of polyether, the ratio of end allyl group to end isopropylene group increases. For example, the functionality of polypropylene oxide diols with a molecular weight of 2,000 is 1.96 instead of 2. In order to make the hydroxyl groups at the end of polyether conform to the theoretical values, special catalysts are needed. TPPU with excellent properties can be prepared by using this polyether.

Polytetrahydrofuran diol is synthesized by cationic ring-opening polymerization of tetrahydrofuran with a functionality of about 2.

Table 9.2 lists the main soft segments and the corresponding properties of polyurethane.

The low-temperature properties of TPPU depend on the initial glass transition temperature T_e of the soft segment and the corresponding temperature of the complete melting of the soft segment. For the TPPU with medium and low hardness, the initial glass transition temperature of the soft segment is usually 20–30 °C above the corresponding temperature of the pure soft segment (the glass transition temperature of the pure soft segment). The range of degree transition is very narrow.

Tab. 9.2: Important polyether and polyester diol[1] and properties of corresponding polyurethane elastomers.[2]

Polyether or polyester diols	Polyether or polyester diols		Elastomer	Hydrolysis stability
	T_e^3 (°C)	T_m^4 (°C)	T_e (°C)	
Polyethylene glycol adipate	−46	52	−25	Passable
1,4-Butanediol polyadipate	−71	56	−40	Good
Polyethylene adipate and 1,4-butanediol mixed esters	−60	17	−30	Passable/good
Polyethylene adipate and 2,2-dimethylpropylene glycol mixed esters	−57	27	−40	Good
Polycaprolactone	−72	59	−30	Good
Polyethylene glycol adipate	−53	–	−30	Bad
Poly(1,6-hexanediol carbonate)	−62	49	−30	Very good
Polytetrahydrofuran diol	−100	32	−80	Very good

Note: [1]Relative molecular mass 2,000; [2]hardness 85; [3]initial temperature of glass transition; [4]melting point.

The glass transition temperature range of soft segment in TPPU depends on the content of hard segment and the phase separation degree of hard segment and soft segment. With the increase of hard segment content and the decrease of phase separation, the glass transition range of soft segment will be widened correspondingly, which will lead to poor cryogenic performance. If polyether with poor compatibility with hard segment is used as soft segment, the low-temperature flexibility of TPPU can be improved. When the relative molecular weight of the soft segment increases or the TPPU is annealed, the incompatibility between the soft segment and the hard segment will also increase.

9.2.3.3 Hard segment in TPPU

1. Polyisocyanate

Of the commercially available polyisocyanates, only a few are suitable for the preparation of TPPU. The most important diisocyanate is MDI. In addition, 4,4′-dicyclohexylmethane diisocyanate, 1,4-phenyldiisocyanate, 1,4-cyclohexane diisocyanate (HPDI), NDI and 3,3′-dimethyl-4,4′-diisocyanate biphenyl are also used in the preparation of TPPU elastomers or have interesting application prospects. The structures of these isocyanates are shown in Fig. 9.4.

OCN —⬡— CH₂ —⬡— NCO   OCN —⬡— CH₂ —⬡— NCO

MDI   HMDI

NCO   NCO   NCO   CH₃

                                                    OCN —⬡—⬡— NCO

NCO   NCO   NCO   CH₃

PDI   HPDI   NDI   TODI

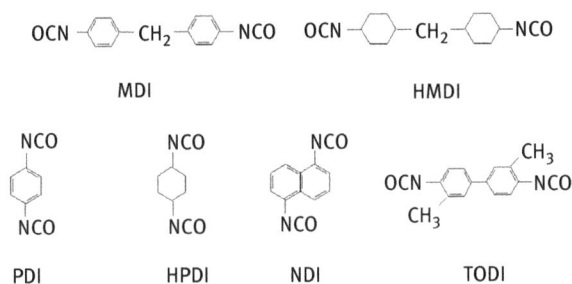

Fig. 9.4: Structure of partial polyisocyanates.

2. Chain extender

Chain extender, as a part of the hard segment of block polyurethane elastomer, also largely determines the properties of block polyurethane.

The most important chain extenders of TPPU are linear diols, such as ethylene glycol, 1,4-butanediol, 1,6-hexanediol, 1,4-di(hydroxyethoxy) benzene and so on. The polyurethane segments formed by the reaction of these diols with diisocyanate almost completely crystallized at room temperature and did not decompose during melting. However, in the preparation of thermoplastic polychloride esters with high hard segment content, glycol should not be used as chain extender, otherwise, the thermal stability at high temperature is poor.

1,4-Butanediol and 1,4-di(hydroxyethoxy) benzene are the most suitable chain extenders for the preparation of thermoplastic polychloride, especially the polyurethane obtained from the latter has better properties at high and low temperatures, and has smaller compression set.

Generally, nonlinear diols are not suitable as chain extenders for TPPU, because the hard segment crystallization of the obtained polyurethane is not complete, resulting in poor high and low-temperature properties of TPPU. It is sometimes recommended to use the mixture of straight-chain diols as chain extender to obtain hard segments with lower order, which is very important for the preparation of extrusion grade TPPU which can adapt to wider processing conditions.

With the increase of hard segment content, the hardness of TPPU increases, and the glass transition temperature of elastic phase also increases. When the mass fraction of hard segment reaches 60%–70%, the reversal occurs, and the polyurethane transforms from elastic polymer to brittle high modulus plastic.

Diamines are also good chain extenders for polyurethane, but they are usually not used to prepare TPPUs because the urea group formed by the reaction of amino group and isocyanate group is decomposed in melting. However, dibasic amines with larger steric hindrance, such as 1-amino-3-aminomethyl-3,5,5-trimethylcyclohexane (isophorone diamine), can be used to react with aromatic or aliphatic diisocyanates. If the mixture of diamine and diol with steric hindrance effect is used as chain extender, the TPPU with poor crystallinity in hard segment can be obtained. This TPPU has poor elasticity but good damping property, so it can be used as a damping material.

The first chain extender used in polyurethane elastomers was water. Two isocyanate groups formed a urea group with one molecule of water. It was found that the content of urea in hard segment was less than that in dibasic amine when water was used as chain extender, so the TPPU could be melted.

9.2.3.4 Synthesis of TPPU

The synthesis of block polyurethane is usually carried out at above 80 °C. The optimum ratio of isocyanate group to hydroxyl group should be close to 1:1. If the ratio is less than 0.96, the relative molecular mass of the polymer is too low; on the contrary, when the ratio is higher than 1.1, the cross-linking reaction will lead to the difficulty of thermoplastic processing. When the ratio is 0.98 or slightly higher, the average-molecular weight of the block polyurethane is about 40,000. Currently, the performance of the block polyurethane is better.

There are many ways to synthesize block polyurethane. The "one-step method" is to mix all the components before reacting. In prepolymer method, polyester or polyether diol reacts with diisocyanate to form isocyanate-terminated prepolymer, and then extend the chain with chain extender. The reaction can be carried out intermittently or continuously in a reactor or a reactive extruder.

In large-scale industrial production, conveyor belt method and reactive extrusion method are usually used. In the former method, all raw materials are mixed and heated to a liquid state, dumped on the conveyor belt, solidified on the conveyor belt and then crushed into granules. This granular product can be used directly, but usually it is extruded into more uniform granules. When the reaction extrusion method is adopted, the uniform granules can be obtained directly by nearly complete polymerization at the exit of the extruder.

Because the separation between hard and soft phases of TPPU is closely related to temperature, the thermal history in the production process is extremely important. Therefore, even if the same raw material is used, if the thermal history is different, the performance of the final product will be very different.

9.3 Structure and properties of styrene block copolymers

9.3.1 Morphological structure of styrene block copolymers

Among all kinds of thermoplastic elastomers, styrene block copolymer thermoplastic elastomer is one of the earliest thermoplastic elastomers studied, and its output ranks first at present. Among them, the morphology of styrene, diene and styrene triblock copolymers is the most interesting. The biggest difference between them and random copolymers is that they are phase separation systems. Polystyrene and polydiene retain many properties of their homopolymers. For example, these block copolymers have two glass transition temperatures corresponding to their respective homopolymers, while random copolymers have only one glass transition temperature, as shown in Fig. 9.5. This indicates that polystyrene phase is strong and hard, while polydiene phase is soft and elastic in SBS and SIS block copolymers at room temperature.

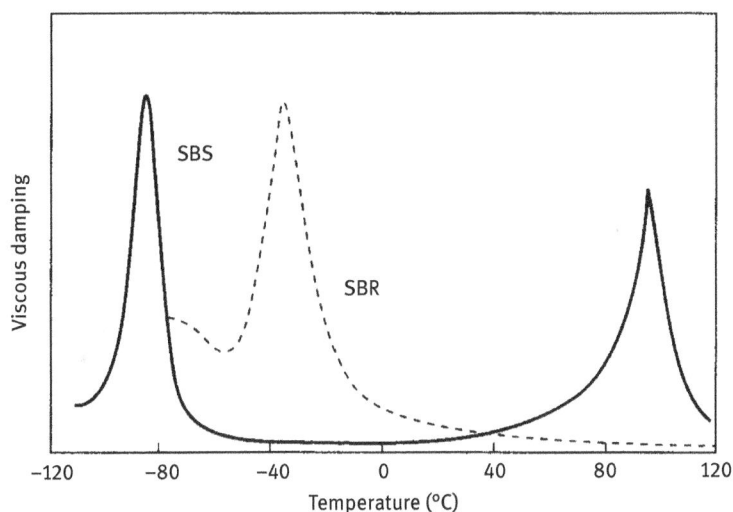

Fig. 9.5: Viscous damping comparison of block copolymer SBS and random copolymer SBR.

If the polystyrene phase only accounts for a small part of the total volume, it is reasonable to assume the phase structure shown in Fig. 9.6. In this phase structure, the polystyrene phase is a separated spherical region. Because the ends of each polydiene molecular chain are terminated by polystyrene segments, these hard polystyrene domains as physical joints become the cross-linked network structure of the whole system, which is similar to conventional vulcanized rubber in many aspects. However, this cross-linking is formed by physical rather than chemical processes, so it is unstable. At room temperature, this type of block copolymer has many properties of vulcanized rubber, but after heating, the polystyrene domain softens, the strength of the cross-linking network decreases, and finally the block copolymer can flow. When the heated block copolymer is cooled, the polystyrene phase domain hardens again and the original properties are restored. Similarly, the block copolymers can also be soluble in solvents (i.e., solvents that can simultaneously dissolve their homopolymers), and the same phase structure can be obtained when the solvents are volatilized.

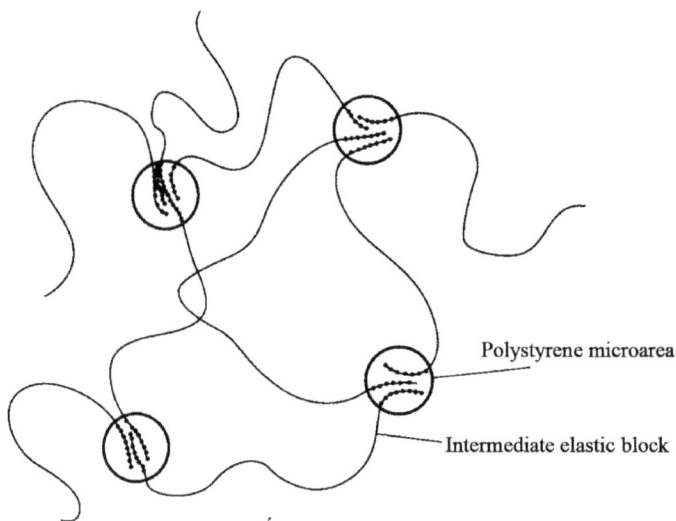

Polystyrene microarea

Intermediate elastic block

Fig. 9.6: Diagram of phase structure of block copolymer SBS.

The ideal phase structure shown in Fig. 9.6 is inferred from the mechanical properties and rheological behavior of SIS and SBS block copolymers, and the phase structure changes with the proportion of two blocks. However, at that time, the phase structure was only speculated, and was directly supported by the electron microscopy photos. This is because the size of polystyrene microarea cannot be observed in the visible range, which is why these block copolymers are transparent. Later, the development of osmium tetroxide dyeing technology enabled people to observe the micromorphology of triblock copolymers directly through electron microscopy, which proved that the above assumption was at least conceptually correct.

Later, a more detailed model of the morphology of block copolymers varying with composition was proposed, as shown in Fig. 9.7. With the increase of polystyrene content, the polystyrene phase changes from spherical to cylindrical. When the volume fraction of rubber phase and polystyrene phase is about the same, sandwich structure is formed. When the content of polystyrene continues to increase, polystyrene forms continuous phase, while rubber phase is dispersed in cylindrical or spherical shape. The shapes shown in Fig. 9.7 are supported by electron microscopic photographs. When the polystyrene content is about 50%, the block copolymer has regular sandwich structure. When the content of polystyrene is about 30%, the polystyrene phase domains are dispersed in the elastic continuous phase with a very regular hexagonal column structure. When the processing conditions are strictly controlled, the extruded bar has similar structure and the small angle X-ray scattering proves that it is almost a complete "single crystal." The morphology and

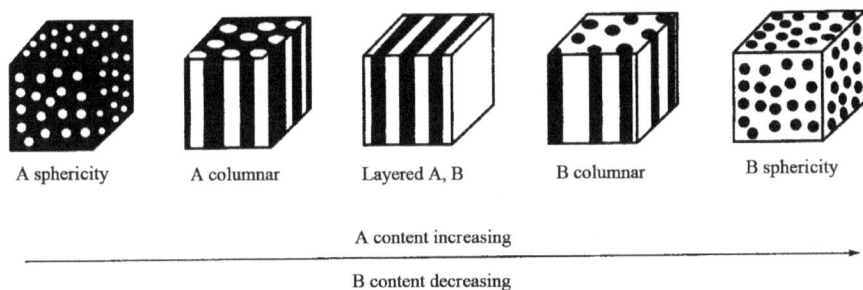

| A sphericity | A columnar | Layered A, B | B columnar | B sphericity |

A content increasing

B content decreasing

Fig. 9.7: The morphology of ABA block copolymer changes with composition.

structure of the extruded bar are regular and the phase domain of the column poly-styrene is infinite. In addition, it was found that block copolymers can be arranged more effectively into compact hexagonal columnar or sandwich structures under oscillating shear field.

Although the above explanations are for SIS and SBS block copolymers, they are also suitable for other block copolymers, such as $(SB)_n$ and $(SB)_nX$ (X denotes multifunctional junctions). This is because they can also form a network structure like that shown in Fig. 9.6. However, block copolymers such as SB and BSB cannot form this phase structure because only one end of each polydiene molecular chain is associated with the polystyrene chain. Because of this, the properties of SB, BSB and other block copolymers are completely different from those of conventional vulcanized rubber, and their strength is often very low.

9.3.2 Conditions for phase domain formation of styrene block copolymers

In block copolymers, if the mixed free energy G_m is negative, the two blocks are completely compatible. The free energy G_m can be expressed as follows:

$$\Delta G_m \; \Delta H_m - T\Delta S_m \tag{9.1}$$

In the formula, ΔH_m and $T\Delta S_m$ are the enthalpy of mixing and the entropy of mixing, respectively, and T is the absolute temperature. Therefore, the conditions for the formation of phase domains are as follows:

$$\Delta H_m > T\Delta S_m \tag{9.2}$$

For hydrocarbon polymers, because there is no strong interaction group, the ΔH_m value is usually positive, and the greater the difference between the two blocks, the higher the ΔH_m value. $T\Delta S_m$ is always positive, but when the relative molecular mass of the block is large, $T\Delta S_m$ tends to zero. Therefore, it can be predicted that the following factors will contribute to the microphase separation (formation of phase domains):

1. The larger the structural difference between the two blocks, the better the microphase separation.
2. The higher the relative molecular mass of the block, the better the microphase separation.
3. The lower the temperature, the better the microphase separation.

The theory of microphase separation is quite mature and can be quantified. In the experimental aspect, the influence of the difference of two block structures on phase separation is studied. It is found that the larger the difference of structure is, the more obvious the phase separation phenomenon is. For example, E-EB-E block copolymers are not separated in melting state, while S-EB-S block copolymers have serious phase separation. In addition, the effects of molecular weight and temperature on phase separation were studied. For example, when the relative molecular mass of the polystyrene block at the end of SBS is 7,000, both steady-state and dynamic viscosity measurements show that the critical temperature for phase separation of the polymer is 150 °C; when the relative molecular mass of the polystyrene block at the end of SBS is 8,000, it shows Newtonian flow behavior at 160 °C, so it can be considered that there is no phase separation at this temperature. When the relative molecular ester content of the terminal polystyrene block is more than 10,000, it is still phase separated above 200 °C. The above experiments are in good agreement with the theoretical calculations.

9.3.3 Properties of styrene block copolymers

9.3.3.1 Mechanical property

From the application point of view, the most interesting aspect of styrene triblock copolymers is the elasticity of rubber at room temperature, that is, its properties are like those of vulcanized rubber. Early studies have proved this point. The stress–strain behavior of SBS was compared with vulcanized natural rubber and vulcanized styrene-butadiene rubber (SBR) (see Fig. 9.8). The tensile strength of SBS and similar block copolymers can reach up to 30 MPa and the elongation can reach 800%. These indexes (especially the tensile strength) are superior to those of vulcanized SBR or *cis*-butadiene rubber.

In order to explain this abnormal phenomenon, two mechanisms have been proposed. The first mechanism assumes that hard polystyrene domains play a similar role to conventional vulcanized rubber reinforcing fillers such as carbon black. This mechanism is supported by the fact that the size of polystyrene phase domain (about 3×10^{-8} m) is the same as that of typical reinforcing filler particles, which can be well dispersed in the rubber phase and firmly bonded with the rubber

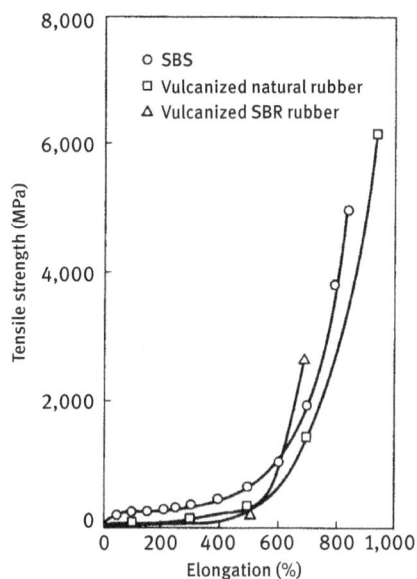

Fig. 9.8: Comparison of stress–strain properties of SBS, vulcanized natural rubber and vulcanized SBR rubber.

phase. The second mechanism is that the entanglement of elastic chains results in an increase in strength. Experiments show that the tensile modulus and tensile strength of SBS and SIS are independent of the molecular weight of polystyrene block if the molecular weight of polystyrene block is high enough to form a strong and phase-separated phase domain when the content of polystyrene is constant.

Another interesting aspect of styrene triblock copolymer is that their modulus of elasticity is unusually high and does not vary with molecular weight. Some people attribute this phenomenon to the intertwining between elastic blocks, which plays a cross-linking role. If so, the molecular weight between the entanglement points is an important parameter. For styrene thermoplastic elastomers, the following semi-quantitative formulas can be obtained by dealing with the effects of rubber elasticity and filler on elastic modulus:

$$f = (\rho RT/Mc + 2C_2/\lambda)\ (\lambda - 1/\lambda^{-2})\ (1\ +\ 2.5\varphi_s\ +\ 14.1\,\varphi_s{}^2) \qquad (9.3)$$

In the formula, f is the tensile stress acting on the sample; λ is the elongation ratio; ρ is the density of the sample; R is the gas constant; T is the absolute temperature; M_c is the molecular weight between the entanglement points in the elastic phase; C_2 is the constant deviating from the ideal elastic behavior; and φ_s is the volume fraction of the polystyrene phase domain.

According to the above formula, a straight line can be obtained when $f/(\lambda - 1/\lambda^{-2})$ $(1 + 2.5\varphi_s + 14.1\varphi_s{}^2)$ is used to plot $1/\lambda$. This is verified by SIS triblock copolymers (see Fig. 9.9), and M_c is calculated to be about 20,000 according to the intercept, which is similar to the molecular weight of the polyisoprene entanglement points measured by viscometry.

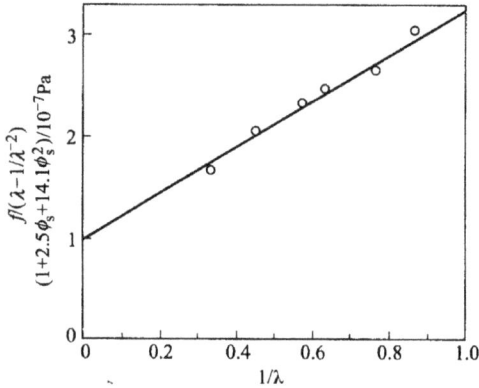

Fig. 9.9: Stress–strain behavior of SIS.

According to M_c, the elastic modulus and swelling degree of styrene triblock copolymers in solvents can be calculated. The results obtained are in good agreement with the experimental values. Table 9.3 lists the M_c values of polyisobutylene, polyisoprene, polybutadiene and ethylene/butene copolymers. According to the data in the table, it can be inferred that when the content of styrene is the same, the modulus of elasticity of block copolymers increases in the following order: SIBS, SIS, SBS and SEBS. However, the above four block copolymers, with clear structure, have not been used to validate this conclusion. However, the elastic modulus of SBS and SEBS block copolymers conforms to the above rules, and SIBS block copolymers are indeed very soft.

There are three possible mechanisms for the tensile failure of styrene block copolymers, which are still controversial.
1. Plastic failure of polystyrene phase domains;
2. Brittle failure of polystyrene phase domains;
3. Elastic failure of intermediate block phase in polydiene.

Tab. 9.3: Relative molecular mass M_c between entanglement points of different block copolymers.

Elastomer	M_c^1	M_c^2	Elastomer	M_c^1	M_c^2
Ethylene/butene copolymer	1,660	–	Polyisoprene	6,100	14,000
Polybutadiene	1,700	5,600	Polyisobutylene	8,900	15,200

[1]Measured by viscoelastic spectrometer; [2]measured by viscometry.

The first mechanism is supported by the fact that the tensile strength and modulus almost doubled when the end block of polystyrene was transformed into poly-α-methylstyrene. This indicates that the glass transition temperature of hard microregion is the controlling factor (the glass transition temperature of poly(α-methylstyrene) is 70 °C higher than that of polystyrene). However, the molecular weight, which must have a great influence on the plastic flow, has no effect on the tensile properties, which raises doubts about the plastic failure of polystyrene. In fact, there are two forms of plastic flow and brittle failure in polystyrene phase domains observed in electron microscopic photographs. Some people have measured the fracture point trajectories of SBS block copolymers. It is found that the transition occurs at about 40 °C, so it can be inferred that the damage occurs in polystyrene phase domains. The failure mode changes from brittle failure to plastic failure at about 40 °C. The experimental basis for failure occurring in elastic phase is that the tensile strength and modulus of SBS, SIS or SEBS triblock copolymers decrease when SB, SI or SEB triblock copolymers are added respectively.

In fact, the dominant failure mechanism depends on the experimental conditions. At high temperature, the polystyrene domain softens, and the polystyrene domain is the weak-cross-linking point in the molecular chain. At this time, plastic damage plays a dominant role; when the experiment lasts a long time, so does the case; at low temperature or for a short time, the other two failure mechanisms play a dominant role. However, when more diblock copolymers are added, the elastic phase transition of polydiene is weak, and it becomes the breaking point.

9.3.3.2 Viscosity and viscoelastic properties

A remarkable feature of SBS and SIS block copolymers is their abnormal melting viscosity. When the shear rate is low, the viscosity is much higher than that of random copolymers of polybutadiene, polyisoprene or styrene-butadiene with the same molecular weight. The viscosities of SBS and polybutadiene with molecular weight of 75,000 are compared in Fig. 9.10. In addition, the block copolymers exhibit non-Newtonian flow behavior, that is, the viscosity increases with the decrease of shear rate, and tends to infinite at zero shear rate. Figures 9.11 and 9.12 are the steady and dynamic viscosities of SBS triblock copolymers, respectively. The reason for this phenomenon is that the block copolymer maintains the two-phase structure shown in Fig. 9.6 in the melting state, and the flow occurs only when the end block of polystyrene is pulled out of the phase domain under the action of shear force. When the molecular weight of polystyrene block is larger than a certain critical value, phase separation occurs and polystyrene phase domain is formed. Therefore, even if the glass transition temperature of polystyrene is higher than that of polystyrene, some energy is needed to make the polystyrene block

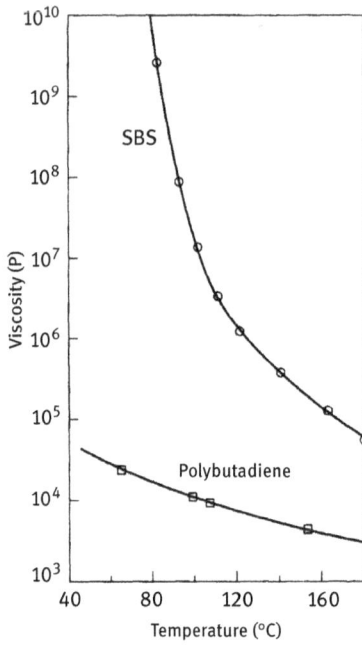

Fig. 9.10: Viscosity of SBS and polybutadiene at constant shear stress.

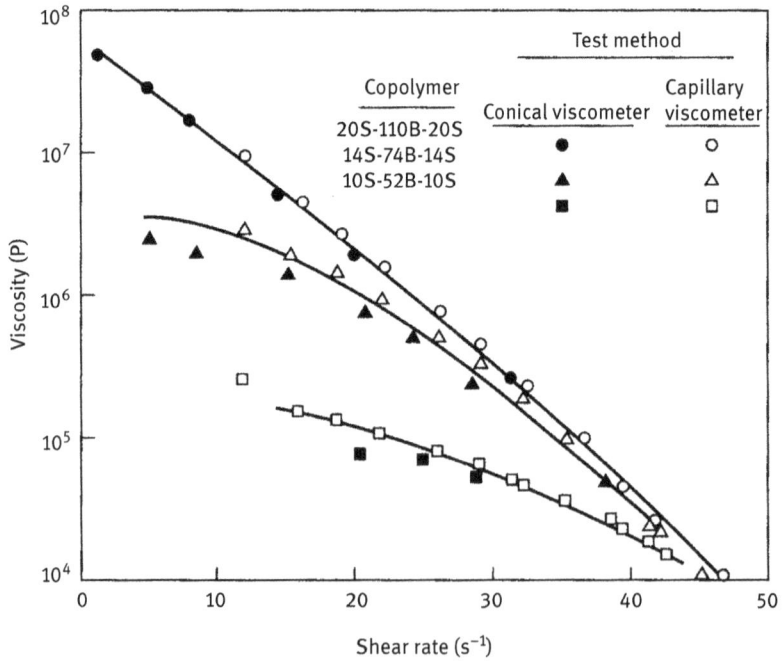

Fig. 9.11: Steady viscosity of SBS at 175 °C.

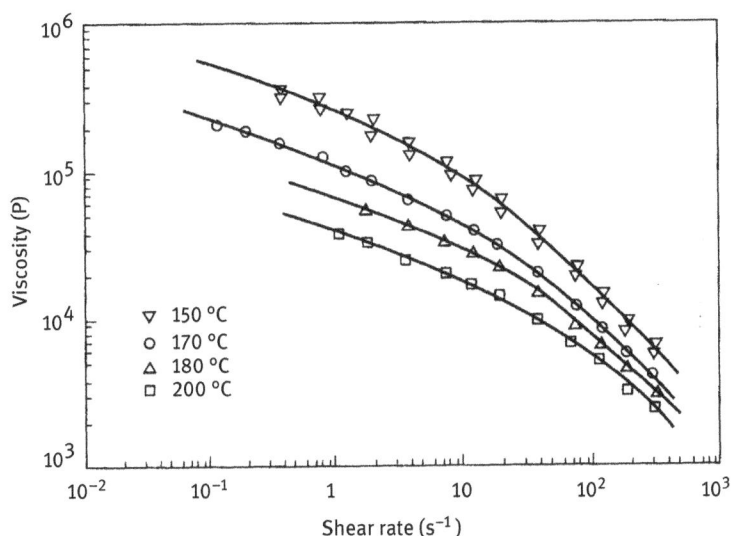

Fig. 9.12: Dynamic viscosity of SBS at different temperatures.

enter into the elastic phase (i.e., flow), which shows an increase in viscosity. Obviously, the required energy increases with the increase of the incompatibility between the middle block and the end block, then the viscosity also increases. For SEBS block copolymers, because of the poor compatibility between the two blocks, their viscosity is very high (non-Newtonian behavior is serious). On the contrary, when polyethylene is used instead of polystyrene as the end block, namely EEBE, because the end block and the middle block are mutually soluble, there is no phase separation in the melting state. It is for this reason that the melting viscosity of EEBE block copolymer is very low.

The dynamic mechanical properties of polystyrene triblock copolymers are also unique. For most polymers, the Williams-Landel-Ferry (WLF) method can be used to describe the dynamic mechanical properties, that is, to measure the dynamic mechanical properties within a certain frequency range at different temperatures. According to the temperature difference between the measured temperature and the reference temperature, the appropriate displacement factor is selected and the results are plotted. The reference temperature is related to the glass transition temperature of the polymer. However, as mentioned earlier, styrene triblock copolymers have two glass transition temperatures, so the above methods cannot be used directly. Several improvement methods have been put forward. One method is that the displacement factors of these block copolymers can be calculated by the reference temperature of polybutadiene at low temperatures, polystyrene at high temperatures and an "adjustable" intermediate glass transition temperature between two glass transition temperatures at intermediate temperatures. The WLF equation can be used to transform the viscoelastic behavior at different temperatures.

9.3.4 Other styrene block copolymers

In addition to the triblock copolymers with polystyrene block at the end and rubber block at the middle, many other types of styrene block copolymers have been developed. It has been reported in the literature that the total number of blocks of $(SI)_n$ linear multiblock copolymers can reach nine. The rheological properties of SBS, $(SB)_3X$, $(SB)_4X$, BSB and $(BS)_3X$ block copolymers with polystyrene content of 30% were also compared. When the total molecular weight is the same, the viscosity of the first three $(SI)_n$ lock copolymers (the end block is polystyrene) is much higher than that of the last two (the end block is polybutadiene), and the viscosity of the linear block copolymers SBS and BSB is much higher than that of the star block copolymers. However, if linear block copolymers are also considered as star-shaped structures, that is, $(SB)_2X$, $(BS)_2X$, then there is no difference in viscosity between linear block copolymers and star block copolymers when arm lengths are equal. The solution viscosity and melting fluidity of SBS, $(SB)_3X$, and $(SB)_4X$ with 25% styrene also have the same rule, that is, if the arm length is the same, the solution viscosity and melting fluidity of these polymers are similar.

As mentioned above, for SBS, $(SB)_3X$, $(SB)_4X$, BSB and $(BS)_3X$ block copolymers, when the total molecular weight is the same, the viscosity of the first three block copolymers (the end block is polystyrene) is much higher than that of the last two (the end block is polybutadiene), which indicates that for SBS or $(SB)_nX$, the flow is achieved by the destruction of dispersed polystyrene phase. Additional energy is required, so the viscosity is very high, and the magnitude of the added energy depends on the molecular weight of the styrene block, while polybutadiene has little effect. On the contrary, for BSB and $(BS)_nX$, the flow occurs in the continuous phase of polybutadiene without destroying the polystyrene domain, so the energy required is greatly reduced and the viscosity of macroscopic performance is low.

9.4 Structure and properties of TPPU

9.4.1 Morphological structure of TPPU

The morphology of the multiphase system plays a decisive role in the final properties of the product. By controlling the morphological changes, the desired properties can be obtained. Therefore, in order to understand the relationship between structure and properties, it is necessary to study the morphology of multiphase systems. However, it is very difficult to study the morphology and structure of TPPU copolymers because of a series of effect elements to properties such as crystallization, phase mixing, hydrogen bonding and thermal history.

In theory, phase separation is caused by the thermodynamic incompatibility between two phases. For a certain length of polyurethane block copolymer, with the increase of the number of blocks, the length of segments decreases, and phase separation becomes more and more difficult. On the other hand, for a block copolymer with certain component, phase separation becomes easier with the increase of total molecular weight and the number of blocks in each molecule. In addition, the soft and hard phases of TPPU can be amorphous or partially crystallized. Generally, when one block of block copolymers is a crystalline polymer, the degree of phase separation is higher.

9.4.1.1 Effect of hard segment structure on morphological structure of TPPU

In order to elucidate the nature of hard-phase microzone in TPPU, a lot of research work has found that the hard segment formed by the reaction of linear diol and MDI is crystallizable, but wide-angle X-ray diffraction studies show that crystallization seems to be inhibited under normal conditions. That is to say, the ordered state of the hard-phase microzone of TPPU is semicrystalline. After heat treatment of TPPU, the hard microregion can be changed from semicrystalline to crystalline state. It is considered that in order to obtain higher crystallinity of hard segment, it is necessary to anneal the TPPU at high temperature for a long time (190 °C, 12 h).

Blackwell et al. studied the structure of MDI-diol-polybutylene adipate polyurethane hard segment by small angle X-ray scattering. They used BDO, propylene glycol (1,3-propane diol, PDO) and ethylene glycol (1,2-ethane diol, EDO) as chain extenders and found that the crystallinity of the hard segment of MDI-BDO was the best. According to conformation analysis and model compound study, it is found that the hard segment of MDI-BDO with fully extended chain conformation can form a hydrogen bond network on the plane perpendicular to the axis of the molecular chain. The hard segments of MDI-PDO and MDI-EDO crystallize in folded chain conformation. Eisenbach and Guenther strictly controlled the reaction conditions, synthesized and analyzed a series of soft and hard blocks with clear molecular structure, which confirmed the above findings. Small-angle X-ray scattering analysis shows that the oligomers formed by the reaction of MDI and BDO crystallize in an extended chain rather than in a folded chain. Recently, polycrystalline hard segments have been found in MDI-diol polyurethane elastomers. The crystallization of MDI-polyester polyurethane was studied by changing the chain length of chain extender [from ethylene glycol EDO to hexanediol (HDO)]. Experiments show that polycrystalline polyurethane is obtained by using HDO and BDO as chain extenders. Differential scanning calorimetry (DSC) tests also confirm the above conclusions. The appearance of polycrystalline phenomena in DSC spectra results in several melting point peaks. The crystallinity of TPPU obtained with HDO as chain extender is not high when melt pressed, but it can reach a high crystallinity after stretching or annealing. On the contrary, there is no polycrystalline phenomena in the TPPU obtained by using EDO as chain extender. Based on the small-angle X-ray scattering analysis of single crystals of MDI-methanol dicarbamate model compounds, a molecular chain alignment model in the hard microregion of MDI-BDO polyurethane was proposed, as shown in Fig. 9.13.

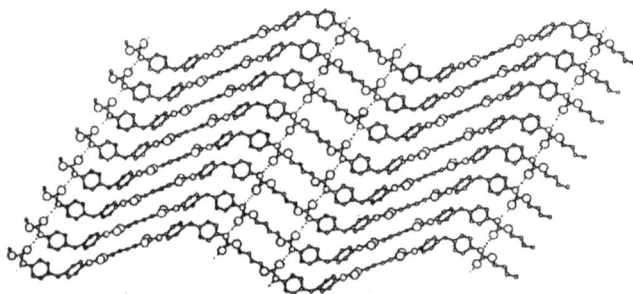

Fig. 9.13: MDI-BDO hard segment molecular chain arrangement model.

Similar studies have been carried out on the model compounds prepared by the reaction of 4-isocyanate-diphenylmethane with $HO(CH_2)_nOH$ ($n = 2$–6). The small-angle X-ray scattering analysis of these model compounds shows that the hydrogen bonds between adjacent molecules are relatively stable when n is even, while the stability of physical cross-linking systems is poor when n is odd.

9.4.1.2 Thermal transition behavior of TPPU

DSC is a common method to characterize the phase separation, glass transition, crystallization and melting of materials. Figure 9.14 is a thermal transition process of TPPU determined by DSC. The samples used are all made from polyester diols, BDO and MDI, but their ratios are different (see Tab. 9.4).

As shown in Fig. 9.14, with the increase of hard segment content in polymer, the glass transition zone widens and moves toward high temperature. This movement can be explained by the increase of hard-phase microregion in the system, and is also related to the length of hard-phase microregion. It can be seen from Tab. 9.4 that the increase of chain extender dosage not only increases the content of hard-phase microregion, but also increases the length of hard-phase microregion. The length of hard-phase microregion is the upper limit of the size of hard segment crystallization in the direction of molecular chain. Therefore, it can be said that the length of hard-phase microregion determines the melting point and stability of TPPU. It can also be seen from Fig. 9.14 that with the increase of hard-phase microregion content, the temperature region of hard segment crystallization melting

Endothermic

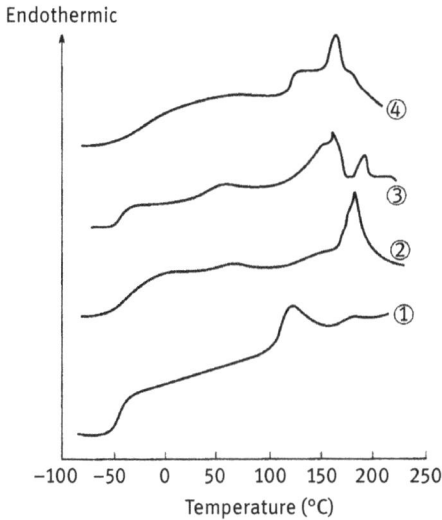

Fig. 9.14: DSC curves of thermoplastic block polyurethanes with different hard segment contents.

Tab. 9.4: Thermoplastic block polyurethane samples with different hard segment contents.

Sample number	Component			Hard segment content (%)	Average length of hard segment (μm)
	Polyester[1]	BDO	MDI		
1	1	1.77	2.8	31	5.0
2	1	3.55	4.6	40	8.6
3	1	5.55	6.6	52	12.7
4	1	10.0	11	65	21.7

[1]Poly(ethylene adipate glycol and butylene glycol mixed ester)diol, $\overline{M}^n$ = 2,000.

(i.e., the maximum value in DSC curve) moves toward high temperature, reaching the maximum at about 190 °C. Moreover, the melting temperature range of crystallization is widened, indicating that the thickness of hard-phase microregion also has a wider distribution.

In addition, the size distribution of the crystalline region produced by TPPU during melt cooling is not only related to the length distribution of the hard segment, but also to the crystallization kinetics. The curves in Fig. 9.15 are DSC curves of TPPU prepared from a mixture of MDI, BDO, polyethylene adipate and butylene glycol (molar ratio 6:5:1). There are many maxima in the curve, especially the endothermic peak at 205 °C. The samples were heat treated at 118, 135, 180 and 205 °C

Endothermic

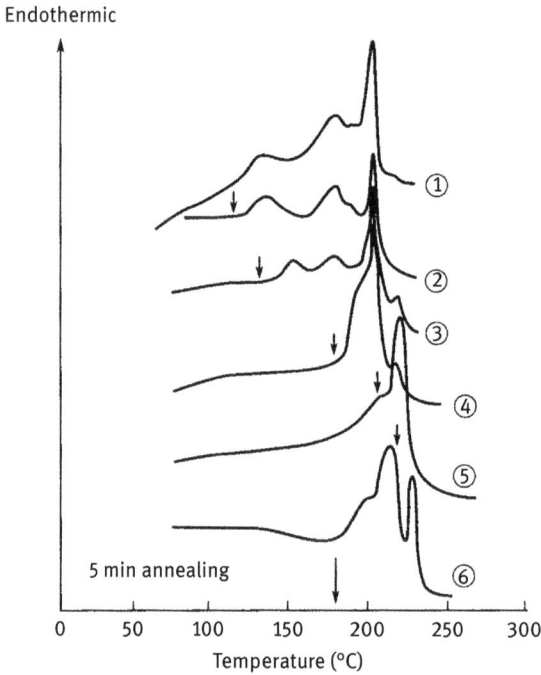

Fig. 9.15: Effect of annealing treatment on thermal transition of thermoplastic block polyurethane (DSC).

respectively for 5 min (curves ②–⑤), and the maximum endothermic peak tended to 230 °C.

It can also be seen from Fig. 9.15 that the range of melting temperature gradually narrows with the increase of heat-treatment temperature, which is due to the formation of crystalline zones with higher order and larger volume when partial crystallization is recrystallized after melting. The specific process can be described as follows: during rapid cooling, the hard segment crystallizes according to its arrangement in the melt, and the thermodynamic effective thickness (d_1) of the crystallization is smaller, while the thickness of the phase interface zone is larger (see Fig. 9.16a); during annealing, the structure is rearranged to obtain thermodynamic favorable arrangement, and the thickness of the crystal zone (d_2) increases (Fig. 9.16b), which ultimately leads to the increase of the melting point. Of course, the improvement of radial crystallization order cannot be ruled out. When the annealing temperature exceeds 21 °C, the melting temperature decreases (Fig. 9.15, curve ⑥). This is because most of the original crystals have been melted, and the hard segment cannot form new crystals in a relatively short period of time. Only after cooling, the crystals with poor orderliness and low melting point can be formed.

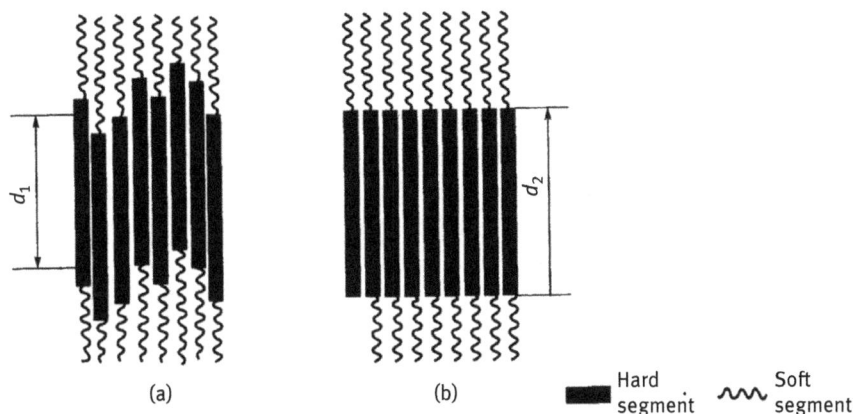

Fig. 9.16: Hard segment arrangement diagram: (a) before annealing and (b) after annealing.

By thermal analysis and small-angle X-ray scattering analysis, it is considered that the interphase mixing is related to the ability of hydrogen bond formation in the soft segment. For binary alcohols with the same molecular weight ($M = 1,000$), the degree of phase separation of polyether-MDI-BDO (molar ratio = 1:2:1) polymer is higher than that of polyester-MDI-BDO (molar ratio = 1:2:1). The latter has smaller grain size and lower crystallinity, resulting in better transparency. On the DSC curves of these two polymers, there are obvious changes at 60–80 °C, which can be attributed to the destruction of hydrogen bonds between carbamate-polyester and carbamate-polyether. Infrared studies show that the dissociation of hydrogen bonds is controlled by the fluidity of hard molecular chains (that is to say, T_g) and hydrogen bonds play a role in increasing the cohesion in hard microregions. In hard microregions, hydrogen bonds determine the arrangement of the groups of hydroformic acid esters. However, some studies have shown that this transformation may not be caused by the dissociation of hydrogen bonds, but by the glass transition of hard phase and the plasticizing effect of soft segment on hard phase. Therefore, the role of hydrogen bonds in TPPU should not be overemphasized, for example, in polyether ester elastomers, there is no hydrogen bond. Witsiepe synthesized polyether ester elastomer without hydrogen bond by condensation reaction of short-chain diols with polytetrahydrofuran polyether diols or other long-chain polyether diols with piperazine. This elastomer has no hydrogen bond formed by carbamate, but its physical properties are still comparable to those of TPPU. In addition, the thermal stability of the thermoplastic elastomer is better than that of the TPPU, and the processing conditions of the elastomer are also wider.

Schneider has studied extensively the polyurethane prepared by TDI. TPPU was synthesized from 2,4- or 2,6-TDI isomers. The structure and thermal transformation of the obtained polyurethane were studied by thermomechanical analysis, DSC, small-angle X-ray scattering and infrared spectroscopy. The results show that the degree of phase separation strongly depends on the type of TDI and the molecular weight of soft segments. The T_g of polyurethane synthesized from 2,4-TDI and $M = 1,000$ polyether is closely related to its composition, indicating the good mixing between soft and hard phases. On the contrary, when 2,4-TDI is replaced by 2,6-TDI, the T_g of the polyurethane is independent of its composition, and the transition temperature is higher (melting in the hard zone of crystallization), which indicates that the structure of the polyurethane is orderly. For polyurethane prepared by 2,4-TDI, phase separation occurs when the relative molecular weight of polyether increases from 1,000 to 2,000, while for polyurethane prepared by 2,6-TDI, the degree of phase separation is improved.

In a word, T_g of soft phase is a very sensitive index to characterize the degree of phase separation. From the application point of view, with the increase of hard segment content, the hardness of TPPU also increases, and the degree of mixing between the two phases also increases. At this time, the low-temperature flexibility will decrease correspondingly. This deficiency can be compensated by using high-molecular-weight soft segment or preextended chain soft segment.

9.4.2 Properties of TPPU

TPPU is the first polymer material with both rubber elasticity and thermoplasticity. Moreover, the properties of TPPU have great variability, which provides great convenience for practical use. For example, Young's modulus can change from 8 to 2,000 MPa with the change of hard segment content in block polyurethane. The rigidity of TPPU can also be improved by adding organic or inorganic additives, especially glass fibers.

9.4.2.1 Mechanical properties of TPPU

TPPU has excellent mechanical properties, such as high tensile strength and elongation (see Fig. 9.17). The chemical structure and Shaw hardness of TPPUs are different, and their tensile strength varies from 25 to 70 MPa. The tensile strength of soft TPPU (Shao's A is 70–85) is lower, while that of hard TPPU (Shao's D is 50–83) is higher. TPPU has flexibility in a wide temperature range, and the soft segments in the molecular chain determine its low-temperature properties (see Tab. 9.5). In addition, the TPPU does not contain any plasticizer and can be used as laminating material with other materials such as ABS and polycarbonate.

One of the main advantages of TPPU is its excellent wear resistance. Many commercial TPPUs are polyester and polyether ester type. The wear resistance, tear resistance and tensile strength of these two types of TPPUs are better than those of polyether-type TPPUs. Therefore, polyester, polyether ester TPPU is often used to manufacture soles and cable sheaths.

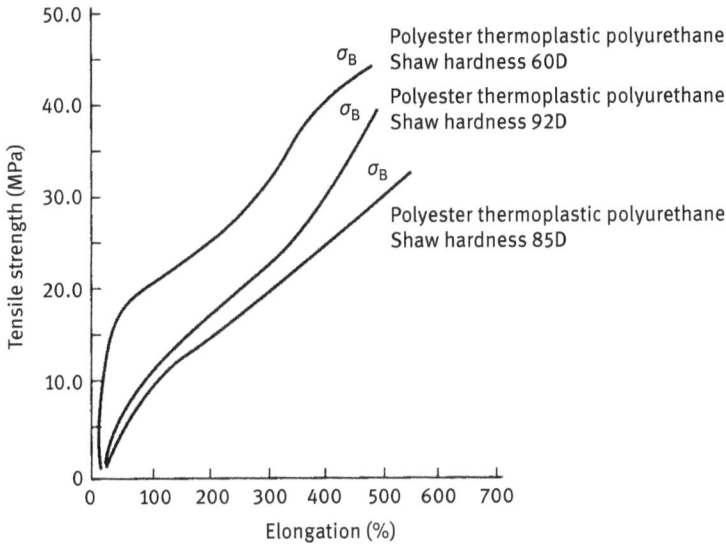

Fig. 9.17: Stress–strain curves of three thermoplastic block polyurethanes with different Young's modulus.

Tab. 9.5: Performance comparison of various types of thermoplastic block polyurethane.

Performance	Polyester	Polyether	Polyether ester
Tensile strength	++	−	+
Wear resistance	++	−	+
Swelling property in oil, fat and water	+	−	+
Weatherability	+	−	+
Crack propagation resistance	+	−	+
Tear strength	++	−	+
Resilience	++ ~ +	++ ~ +	+
Microbial and bacterial degradability	+ ~ −	++	++
Low-temperature impact resistance	+ ~ −	++ ~ +	+
Hydrolysis resistance	−	+	+

Note: ++, excellent; +, good; −, poor.

The swelling of polyester TPPU in oil, grease and water is very small (see Tab. 9.5). Polyether-type TPPU is recommended for high requirements of hydrolysis resistance, microbial degradation resistance, low-temperature resistance and flexibility. In contrast, polyether ester TPPU has more excellent properties. It also has the properties of polyether and polyester TPPU. It can be used to prepare fire water pipes, cable sheaths and films.

9.4.2.2 Thermal properties of TPPU

TPPU has wide using range of temperatures. Most of the products can be used for a long time in the range of −40 to 80 °C. The short-term using temperature can reach 120 °C. The polyurethane prepared from piperazine can even withstand much higher temperatures. At high temperatures, the properties are maintained mainly by hard segments, and the higher the hardness of the product (i.e., the higher the content of diisocyanate and chain extender), the higher the temperature of use. In addition, the high-temperature properties are affected not only by the amount of chain extender, but also by the type of chain extender. For example, using temperature of TPPU with 1,4-di (hydroxyethoxy)benzene as chain extender is higher than that of TPPU with 1,4-butanediol or 1,6-hexanediol as chain extender.

The type of diisocyanate also affects the high-temperature properties of TPPU. As shown in Tab. 9.6, different diisocyanates and chain extenders show different melting points as hard segments of TPPU.

Tab. 9.6: Effect of diisocyanate and chain extender on melting point of thermoplastic block polyurethane.

Diisocyanate	Chain extender	Melting point (°C)	Diisocyanate	Chain extender	Melting point (°C)
2,4-TDI	1,4-Butanediol	209	MDI	1,4-Butanediol	194
2,4-TDI	1,6-Hexanediol	198	MDI	1,6-Hexanediol	185
2,6-TDI	1,4-Butanediol	215	HDI	1,4-Butanediol	182
2,6-TDI	1,6-Hexanediol	202	HDI	1,6-Hexanediol	176

Note: Polyester is hexanediol adipate with molecular weight of 2,000; polyester:diisocyanate: chain extender molar ratio is 1:2:1.

9.4.2.3 Hydrolysis stability of TPPU

At room temperature, TPPU can be used in pure water for several years, and its performance does not change significantly. However, the mechanical properties will change dramatically even if it is immersed in water for only a few weeks to several months at 80 °C (see Fig. 9.18). It can be seen from the figure that the hydrolysis stability of TPPU is related to the structure of soft segment. Polyether ester TPPU and polyether TPPU have excellent hydrolysis resistance at high temperature, while polyester TPPU has poor hydrolysis resistance. The hydrolysis resistance of TPPU prepared by modification of polyester diol with carbodiimide is improved.

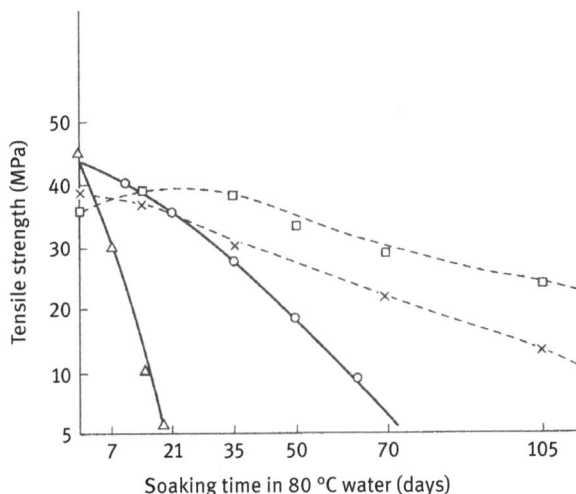

Fig. 9.18: Effect of soaking time of different thermoplastic block polyurethanes in water at 80 °C on tensile strength.
Δ: Polyester type; o: polyester (modified by carbodiimide); ×: tetrahydrofuran polyether; □: mixed polyethylene oxide polyether-polycarbonate.

With the increase of hardness of TPPU, the hydrolysis stability of hard segment becomes better and better because of its hydrophobicity.

TPPU is sensitive to acid and alkali. At room temperature, TPPU can be slowly eroded by dilute acid or alkali. At high temperature, TPPU is not resistant to concentrated acid or alkali. Acids also have different effects on different polyurethanes. Therefore, for each specific use, it is better to undertake immersion experiments under simulated conditions.

9.4.2.4 Oil and solvent resistance of TPPU

Nonpolar solvents such as hexane, heptane and paraffin oil have little effect on the polar TPPU. Even at high temperature, the swelling of polyurethane in nonpolar solvents is very small.

TPPU has excellent resistance to mineral oil, diesel oil and lubricating oil, but for some industrial oils and lubricating oils, it may corrode TPPU at high temperature because of its additives.

TPPU can swell seriously in chlorinated hydrocarbons or aromatic hydrocarbons (such as toluene), and the degree of swelling depends on the structure of polyurethane. The swelling of polyester is smaller than that of polyether, and that of hard is smaller than that of soft.

Some polar solvents, such as tetrahydrofuran, acetone, butanone and dimethylformamide, can partially or completely dissolve thermoplastic block polychloridase. For example, soft TPPU can be dissolved in butanone/acetone mixed solvent and used as an adhesive, while hard TPPU can be used as a surface finishing agent for textiles and leather when dissolved in ketone.

9.4.2.5 Microbial resistance and ultraviolet resistance of TPPU

Soft polyester TPPU can be eroded by microorganisms when it is in contact with moist soil for a long time. Soft or rigid polyether TPPU and polyether ester TPPU is usually not subject to microbial erosion.

Bacteria may discolor the surface of TPPU products, but in the short term, they generally do not damage their mechanical properties.

The TPPU prepared from aromatic isocyanates (such as 2,4-TDI, 2,6-TDI and NDI) yellows under ultraviolet light, but has little effect on the mechanical properties. The yellowing phenomenon can be alleviated by adding some ultraviolet absorbents.

9.4.3 Blending of TPPU with other polymers

TPPU has good blending properties with many other macromolecule materials. Apart from nonpolar resins such as polyethylene and polypropylene, polymers with processing temperature below 280 °C can be blended with thermoplastic front-end polyurethane in various proportions. According to the proportion of TPPU in the blends, the blends can be divided into three types.

1. Blends of TPPU as secondary component

In this case, the amount of TPPU is generally 10%–30%, which mainly acts as modifier. For example, soft TPPU can be used to improve the impact resistance of high modulus plastics (such as unsaturated polyester, epoxy resin, polyformaldehyde and polybutylene terephthalate). The impact strength and low-temperature flexibility of the plastics are improved by adding a small amount of TPPU, but other properties of the plastics are not damaged. TPPU can also be used as nonvolatile long-acting plasticizer for PVC and can be blended with PVC in any proportion.

2. Blends of TPPU and other resins mixed in the same proportion

When TPPU and other thermoplastic plastics are blended in the same proportion, new mechanical properties of the blends are usually obtained. For example, the modulus of TPPU blends with the same amount of polycarbonate increases, and the blends have excellent processing properties, which can be used in the automotive industry. TPPU can be blended with ABS at any ratio. Increasing the proportion of ABS is beneficial to the modulus of the blend, but the wear resistance and tear strength decrease. As ABS is cheaper, the blend has a price advantage. When styrene–maleic anhydride copolymer or styrene–maleimide copolymer is blended with TPPU, the impact strength is significantly improved, and the Vicat softening temperature is still very high. TPPU can also be blended with the following polymers: styrene–methacrylic acid–diene copolymer, styrene–acrylonitrile copolymer or styrene–methacrylate copolymer, styrene–butadiene–acrylonitrile copolymer and styrene–butadiene–methyl acrylate copolymer.

Examples of TPPU ternary blends are TPPU–polycarbonate–ABS ternary blends. The blend has good processability and fuel oil resistance, and the TPPU–polycarbonate–polybutylene terephthalate ternary blend has excellent stress-cracking resistance in solvents.

3. TPPU blends as major components
It has been patented to use ABS as impact modifier of high modulus TPPU and compatibilizer of polyether TPPU. These blends have been commercialized.

Some acrylate polymers can be used as processing aids for thermoplastic polyurethane. For example, ethylene–acrylate copolymers can be used to improve the processing properties of TPPU in blow molding, in which ionic groups play the role of thermoplastic polyurethane and polyethylene compatibilizer.

In addition to blending with other polymers, different types of TPPUs are often blended with each other. For example, the blend of soft TPPU and hard thermoplastic polyurethane can produce TPPU with medium hardness and good processability. The blends of polyester and polyether TPPU have some special properties, such as good wear resistance and excellent impact resistance at low temperature. Blending TPPU with different melting index and hardness can significantly improve the fluidity and demolding of polyurethane in blow molding process.

Exercises

1. What is active anionic polymerization? What are its characteristics?

Answer:
Anionic polymerization is a kind of ionic polymerization. In this kind of reaction, the substituents of vinyl monomers are electron absorbing, which makes the double bonds have a certain positive charge and electrophilic property.

Characteristics:
(1) Polymerization reaction rate is very fast.
(2) Monomers have strong selectivity for initiators.
(3) Chain-free termination reaction.
(4) The molecular weight distribution is very narrow.

2. What are the basic components of polyurethane?

Answer:
(1) Polyester or polyether high-molecular-weight diols
(2) Chain extenders, such as water, short-chain diols or diamines
(3) Diisocyanates such as NDI

3. What are the common preparation methods of triblock copolymers?

Answer:
(1) Sequential polymerization: The first monomer is initiated by a single-active initiator, and then the other monomers are added to the first monomer for sequential polymerization.
(2) Coupling method: The first monomer and the second monomer are polymerized by a single active initiator, and then the active chain is linked by a coupling agent.
(3) Multifunctional initiator method: The first monomer is initiated by initiator with two or more active centers, and then the second monomer is added to continue the polymerization.

4. How to prepare star block copolymers by active anion polymerization?

Answer:
There are two ways to prepare star-shaped copolymers by active anionic polymerization: one is to couple block anions with multifunctional electrophilic reagents, and the other is to add dimeric functional monomers, such as DVB.

5. Discuss the blending of TPPU with other polymers.

Answer:
1. Blends of TPPU as secondary component
In this case, the amount of TPPU is generally 10–30%, which mainly acts as modifier. The impact strength and low-temperature flexibility of the plastics are improved by adding a small amount of TPPU, but other properties of the plastics are not damaged. TPPU can also be used as nonvolatile long-acting plasticizer for PVC and can be blended with PVC in any proportion.

2. Blends of TPPU and other resins mixed in the same proportion

When TPPU and other thermoplastic plastics are blended in the same proportion, new mechanical properties of the blends are usually obtained. For example, the modulus of TPPU blends with the same amount of polycarbonate increases, and the blends have excellent processing properties, which can be used in the automotive industry.

3. TPPU blends as major components

It has been patented to use ABS as impact modifier of high modulus TPPU and compatibilizer of polyether TPPU. These blends have been commercialized.

Chapter 10
Interpenetrating polymer networks

10.1 Summary

Interpenetrating polymer networks (IPN) are systems in which two or more polymers form cross-linked networks through chemical bonds, while two cross-links penetrate each other and cannot be separated by simple physical methods. For example, a typical IPN structure is formed by the interpenetration of epoxy (EP) resin cross-linking system and polyurethane (PU) cross-linking system.

In recent years, the definition of IPN has been further expanded. It is believed that the structure of IPN is not only the interpenetration of chemical networks, but also a multiphase system of complex interpenetrating networks, that is, the interpenetrating structure of physical cross-linking networks composed of macromolecular aggregates is contained in the structure of IPN. The physical cross-linking includes block polymer, hard segment (glass region) of chain in graft copolymer and small crystalline region in semicrystalline polymer. This network polymer with physical cross-linking is also called IPN-like.

In conclusion, the structure of IPN should be based on the different compatibility of polymers, the heterogeneity of polymer blends and the cross-linking of polymer macromolecules. Therefore, IPN is more precisely defined as a network formed by physical and chemical cross-linking of two or more polymers, each network segment interacts with each other to form a multiphase system with different compatible microstructure.

IPN is a new type of composite polymer material developed in the past three or four decades. It is a new field of polymer design and blending modification technology, which opens a new way to prepare polymer materials with special properties. It plays an important role in the theory and practice of polymer modification.

There are usually no chemical bonds between the components of IPN, which is different from the graft copolymers or block copolymers. At the same time, there are cross-linking networks (including chemical cross-linking and physical cross-linking) between the components of IPN, which are different from the physical blend polymers. Figure 10.1 is a schematic diagram of several simple two-component polymer systems, showing the differences and connections between IPN and physical blends, graft copolymers, block copolymers and cross-linking copolymers.

In Fig. 10.1, the structure (a) between the two polymers without any chemical bonds is called a polymer mixture. Structure (b) is a graft copolymer, indicating that side chain polymer II is grafted on the side of polymer I. The position of grafting is often irregular and not limited to one side of the chain. The difference between block copolymers and graft copolymers is that the two chains are head-to-tail. For example, the structure (c) represents a triblock copolymer, which is the basis of thermoplastic

https://doi.org/10.1515/9783110596335-010

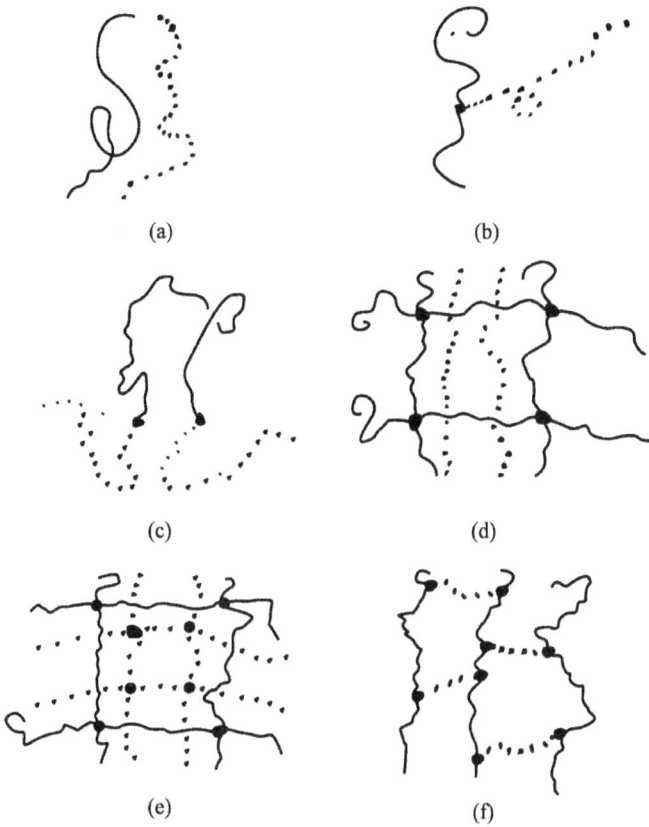

Fig. 10.1: Schematic diagrams of several simple two-component polymer systems.
(a) Polymer blends; (b) graft copolymers; (c) triblock copolymers;
(d) semi-IPN; (e) all-IPN; and (f) cross-linked polymers.

elastomers. Semi-IPN (d) is called when only one polymer is cross-linked and the other polymer is linear and intertwined in the cross-linked polymer. If both polymers are cross-linked and intertwined, the full IPN shown in Fig. 10.1(e) is obtained. Structure (f) is a cross-linked polymer. The two polymers together form a network structure.

10.2 History of IPNs

As with many other aspects of science and engineering, it is difficult to determine the exact time at which IPN concepts originated. Table 10.1 begins with Goodyear's work on vulcanization or cross-linking of rubber, recording the development of IPN, polymer blends, grafted polymers and block polymers.

Tab. 10.1: The historical development of IPN and related materials. IPN, interpenetrating polymer network.

Event	First researcher	Year
Rubber vulcanization	Goodyear	1844
IPN-type structure	Aylsworth	1914
Interpretation of polymer structure	Staudinger	1920
Graft copolymer	Ostromislensky	1927
IPNs	Staudinger and Hutchinson	1951
Block copolymer	Dunn and Melville	1952
HIPS and ABS	Amos, MeCurdy and McIntire	1954
Block copolymer surfactant	Lunsted	1954
Homogeneity IPN, proposing the concept of IPN	Millar	1960
Thermoplastic elastomer	Holden and Milkovich	1966
AB cross-linked copolymer	Bamford, Dyson and Eastmond	1967
Sequential IPN	Sperling and Friedman	1969
Latex IEN	Frisch, Klempner and Frisch	1969
Simultaneous IPN	Sperling and Arnts	1971
IPN naming	Sperling	1974
Thermoplastic IPN	Davison and Gergen	1977

Goodyear put forward the idea of rubber vulcanization in 1844, and began the manufacturing process of natural polymers. The first one to make IPN structure was Aylsworth. In 1914, he prepared a mixture of natural rubber, sulfur and partially reactive phenol–formaldehyde resin, which formed IPN structure when cured. However, the term "polymer" was not used in the patent at that time, because there was no concept of polymer at that time. Until H. Staudinger proposed polymer in 1920, the chain characteristics of polymer materials were unknown, and it was almost impossible to systematically study the polymer topology. Over the next decade, however, Ostromislensky's patents clearly demonstrated an understanding of the structure of the graft copolymer. During the 1940s and 1950s, great progress was made in polymer blends, graft copolymers and block copolymers. At the initiative of Amos and others, polymer blends and graft copolymers were used as rubber toughening plastics, including high-impact PS and acrylonitrile (AN)–butadiene–styrene (St) plastics. Table 10.1 further illustrates that block copolymers become important because Lunted uses block copolymers containing water-soluble and oil-soluble blocks as surfactants. Block copolymers consisting of elastomers and thermoplastic polymers can be used as thermoplastic elastomers.

J. J. P. Staudinger, son of H. Staudinger, began his research work during this period. He applied for his first IPN patent in 1941. Ten years later, the patent was applied in the manufacture of plastics with smooth and transparent surfaces. The name "interpenetrating polymer networks" was first proposed by J. R. Millar in 1960. Inspired by small molecular locks, Millar first studied the homogeneous IPN structure of polystyrene (PS), and made ion-exchange resin with IPN structure. Figure 10.2 is a schematic diagram of the structure of small molecular locks proposed by Millar.

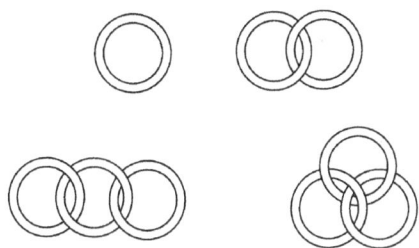

Fig. 10.2: Millar's small molecular locks (unpublished).

Since the 1960s and 1970s, H. Frisch and H. L. Sperling have made important academic contributions to the synthesis, phase structure and properties of IPN.

Frisch's team consists of Harry Frisch at New York University, Albany, and Kurt Frisch brothers at Detroit University. Daniel Klempner was a student of H. Frisch at the time, and later worked at the University of Detroit and became an experienced team member. H. Frisch has long been interested in locking rings. He imagines IPN as a macromolecule similar to locking rings, and based on this, he has successfully prepared many IPN products. K. Frisch is one of the most famous PU scientists in the world. Because of his interest in PU, almost always one component of IPN developed by Frisch team is PU.

Sperling's initial interest was to study how to produce finely dispersed polymer mixtures without requiring any major mechanical mixing equipment. He wanted to use the dual-network structure as a means of restraining gross phase separation. Shortly after Spelling began IPN research, metallurgist David Thomas took part in his work. Thomas brought important concepts of electron microscopy and mechanical properties to this study, and Sperling's contribution was to understand the peculiar behavior of viscoelasticity and network topology and phase structure.

There are two other early researchers that should be introduced. Yury Lipatov of the Kiev Academy of Sciences in Ukraine, the Soviet Union, has a strong interest in the phase boundaries of IPN, which he believes are caused by thermodynamic incompatibility. Dr. Lipatov was the first person to think of IPN as a polymer/polymer complex. He believes that in this complex, the second network is the filler of the first network formed first; Guy Meyer of Louis Baster University in Strasbourg, France, considers the semi-IPN as an adhesive and got the result of IPN. In addition to the numerous scientific papers on IPN and related materials provided by many scientists, the developing patent literature also describes completely different approaches to IPN chemistry and technology.

Since the 1970s, IPN has developed rapidly, and the research and development of IPNs are in the ascendant at home and abroad. The research on IPN in China started late and developed rapidly after 1980s.

It is reported that IPN has been widely used in leather modifier, pressure-sensitive osmotic membrane, ion-exchange resin, sound insulation material, high-impact plastics, pressure-sensitive adhesives, medical materials, rubber reinforcement, damping materials, textile auxiliaries and so on. As we all know, at present almost all the polymer materials involve modification such as blending and compounding of polymers. Therefore, IPN synthesis technology is of great significance in the application of polymer materials.

10.3 Types and representation of IPNs

10.3.1 Types of IPNs

10.3.1.1 Classification by material morphology
According to the morphology of materials, IPN can be divided into ideal IPN, partial IPN and phase separation IPN.

1. Ideal IPN
Ideal IPN refers to IPN in which two polymers forming interpenetrating networks penetrate evenly at the molecular level. Due to the lack of thermodynamic compatibility between most polymers, it is difficult to obtain ideal IPN in practice.

2. Partial IPN
Partial IPN means that the two polymers forming interpenetrating networks are only partially compatible, so it is impossible for the two polymers to penetrate each other completely at the molecular level. There are microphase separation structures in the aggregates. This IPN is characterized by a wide glass transition region. At present, most IPNs belong to this category.

3. Phase separation IPN
Phase separation IPN means that the two polymers forming interpenetrating networks are completely incompatible. Because of the special structure of interpenetrating networks, incompatible polymers can improve their compatibility under the

action of "forced mutual solubility," so they can also produce interacting multicomponent polymers, for example, PS/polyisoprene IPN.

10.3.1.2 Classification by synthesis method

IPNs are physical mixtures prepared by chemical synthesis. According to their preparation methods, IPNs can be divided into four types: step IPN, synchronous IPN, semi-interpenetrating IPN and emulsion IPN.

1. Step-by-step IPN

IPN synthesized by step method is called step IPN. IPNs are formed step by step when IPN is synthesized. This method was first proposed by J. R. Millar in 1960. The step-by-step IPN method is to first synthesize polymer network I through bulk polymerization, and then place polymer network I in a mixed solution consisting of monomer II, initiator and cross-linking agent, so that the substances in the mixed solution infiltrate into polymer network I. After initiation, monomer II and other substances are polymerized and cross-linked in polymer network I, and finally polymer network II is formed. The macromolecule chains of the two polymer networks penetrate each other and form IPNs. Another synthesis method of step IPN is to prepare a mixed solution of two monomers, initiator and cross-linking agent. Polymerization and cross-linking conditions are controlled in the reaction, so that only polymer network I is formed in the first stage of the reaction. After the formation of network I, the second monomer reacts and forms another network. In this way, the two kinds of networks formed successively run through each other to obtain IPNs.

For example, vinyl acetate monomer containing cross-linking agent tetraglycol dimethacrylate and benzoin was initially polymerized to form cross-linked polyvinyl acetate, which was then swelled with the same amount of St monomer containing initiator and cross-linking agent. After equilibrium, the St was polymerized and cross-linked, and white leather-like IPN PVAc/PS was prepared.

2. Synchronized IPN

IPN synthesized by synchronization method is called simultaneous IPN (SIN). In the synthesis reaction, the networks of the two polymers are formed at the same time. This method was proposed by K. Frisch et al. in 1974. When IPN was synthesized by simultaneous synthesis, the two monomers were mixed with the initiator and cross-linking agent needed. The two monomers were copolymerized and cross-linked at the same time by noninterference mechanism. For example, the EP resin/polyacrylate (PA) SIN can be prepared by mixing the components of EP resin and cross-linking acrylate resin to initiate the polymerization of acrylate monomers and cross-linking the EP resin at the same time.

3. Semi-IPN

Semi-IPN is not so much a preparation method as a type. It is an IPN, semi-IPN, which was first made by K. Frisch in 1969. The so-called semi-IPN refers to the fact that only one of the two polymers constituting IPN has a network structure, while the other has a linear structure. In other words, the network and linear macromolecules penetrate each other and form semi-IPN. This is also an important IPN. For example, the toughness of the product is much higher than that of the EP resin when the linear polymer of butyl PA runs through the EP resin network.

4. Latex IPN

The IPN prepared by emulsion method is called latex IPN (LIPN). There are two types of LIPN. The first one is to add monomer B, initiator and cross-linking agent to polymer network latex A, so that monomer B and other substances can penetrate into latex A particles. Through polymerization and cross-linking reaction, monomer B forms a polymer network B interacting with network A in latex A particles. The second kind: polymer latex A, polymer latex B and cross-linking agent are mixed and then heated, cross-linked and solidified, which promotes the two latex particles to penetrate the macromolecular chain to a certain extent at the contact sites. Usually, the average particle size of latex particles is 1–4 μm, and interpenetrating networks only occur on the surface of latex particles.

In recent years, many core shell structural emulsions often contain interpenetrating network structures. Because in the synthetic and shell structured emulsion, the shell monomer B polymerized on the surface of the polymer nucleus A, and then penetrated into the polymer A. If both A and B polymers are cross-linked or one of them is cross-linked, an interpenetrating network structure is formed at the core-shell interface.

10.3.2 Representation of IPN

At present, there is no uniform regulation on the naming and labeling of IPN. The simplest method is to imitate the representation of linear polymer blends. If polymer I is A, polymer II is B and weight ratio is X/Y, the IPN is IPN X/Y A/B. If it is made by synchronous method, it is expressed as SIN X/Y A/B, half IPN X/Y A/B and half SIN X/Y A/B and so on. This representation method is the simplest, if there is no special need, it is mostly used at present.

However, this annotation method does not represent system naming, it cannot represent the relationship between IPN and other related materials.

In terms of preparation methods, IPN is close to graft copolymerization blending method, and mechanical blending method is close to chemical bonding between phases. In order to systematically describe IPN and other blending methods, as well as their differences and connections, Sperling proposed a system naming and annotation method. This naming and labeling is composed of three parts: basic elements (see Tab. 10.2), binary operation symbols (see Tab. 10.3) and subscripts (see Tab. 10.4).

Tab. 10.2: Basic elements.

Symbol	Significance
P	Linear polymer
R	Random copolymer
A	Alternating copolymer
M	Mechanical or physical blends
G	Graft copolymer
C	Cross-linked polymers
B	Block copolymer
S	Star block copolymer
I	Interpenetrating network polymer
U	Reaction mixtures with unknown structures

Tab. 10.3: Binary operators and their represented reactions.

Symbol	Reactions expressed
O_R	Formation of random copolymer
O_A	Formation of alternating copolymers
O_M	Mechanical or physical blending
O_G	Formation of graft blends
O_C	Formation of cross-linked polymers
O_B	Formation of block copolymers
O_S	Formation of star block copolymers
O_I	Formation of interpenetrating network polymers
O_U	Formation of unknown reaction mixtures

Operators are placed between two basic elements to indicate how they are combined. For example, $C_{11}O_1C_{22}$ means that IPN consists of cross-linked polymer 1 and cross-linked polymer 2. Subscripts 1, 2, 3, . . . i, j denote the number of the basic element (polymer). For example, P_1 is for polymer 1, P_j is for polymer j and so on. If there are two or more subscripts, the first subscript represents the first polymer formed, the second subscript represents the second polymer formed and so on. When the time sequence does not work (e.g., in the case of mechanical blends), the order of subscripts denotes the component content or the order of importance. The usage and meaning of the subscript are shown in Tab. 10.4.

Tab. 10.4: Illustration of subscript usage.

P_1	Homopolymer consisting of monomer 1
G_{12}	Polymer 1 is the backbone and polymer 2 is the side-chain graft copolymer
C_{11}	Cross-linked polymer consisting of linear polymer 1
U_{12}	The unknown reaction mixture consisting of polymer 1 and polymer 2, that is, the reaction and the structure of the product are unknown
R_{27}	Random copolymer of monomers 2 and 7
M_{21}	Mechanical or physical blends consisting of polymer 2 and polymer 1

The rules of expression reading stipulate that from left to right, from top to bottom, this order also indicates the order of operation or operation.

Examples are as follows:

1. $P_1O_GP_2=G_{12}$ (graft copolymer with polymer 1 as main chain and polymer 2 as side chain);
2. $P_iO_CP_j=C_{ij}$ (polymers i and j form cross-linked copolymers);
3. $C_{11}O_1P_2=$Half-IPN (semi-IPN consisting of cross-linked polymer 1 and linear polymer 2);
4. $P_1O_MP_2=M_{12}$ (mechanical blends consisting of linear polymer 1 and linear polymer 2, linear polymer 1 being the main component);
5. $(P_1O_MP_2)O_G\ P_3=G_{13}O_MG_{23}$ (the bracketed components in the formula represent an independent process). The system represents the addition of polymer 3 to the mechanical blends of polymer 1 and polymer 2 for graft copolymerization. The final product is the mechanical blends of graft copolymers 1, 3 and 2, 3.

In this naming and annotation system, for the representation of IPN, the synchronization process is represented by square brackets, and the step-by-step process is represented without square brackets. For example,

$$[C_{11}O_1C_{22}]$$

represents a synchronous IPN consisting of a cross-linked polymer L and a cross-linked polymer 2. The corresponding step-by-step 1 PN is expressed as

$$C_{11}O_1C_{22}$$

If the composition of the blend is known, it can be expressed by numerical coefficients. For example, $P_1O_R5P_2$ represents a random copolymer of 1 mol of P_1 and 5 mol of P_2. If the molecular weight is known, it can also be expressed in parentheses after the corresponding elements, such as $P_1(5 \times 10^4 \text{ g/mol}) O_GP_2(3 \times 10^5 \text{ g/mol})$.

For some copolymers with complex structures, vertical symbolic representations can be used, for example:

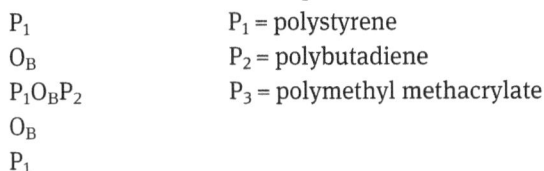

P_1 P_1 = polystyrene
O_B P_2 = polybutadiene
$P_1O_BP_2$ P_3 = polymethyl methacrylate
O_B
P_1

represents a triblock copolymer consisting of polymers 1 and 2 and grafted with polymer 3 on the central block. Vertical symbolic representations are read from top to bottom and from left to right.

10.4 Preparation of IPNs

10.4.1 Preparation of step-by-step polymer networks

10.4.1.1 Preparation of step-by-step polymer networks

The preparation method of step-by-step IPN is to make monomer I polymerize and cross-link to form network I, then swell network I with monomer II containing initiator and cross-linking agent, and polymerize and cross-link in situ to form network II.

The synthesis of step-by-step IPN requires a cross-linking polymer I, which can be either rubber or plastic; linear polymers can be prepared before cross-linking, or both synthesis and cross-linking can be carried out simultaneously; the synthesis and cross-linking of network II can be initiated by initiator heating, or photopolymerization can be initiated by adding photosensitizer. There is no chemical interaction between the polymer chains of the two networks, but only physical interpenetration. The polymer components commonly used in the preparation of step IPN are listed in Tab. 10.5.

Interpenetrating liquid crystal polymer network (ILCPN) can also be prepared by step IPN method. It is difficult for two kinds of liquid crystalline polymers to form stable systems by general mixing method, which can be achieved by using step-by-step IPN technology. First, the side chain liquid crystal elastomer based on PA was used as network I polymerization, and then another side chain liquid crystal elastomer was added to form network II by in situ polymerization, thus the ILCPN was obtained.

In step IPN, if monomer II has not reached the swelling equilibrium, it has completed the polymerization and/or cross-linking, which makes the concentration of polymer II change gradiently in polymer I, then it is called gradient IPNs.

There is also a step-by-step IPN called Millar IPN, which is an interpenetrating network composed of two chemically identical cross-linked polymers. For example, the PS/PS interpenetrating network first made by Millar in 1960 has different properties from the general cross-linked PS. Under the same degree of cross-linking, the swelling property of IPN PS/PS is smaller, while the mechanical properties are better.

The step-by-step IPN is of great significance to the modification of existing polymers, especially cross-linked polymers. The main disadvantage of IPN is that the cross-linking structure of network I brings great difficulties to molding and processing, and the heat dissipation of network II polymerization is also difficult.

10.4.1.2 Examples of preparation of step-by-step polymer networks

There are many examples of step IPN, such as polybutadiene rubber or St–butadiene rubber (SBR)/PS, butyl polyacrylate/PS, ethyl polyacrylate/polymethyl methacrylate, ethyl polyacrylate/poly(St–methyl methacrylate), St–butadiene–St triblock copolymer (SBS)/PS, PU, ester/polymethyl methacrylate, castor oil PU/PS, castor oil PU/poly(methyl) acrylate and polysiloxane/PS. Here are three examples.

1. SBR/PS IPN

IPN or semi-IPN can be prepared from SBR as polymer I and PS as polymer II. First, SBR was dissolved in benzene, and dicumyl peroxide (DCP) was added as cross-linking agent. The cross-linking reaction was carried out at about 165 °C. The pressure was 280–350 kPa and the reaction time was 45 min. Then, the cross-linked SBR was placed in a closed vessel containing cross-linking agent divinylbenzene (DVB) and initiator DCP. After swelling for 12 h, the SBR was polymerized at 50 °C for 96 h and at 100 °C for 1 h. Finally, IPN SBR/PS was obtained by vacuum drying to remove unreacted monomers. If one of the polymers is not cross-linked, it is called semi-IPN.

2. Ethyl polyacrylate/PS IPN

Ethyl acrylate (EA) dissolved in diphenylethanone and cross-linking agent tetraethylene dimethacrylate (TEGDM) was initiated by ultraviolet light, then dried in vacuum to a constant weight, so that the content of unreacted monomers was less than 2%. The above products were swelled with St-containing initiator and cross-linking agent (see Tab. 10.5), then the St was polymerized in situ and then dried. The extractable substance in the product was less than 2%. The ratio of EA to St can vary from 75:25 to 25:75. When the composition changes, the properties of the product also change accordingly, and it can be changed from elastomer to plastic product.

Tab. 10.5: Common polymer components in step IPN.

Monomer	Cross-linking agent	Initiator	Polymerization method
Polymer I (generally elastomer)			
Ethyl acrylate	TEGDM	Diphenylethanone	Bulk polymerization initiated by ultraviolet light
Butadiene	Peroxide	Potassium persulfate	Emulsion polymerization
Castor oil	TDI	None	Bulk polymerization
Polymer II (generally plastics)			
Styrene	DVB	Diphenylethanone	Bulk polymerization initiated by ultraviolet light
Methyl methacrylate	TEGDM	Diphenylethanone	Ultraviolet-induced Polymerization

In the above process, if only one polymer component is cross-linked, the corresponding semi-IPN can be obtained.

3. Castor oil/PS IPN

The molecular structure of castor oil is shown in Fig. 10.3.

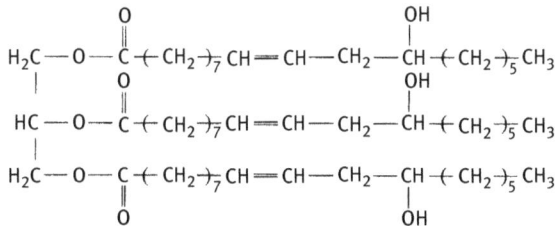

Fig. 10.3: Molecular structure of castor oil.

Because castor oil contains three hydroxyl groups, it can react with diisocyanates such as toluene diisocyanate (TDI) to form PU. The reaction between castor oil and

TDI is exothermic. Because castor oil contains a small amount of volatile substances, a step-by-step reaction can be adopted to prevent bubbles in the product. The first step is reaction of a known amount of castor oil with an excess of TDI (NCO/OH > 1) to form a prepolymer with isocyanate end groups. The reaction was carried out at room temperature, and then the reaction mixture was stirred strongly for more than 1 h, and the bubbles were removed in vacuum to obtain transparent viscous liquid. The second step is to add the amount of castor oil to achieve the required NCO/OH ratio for cross-linking reaction.

IPN castor oil/PS was prepared by swelling the prepared cross-linked PU in STT containing 0.4% diphenylethanone (initiator) and 1% DVB (cross-linker). IPN castor oil/PS was synthesized by polymerization at room temperature for 24 h.

10.4.2 Preparation of synchronized polymer networks

SIN is prepared by mixing two monomers or prepolymers with initiators, cross-linkers and so on evenly. Then two noninterference parallel reactions (such as condensation reaction and free-radical polymerization) are carried out simultaneously according to different reaction mechanisms to obtain two IPNs, that is, SIN.

The main advantage of SIN is that its initial viscosity is small and it is easy to form. Therefore, compared with step-by-step IPN, people pay more attention to the research and application of SIN.

SIN with PU as component is the most studied in SIN. About one third of the published literature on IPN is related to PU, most of which are SIN. The PUs used include polyether, polyester, castor oil and polybutadiene. The second network includes polymethyl methacrylate, PS, PA, EP resin, unsaturated polyester, polysiloxane, polyacrylamide, polybutadiene–AN, poly(methyl methacrylate-methacrylic acid) and so on.

Following is a brief description of the preparation method of synchronous IPN with PU and PMMA as the main components. The raw materials and requirements are listed in Tab. 10.6.

Preparation of PU and its prepolymer: PU prepolymer with isocyanate as terminal group was synthesized by reaction of 2 mol MDI and 1 mol poly(ε-caprolactone) at 60 °C. The prepolymer is sensitive to moisture and should be stored in vacuum for no more than 2 days. The prepolymer of 1 mol was heated to 80 °C and stirred with the mixture of 1,4-butanediol (1,4-BD) and trimethylolpropane (TMP) of 1 mol (mol ratio 4:1). The air was removed in vacuum for 5–10 min. The reaction mixture was injected into the stainless steel mold, and the reaction time was 16 h at 80 °C and 4 h at 110 °C. The molding pressure was about 24 MPa.

The preparation of SIN: 1 mol PU prepolymer heated to 80 °C, mixed with 1 mol 1,4-BD and TMP mixture (mol ratio of 1,4-BD to TMP is 4:1). The mixture of MMA and TMPTMA was added to mix evenly, and the air was removed by vacuum for 5–10 min.

Tab. 10.6: SIN raw materials based on polyurethane.

Name	Abbreviations or codes	Requirement
Poly(ε-caprolactone)	Niax D-560	Molecular weight: 1,978. The hydroxyl value was 56.7, and it was dried at 60 °C for 5 h in 266.6 Pa vacuum.
Trimethylolpropane	TMP	Dry for 5 h under 266.6 Pa vacuum at 60 °C.
1,4-Butanediol	1,4-BD	Dry for 5 h under 266.6 Pa vacuum at 60 °C.
4,4'-Diphenylmethane diisocyanate	MDI	–
Methacrylate	MMA	Wash with 5% KOH aqueous solution. Molecular sieve was dried and the inhibitor was removed by distillation in 266.6 Pa vacuum at 40 °C.
Trimethylolpropane trimethacrylate	TMPTMA	Distillation was carried out in 266.6 Pa vacuum at 40 °C.
Benzoyl peroxide	BPO	–

The mixture was injected into the mold, reacted at 80 °C for 16 h, then reacted at 110 for 4 h. The molding pressure was about 24 MPa, that is, step-by-step PU SIN.

Similar methods can be used to prepare corresponding semi-SIN.

10.4.3 Preparation of latex polymer network

10.4.3.1 Step-by-step latex polymer network

The preparation process of LIPN is to use latex-type cross-linked polymer as seed latex, add monomers, cross-linkers and initiators to polymerize to form core–shell structure LIPN, so it is called step-by-step LIPN, also known as two-stage LIPN, which is essentially a step-by-step IPN formed on latex particles. It overcomes the difficulty of step-by-step IPN forming and has great practical value. Here are two examples to illustrate.

1. Polyvinyl chloride (PVC)/nitrile butadiene rubber (NBR) LIPN

PVC/NBR was prepared by two-step emulsion polymerization. The "seed" latex consisting of cross-linked PVC was prepared, and then the mixture of butadiene, AN and cross-linking agent was added to the latex for the second step of polymerization, that is, LIPN. The formula of seed latex is as follows:

Deionized water	115 mL
Sodium dodecyl sulfate	0.3 g
Vinyl chloride monomer	20 g
Potassium persulfate	0.25 g
TEGDM	0.4% of monomer mass

The prepared PVC latex was filtered to remove a small amount of polymer aggregates. The initiator potassium persulfate of 0.25 g (in the form of aqueous solution) and 0.4% cross-linker TEGDM were added under stirring and washed with nitrogen for 5 min. Then AN and butadiene were added. The reaction vessel was sealed and stirred at 40 °C for 12 h to complete the reaction. The product was agglomerated (saturated salt water) and molded at 190 °C.

2. SBR/PA LIPN

SBR latex is used as seed LIPN, which is often used as impact modifier of PVC. This is a kind of LPIN with multilayer cross-linking structure. Its characteristics are that the T_g of the outer cross-linked polymer is below 0 °C and that of the subouter polymer is above 60 °C. The specific synthesis methods are as follows:

Preparation of seed latex (first layer): According to the proportion given in Tab. 10.7 (by mass), the raw materials were added into the autoclave, and the average particle size of the latex particles was 0.16 μm after 5 h of polymerization at 50 °C.

Tab. 10.7: Raw material ratio of styrene butadiene seed latex.

Name of raw material	Dosage/portion	Name of raw material	Dosage/portion
Butadiene	75	Ferrous sulfate	0.01
Styrene	25	Glucose	1.0
Cumene hydrogen peroxide	0.2	Potassium oleate	0.5
Sodium pyrophosphate	0.5	Water	200

Preparation of the second and third layers of polymers: Thirty pieces of seed emulsion (YISHION polymer), 0.6 potassium stearate, 0.2 pieces of white powder (formaldehyde sodium bisulfate) and dissolved in L20 water, stirring evenly, heating up to 70 °C. 39.8 copies of MMA, 0.2 copies of triallyl isocyanurate (cross-linking agent) and 0.15 copies of isopropylbenzene hydrogen peroxide (PBP) were mixed and dripped into the reactor within 2.5 h, then kept for 1 h, the conversion was 99.8%. The obtained latex was mixed with 0.3 phr potassium stearate, 0.2 phr carving powder, 29.8 phr butyl acrylate (BA), 0.2 phr Triallyl isocyanurate and 0.15 phr PBP, and dripped in 2.5 h, then kept for 1 h. The conversion was 99.6% at 70 °C. The average particle size of latex is 0.21 μm. Finally, 1 portion of stabilizer (2,6-di-tert-butyl-4-cresol) and 0.5 portion of di-cinnamyl thiodipropionate were added to the latex and condensed in 1% aqueous solution. White powder is obtained after washing and drying.

10.4.3.2 Interpenetrating elastomer network

Interpenetrating elastomer network (IEN) is also prepared by emulsion polymerization. The specific method is to mix two kinds of linear elastomer emulsions with cross-linking agents, accelerators or catalysts and co-coagulate, and then heat them to cross-link, so that IEN can be obtained. For example, PU urea emulsion was prepared from TDI, double hydroxyl terminated polypropene and ethylenediamine. In addition, BA, St and DVB (divinyl benzene) were used as raw materials to prepare St acrylic elastomer emulsion. The above two emulsions were then mixed with sulfur suspension, zinc oxide suspension, two zinc dithiocarbamate aqueous solution and butyl bisphenol A suspension and were stirred vigorously for 1 h. After the bubbles were disappearing, they were cast into thin films and heated to cross-link. PU-urea is cross-linked by the reaction between isocyanate and hydroxyl group, while PA is cross-linked by the free-radical reaction of sulfur and double bond.

10.4.4 Preparation of thermoplastic polymer networks

Thermoplastic IPN is an interpenetrating system of physical cross-linking networks. Its high-temperature behavior is like that of thermoplastic, which can be melted and processed. At room temperature, it can form physical cross-linking networks through glass, ionic or crystalline microregions and penetrate each other.

Thermoplastic IPN can be prepared by two methods: (1) mechanical blending in melting state, which is basically the same as mechanical blending in general linear polymers; (2) chemical blending, swell monomer II into polymer I or dissolve polymer I in monomer II, and in situ polymerization to form polymer II.

Thermoplastic IPN SEBS/poly(styrene-co-sodium methacrylate) was prepared by chemical blending method by Seigfrid et al. SEBS is a triblock copolymer and a partial hydrogenation product of SBS. SEBS was swelled with a mixture of St and methacrylic acid and polymerized with peroxide. After the reaction, sodium hydroxide is added to neutralize acrylic radical to form sodium salt. Then thermoplastic IPN is obtained by removing water.

Thermoplastic IPN can also be prepared from block copolymer SEBS and polyamide or thermoplastic saturated polyester.

A method of synthesizing thermoplastic elastomers by synchronous interpenetrating network technology was proposed, in which PU prepolymer, chain extender, related monomers and initiator were mixed evenly and reacted with two different mechanisms simultaneously. Semi-IPN polyether PU/poly(styrene–AN) (PU/P (St–AN)) and semi-IPN polyester PU/poly(methyl methacrylate–styrene) were prepared. By adjusting the chemical composition of the two components and appropriate reaction conditions, controlling the compatibility and phase separation of the system, a semi-IPN thermoplastic elastomer with similar morphology and properties of dynamic vulcanized thermoplastic elastomer was obtained.

These two kinds of semi-IPN materials have good thermoplasticity in the temperature range of 130–150 °C and can be processed by thermoplastic forming. Tables 10.8 and 10.9 show that semi-IPN PU/P (St–AN) and semi-IPN PU/P (MMA–St) have the mechanical properties of elastomers in the range of 60/40–80/20 mass ratio of the two components.

Tab. 10.8: Mechanical properties of semi-IPN PU/P (St–AN).

PU/P (St–AN) mass ratio	100% constant tensile stress (MPa)	Tensile strength (MPa)	Elongation at break (%)	Permanent deformation (%)	Shaw hardness (A)
50/50	–	10.6	90	17	92
60/40	5.0	9.3	320	20	80
70/30	3.1	10.2	260	10	69
80/20	2.9	8.6	280	15	66

Tab. 10.9: Mechanical properties of semi-IPN PU/P(MMA-St).

PU/P(MMA-St) mass ratio	100% constant tensile stress (MPa)	Tensile strength (MPa)	Elongation at break (%)	Permanent deformation (%)	Shaw hardness (A)
50/50	10.4	13.4	120	12	85
60/40	12.3	18.2	280	8	80
70/30	5.4	14.1	340	4	70
80/20	3.5	11.8	420	6	64

10.5 Performance of IPNs

IPN is the closest combination of two polymers formed by the interpenetrating entanglement of cross-linking networks. Their unique forced compatibility can make two or more polymers with different properties form stable polymer blends, thus realizing the complementarity of performance or function between components. At the same time, IPN's special cellular structure, interpenetrating interface, dual continuity and other morphological characteristics, as well as the resulting strong interface combination, make them produce special synergistic effect on macroperformance. In addition, some IPN materials have the characteristics of reactive processing, low-processing viscosity, good fluidity and excellent processing performance. The synergistic effect of IPN materials on mechanical properties is emphasized further.

10.5.1 Enhancement performance

Good interfacial bonding in IPN materials is conducive to transfer and disperse stresses, reduces the possibility of interfacial defects and damage. And the interpenetrating entanglement of networks leads to the increase of the actual cross-linking density of the system, so the modulus and strength of IPN materials are often higher than the arithmetic average of the corresponding performance indicators of its component polymers, showing significant synergy. When used, sometimes even elongation at break also shows synergistic effect.

Table 10.10 shows the results of mechanical properties of IPN polybutadiene PU/PS. From the data in the table, with the increase of PS content in the system, the tensile stress, tensile strength and hardness of IPN materials gradually increase, showing a significant reinforcement effect. When 1,4-BD is used as chain extender,

Tab. 10.10: Effect of component ratio on mechanical properties of IPN polybutadiene PU/PS system.

PS mass fraction (%)	300% tensile stress (MPa)	Tensile strength (MPa)	Elongation at break (%)	Permanent deformation (%)	Shaw hardness (A)
0	1.64	2.84	600	2.8	60
20	2.94	4.24	430	–	74
30	3.11	4.57	515	5.0	80
40	3.50	11.76	815	6.6	82
50	4.22	–	570	13.0	85
60	14.60	18.02	450	26.0	87

the 300% tensile stress, tensile strength and elongation at break of pure polybutadiene PU are 1.64 MPa, 2.84 MPa and 600%, respectively, while the corresponding performance values of IPN material reach 3.50 MPa, 11.76 MPa and 815% respectively, when the mass fraction of PS is 40%.

10.5.2 Toughening properties

For IPN composed of plastic and rubber elastomer, when rubber is the main component, it often shows the characteristics of reinforcing elastomer, but when plastic is the main component, rubber often shows prominent toughening effect. This is because the interpenetration of the network results in a good interface bonding between the two phases, and the continuous cellular structure fills the whole material, and the size of the rubber phase is in the range conducive to toughening, so it can produce a very significant toughening effect. The stress–strain curve shown in Fig. 10.4 shows that when the mass ratio of IPN PU/P (St–AN) is greater than 30/70, the material shows the characteristics of toughened plastics.

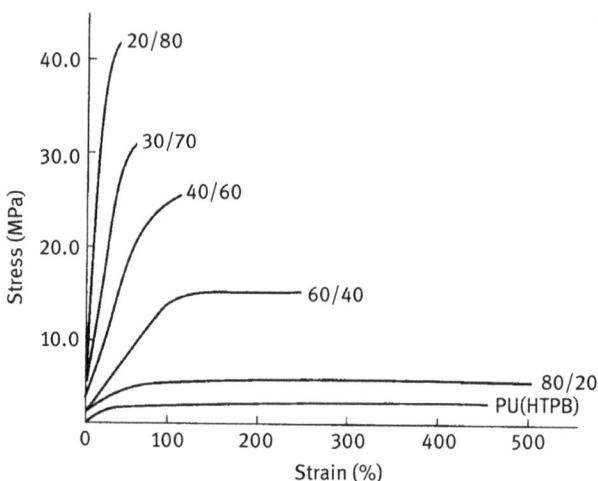

Fig. 10.4: Stress–strain curve of IPN polybutadiene PU/P(St–AN).
(Figure number is PU/P (St–AN) mass ratio.)

10.5.3 Damping performance

In some semicompatible IPN systems consisting of low T_g and high T_g polymers, when the molecular structure of the polymer is reasonably selected and designed and the network has great interpenetration, due to the internal displacement of the

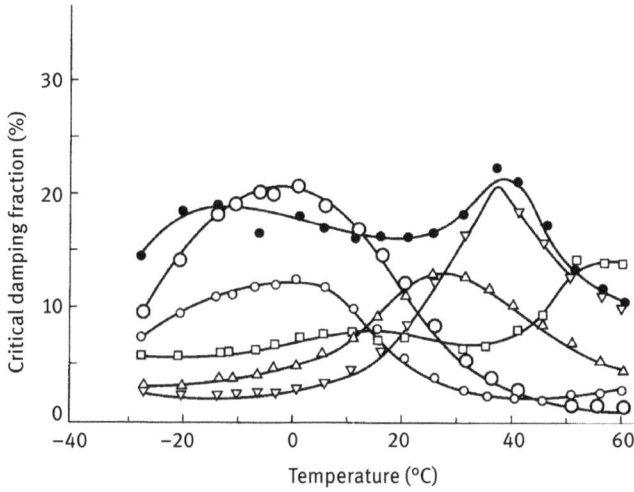

Fig. 10.5: Relation between damping and temperature of IPN and other materials. A, IPN materials; B–F, other materials (● A; ▽ B; △ C; □ D; □ E; ○ F).

two T_g and the overlap of the loss curves, strong mechanical losses can occur in a wide temperature range near room temperature. Therefore, IPN technology is used to prepare materials with high sound and vibration damping properties. It provides a unique way.

Figure 10.5 shows the relationship between the damping and temperature of IPN polyethyl methacrylate/n-butyl polyacrylate (PEMA/PnBA, mass ratio 75/25) and several commercial damping materials. IPN damping materials have high damping in the range of −30 to 45 °C, while other damping materials have narrow loss peaks or very small damping values. The first IPN patent obtained by Sperling et al. in 1974 for noise damping coating was the PEMA/PBA LIPN coating. Klempner et al. reported that PU/EP IPN vibration damping materials were synthesized by synchronous method, and the mechanical loss tan δ delta was as high as 1.75 in the range of 5–35 °C.

10.5.4 Adhesive property

IPN technology can not only improve the compatibility of different polymer materials, thus significantly improving the bulk strength, toughness, environmental resistance of the adhesive, but also improve the wettability of the adhesive to the surface of the adhesives and interface bonding, thereby reducing interface damage. Both effects lead to significant synergistic effect of bonding strength.

EPDM vulcanized rubber and soft PVC plastics were bonded with SIN polyether PU/PMMA as binder. The peeling strength of EPDM vulcanized rubber and soft PVC

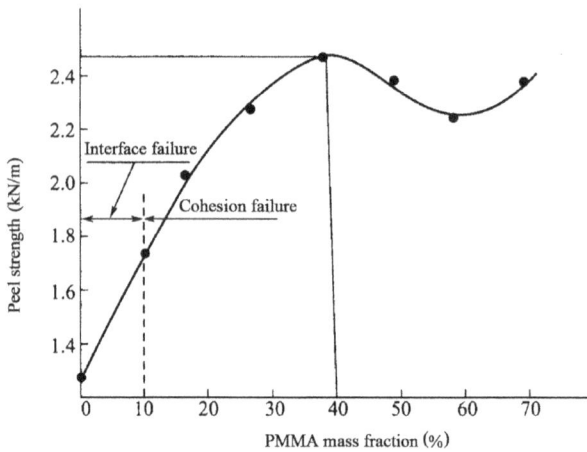

Fig. 10.6: Peeling strength of SIN PU/PMMA adhesive on EPDM/PVC adhesive.

plastics varied with PMMA content as shown in Fig. 10.6. The peeling strength of pure PU is 1.28 kN/m, while that of IPN adhesives is higher than that of pure PU. The maximum peeling strength of IPN adhesives is 2.44 kN/m when the mass ratio of SINPU/PMMA is 60/40. When the mass fraction of PMMA is less than 10%, the failure mode of the specimens is interfacial adhesion failure. When the mass fraction of PMMA is more than 10%, all the bonded specimens are cohesive failure. The reasons for the increase of bonding strength can be summarized as follows: first, the addition of MMA monomers reduces the viscosity and surface tension of the adhesives and improves the surface wettability; second, the PMMA component in IPN adhesives has a reinforcing effect on the body of the adhesives; third, some monomers MMA and cross-linking agent EGDMA in the adhesives penetrate into the surface layer of EPDM and PVC and polymerize in situ. The interfacial conjugate interpenetrating structure greatly enhances the interfacial bonding.

10.5.5 Machinability

SIN has low viscosity before curing and is suitable for pouring and reactive injection molding (RIM). RIM is a reactive processing technology using liquid polymers, which has been widely used in automotive industry, shoemaking industry and other fields. In order to improve the modulus and strength of RIM products, short fibers or reinforcing fillers are usually added to reinforce RIM products, that is, reinforced RIM (RRIM). However, the addition of fibers or fillers significantly increases the initial viscosity of the system, while reducing the stability of the system, especially increasing the wear of equipment. Combining RIM technology with IPN technology can not only achieve the same reinforcing effect as adding fibers or fillers, but also significantly reduce the system viscosity, equipment wears and energy consumption. It can also give some new properties, such as improving heat resistance. IPN-RIM systems studied include PU/EP resin, PU/unsaturated polyester and so on. For example, when the mass ratio of PU/EP is 80/20, the main performance of IPN can reach the level of 10% glass fiber reinforced PU (see Tab. 10.11). IPN with the mass ratio of PU/unsaturated polyester 80/20 has similar performance.

Tab. 10.11: Performance comparison between IPN-RIM and RRIM.

Performance	PU	PU/EP (mass ratio 80/20)	10% Fiber-reinforced PU
Shaw hardness (D)	78	85	83
Tensile strength (MPa)	44.8	49.8	50.5
Elongation at break (%)	30	20	15
Flexural modulus (MPa)	1,007	1,593	1,610

In addition, the processing performance of LIPN is also very good. Thermoplastic IPN can endow the material with precious thermoplastic processing properties. Therefore, the effect of IPN on improving the processability of polymer materials is very significant.

10.6 Application of IPNs

10.6.1 Elastomer reinforcement

There are many reports about the reinforcement of PU elastomer by IPN technology. In 1970s, Kim et al. studied the structure and properties of SIN PU/PMMA system in detail. It was found that when the mass ratio of PU/PMMA was 85/15, the tensile strength, elongation at break and tear strength were 42.6 MPa, 767% and 51.4 kN/m, respectively, while the corresponding values of pure PU were 36.1 MPa, 780% and 44.1 kN/m, respectively. Akay et al. studied the mechanical properties of semi-IPN PU/PMMA by step method. It was found that the modulus of elasticity, tensile strength and hardness of semi-IPNPU/PMMA increased with the increase of volume fraction of PMMA, and the maximum elongation at break appeared near 40% of PMMA content. IPN castor oil PU/PS and castor oil PU/PMMA were synthesized step by step by Sperling et al. The two materials were used as soles of sports shoes. They crossed the Americas without breakage.

IPN technology has special effect on reinforcing PU or claw liquid rubber. The reinforcing effect of carbon black on liquid rubber is often unsatisfactory, which is mainly due to the shielding effect of carbon black on hydrogen bonds. At the same time, the addition of carbon black causes a significant increase in the viscosity of the system, which is not conducive to pouring or reactive injection molding. Some researchers compared the effects of IPN technology and carbon black reinforcement, as shown in Tab. 10.12. It can be seen that the mechanical properties of polybutadiene PU reinforced by IPN technology are better than those reinforced by carbon black (20 g/100 g PU), and the initial viscosity of the former is significantly reduced due to the dilution of monomers, which is very conducive to operation, while the viscosity of the latter increases rapidly with the addition of carbon black, and cannot be poured when more than 20 phr of carbon black is added. The mechanism of reinforcing liquid PU by IPN technology is completely different from that of carbon black. The former is mainly due to the interpenetration of rigid organic particles through networks, especially the interpenetration of interface networks. At the same time, the interpenetration of such networks mainly occurs between rigid networks and rigid segments of PU. The effect of IPN technology is equivalent to increasing the cross-linking density of rigid segments, thus increasing the strength of PU. Modulus and other properties play an active role.

Tab. 10.12: Comparison of IPN technology and carbon black for polybutadiene PU reinforcement.

Performance	IPN PU/PS (mass ratio 65/35)	Carbon black (20 g/100 g PU)
Density (g/cm^3)	1.20	1.08
100% constant tensile stress (MPa)	8.4	3.5
Tensile strength (MPa)	11.4	3.9
Elongation at break (%)	200	170
Permanent deformation (%)	18	3
Tear strength (kN/m)	44.2	32.9
Shaw hardness (A)	76	76
Akron abrasion (cm^3/1.63 km)	0.284	0.379
Volumetric resistivity ($\Omega \cdot$ cm)	1.47×10^{16}	2.91×10^{12}
Dielectric constant (F/m)	4.06	10.03
Dielectric loss (tan δ)	0.06	0.05

The application of IPN in elastomer reinforcement is not only limited to PU and liquid rubber, but also applicable to other rubber systems. For example, the tensile strength and elongation at break of IPN BR/PS system are 16.6 MPa and 190%, respectively, when the mass ratio of IPN BR/PS is 50/50, while the corresponding values of pure BR are only 1.3 MPa and 70%. The stress–strain curves of the system are shown in Fig. 10.7.

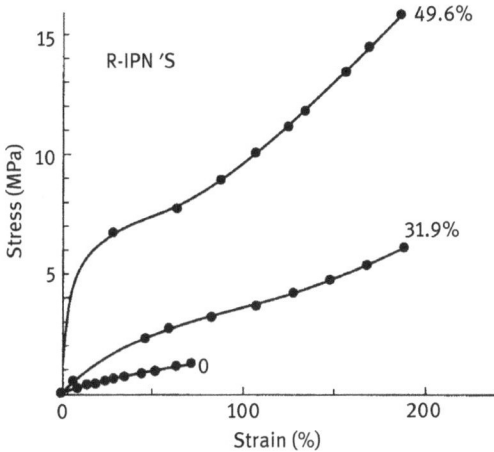

Fig. 10.7: Stress–strain curve of IPN *cis*-butadiene rubber/PS. (The figure in Fig. 10.7 is PS mass fraction.)

10.6.2 Toughened plastics

The toughness of IPN materials was discussed, and the application of IPN technology in toughening plastics involves not only thermoplastics, but also thermosetting plastics and engineering plastics, including some industrialized IPN materials.

Sheet molding compound (SMC) of ICI American company, trade name ITP, is a typical IPN-toughened thermosetting plastic. It is a blend of slightly cross-linked PU with unsaturated polyester and St. After sheeting, St polymerizes and causes cross-linking of unsaturated polyester. In the postcuring stage, PU is further cross-linked to form fully cross-linked SIN, which becomes a tough and impact-resistant sheet material.

Another IPN-based SMC material is Freeman Chemical's product, which is called Acpol. It is a kind of IPN-toughened thermosetting PA material. The material is composed of PA, PS and cross-linking PU. After curing for 24 h, the impact strength of the material can reach twice that of ordinary SMC.

A two-phase IPN was prepared by copolymerization of BA with butylene dimethacrylate and copolymers of St, AN and DVB. The 5 phr of IPN and 100 phr of bisphenol A polycarbonate were extruded and molded. The notched impact strength of the prepared product is 715 kJ/m^2, while that of pure polycarbonate is 117 kJ/m^2.

10.6.3 Thermoplastic IPN

Since the 1980s, thermoplastic IPN has developed rapidly, and many patents and industrial products have appeared. They can be used alone as thermoplastic elastomers and toughening plastics as toughening agents. They have important and wide application prospects.

The thermoplastic IPN based on the thermoplastic elastomer of St–butadiene block copolymer produced by Shell Chemical Company, USA, is named Kraton IPN. They take SEBS (partially hydrogenated SBS, relative molecular weight 9,000–47,000) as a component and nylon, polybutylene terephthalate, polycarbonate, polyformaldehyde and other engineering plastics as another component. Kraton IPN series is melted and blended at 400 °C, and can be injected at 260 °C. It is mainly used to produce beautiful and conspicuous automobile exterior parts. Table 10.13 lists the performance indicators of Kraton IPN GX7590 and GX7595.

Tab. 10.13: Performance of typical Kraton IPN.

Performance index	Kraton IPN GX7590	Kraton IPN GX7595
Relative density	1.03	1.04
Shaw hardness (A)	60	60
Tensile strength (MPa)	19.6	18.2
100% constant tensile stress (MPa)	12.6	15.4
Elongation at break (%)	300	300
Tear strength (kN/m)	96	114
Bending modulus of elasticity, 30 °C	770	1,155
Bending modulus of elasticity, 23 °C	700	1,050
Bending modulus of elasticity, 70 °C	315	455
Impact strength (kJ/m)	10.6	8.9

The thermoplastic semi-IPN products produced by Petrarch System Company in the United States include polysiloxane/PU, polysiloxane/polyamide, polysiloxane/polycarbonate, polysiloxane/polyester, polysiloxane/polysulfone and polysiloxane/SEBS with the mass fraction of 8%–20%. They are prepared by reacting polysiloxane oligomers with functional groups with thermoplastic polymers in melts. For example, in the extrusion or injection process, organosilicon oligomers with active hydrogen and vinyl groups are tightly melted and blended with thermoplastic PU or nylon. Polysiloxane is chemically cross-linked by vinyl addition reaction catalyzed by platinum, while resin is physically cross-linked, thus forming thermoplastic semi-IPN materials. The material exhibits high tensile strength and flexural strength, low relative density, low wear, tear strength more than three times as high as polysiloxane, low oxygen and moisture permeability, good demolding performance, improved dielectric and creep resistance, as well as good high-temperature stability, chemical inertia and biocompatibility. This kind of thermoplastic semi-IPN is mainly used for medical materials, electronic and electrical materials and so on.

10.6.4 Dynamic vulcanization thermoplastic IPN

In the process of melt blending of rubber and plastics, the process of dispersing and vulcanizing rubber in situ is called dynamic vulcanization. As a result, thermoplastic elastomers with thermoplastic matrix and vulcanized rubber particles as dispersing phase are obtained, which are both thermoplastic and highly elastic. The phase interface of this dynamic vulcanized thermoplastic elastomer has network interpenetrating structure. It is the most productive thermoplastic IPN product at present, it mainly includes EPDM/PP, NR/PP, NBR/PP and IIR/PP.

The preparation method of EPDM/PP dynamic vulcanization thermoplastic IPN is that EPDM and PP are melted in twin-screw extruder or internal mixer, and the melting temperature is 20–30 °C higher than that of PP. The vulcanization accelerator and cross-linking agent were added to PP 2–3 min after melting. According to the increase of material viscosity or torque, the optimum vulcanization degree was judged and then blended for 2–3 min. Finally, the material was discharged, pressed or granulated. EPDM can be vulcanized by sulfur, alkyl phenolic resin or bismaleimide system.

Figure 10.8 shows the relationship between EPDM particle size and stress–strain curve. The tensile strength and elongation at break of vulcanized rubber were obtained when the average particle size was about 1 μm.

Figure 10.9 shows the effect of cross-linking on the properties of EPDM/PP elastomers. It can be seen from the figure that with the increase of cross-linking density, the tensile strength increases and the permanent deformation decreases. Table 10.14 further compares the properties of EPDM/PP elastomers without cross-linking and fully cross-linking. Adequate cross-linking of rubber particles and interpenetration of interface networks (interpenetration of EPDM chemical cross-linking network with partially crystallized PP physical cross-linking network) usually lead to the general improvement of the strength, elasticity, fatigue resistance, solvent resistance, high-temperature resistance and melt strength of such thermoplastic IPN elastomers.

Dynamic vulcanized TPE has good thermoplastic processing properties. It can be processed by extrusion, injection, blow molding, calendering, hot forming and hot welding. The processing process is simple and does not need complicated processes

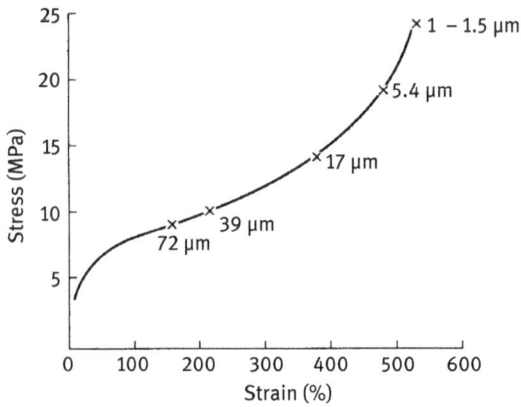

Fig. 10.8: The relationship between EPDM particle size and stress–strain curve.

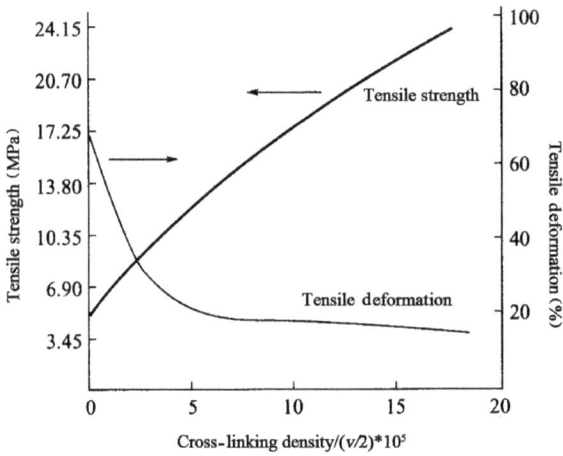

Fig. 10.9: Effect of cross-linking on properties of EPDM/PP elastomer.

Tab. 10.14: Performance comparison of uncross-linked and fully cross-linked IPN EPDM/PP.

Performance	Uncross-linked system	Fully cross-linking system
Shaw hardness (A)	81	84
Tensile strength (MPa)	4.02	13.1
Elongation at break (%)	630	430
100% tensile stress (MPa)	284	50

Tab. 10.14 (continued)

Performance	Uncross-linked system	Fully cross-linking system
Compression deformation (%)	78	31
Tensile deformation (%)	52	14
Swelling rate (in ASTM 3 oil)	162	52

such as batching, mixing, forming and vulcanization in the processing of thermosetting vulcanized rubber. The process cycle is short, energy consumption is low, product quality is easy to control, automatic operation can be adopted and its corner wastes can be recycled to reduce environmental pollution.

10.6.5 Compound material

The SIN formed by EP resin/PU is not only significantly higher than that of ordinary EP resin in toughness, tensile strength, shear strength and hardness, but also can improve thermal stability by about 100 °C.

Freeman Chemical Company has developed two new IPN EP resin/unsaturated polyester resin systems, which have the properties of EP resin and the processability of polyester resin. They can be used in the pultrusion and preimpregnation of advanced composites, and are very easy to wet polyaromatic amide fibers, glass fibers and graphite fibers. Their antiaging properties are similar to those of EP resin and vinyl polyester resin systems. Formed parts are as easy to manufacture as polyester systems.

Poussin Didier of France and others made carbon fiber composites with IPN EP resin/polyester resin system, which had good impact resistance, aging resistance and easy processing properties.

The toughness of semi-IPN formed by EP resin/aromatic thermoplastics is higher than that of two component polymers, showing typical synergistic effect. Excellent resin toughness can significantly improve the damage tolerance of fiber-reinforced composites, and will not damage the heat and moisture resistance of composites.

Thermosetting IPN is formed from brittle thermosetting polyimide (PI) and tough thermoplastic PI. There is no cross-reaction between the two resins. Compared with the pure thermosetting PI, the toughness of graphite fiber reinforced PI IPN increased 3.32 times, and the crack resistance increased significantly. The T_g increased from 339 to 369 °C.

In a 1986 patent, LNP Corporation of the United States reported that high modulus thermoplastics (bending modulus > 1.305×10^{13} Pa) and polydimethylsiloxane and glass fibers containing vinyl and Si–H bonds formed IPN composites during melt injection. Reinforcement fibers can resist warpage of shrinkage and can be used to manufacture high precision parts such as gears and thin-walled parts. Their flow direction and transverse shrinkage are only 0.003% and 0.004%, respectively.

Carbon fiber reinforced semi-IPN system developed by United Carbide Company of the USA is made of thermosetting modified phenolic resin and thermoplastic polycarbonate, polysulfone, polyester and so on. It has good adhesion to carbon fibers, low melt viscosity and easy processing. The properties of composites reach or exceed those of EP-based composites. At the same time, it has high-impact strength, low water absorption and good heat resistance (250 °C). It has the advantages of long service life, short curing time and small shrinkage.

A series of semi-IPN can be synthesized by the combination of thermoplastic plastics such as PET, polycarbonate, polysulfone and aromatic dicyanate. It can combine the strength, toughness, workability and high temperature of thermosetting plastics. Aromatic dicyanoates are heated to form cross-linked polymers with triazine ring structure, which can make the materials have higher softening temperature. It has been reported that the fracture toughness of semi-IPN prepared by the polymerization and cross-linking of bisphenol A dicyanate in graphite fiber reinforced polycarbonate matrix is five times that of graphite/EP resin composite, and its bending property is similar to that of EP resin composite. From the fracture surface analysis, the bond between the fibers and the matrix is good.

10.6.6 Functional materials

IPN can make two or more polymers with different functions form a close and stable combination through interpenetration and entanglement of networks. Its unique structural characteristics such as forced compatibility, interpenetration of interfaces, dual continuity and so on can promote their functional compositions, produce synergies and even bring about new functions. Therefore, IPN has been paid attention to the field of functional materials, and many practical products have been developed. At present, there are more and more reports on IPN functional materials, showing attractive application prospects.

1. Sensitive hydrogels

Sensitive hydrogel is a hydrophilic but water-insoluble polymer cross-linking network. It can perceive the subtle changes of the environment, such as pH, ionic strength, temperature, light, electric field and so on, and respond to these external stimuli through expansion or contraction of gel volume. At the same time, the hydrogel has the characteristics like biological materials after absorbing water. Sensitive hydrogels are the basis for the design of bionic materials, chemical mechanical systems, drug delivery systems and chemical switches.

IPN is a new technology for preparing sensitive hydrogels. It introduces cross-linking through interpolymer complex, which makes two polymers form cross-linking network at the same time of association, or one polymer fills in another polymer to form half-IPN. Subvalent bonds in IPN can be reversibly formed or destroyed with environmental changes, resulting in discontinuous changes in hydrogel volume. A semi-IPN hydrogel membrane based on polyelectrolyte complex CS/PAA was prepared by using chitosan and polyacrylic acid (PAA) as raw materials. It is not only sensitive to the change of pH value, but also sensitive to ions, which provides a possibility for its application in drug release system and separation. Some researchers synthesized a temperature and pH sensitive IPN hydrogel membrane. The first step was to prepare PU/poly(N-isopropylacrylamide) IPN and then to form a thermosensitive hydrogel with PAA. The content of PAA in IPN hydrogel is low, and the swelling rate decreases with the increase of temperature, and the degree of thermal sensitivity is affected by the pH value. Polyvinyl alcohol/PAA IPN can also be used as hydrogel membrane sensitive to pH and temperature.

2. IPN conductive materials

IPN conductive materials belong to IPN fast ionic conductors. IPN fast ionic conductors have high ionic conductivity at room temperature and good mechanical properties, and are potential solid electrolyte materials.

The interpenetrating network fast ionic conductor formed by polyethylene oxide (PEO) containing $LiClO_4$ and EP resin is a widely studied system. Some researchers have made amorphous electrodes with high room temperature conductivity which can be used for lithium batteries. The effect of EP resin network on the conductivity and strength of the system has been studied. The results show that when the mass fraction of EP resin is 30%, its comprehensive performance is the best, and its conductivity is increased by an order of magnitude to 10^{-3} S/cm compared with the single component system ($LiClO_4$–PEO).

Some researchers have prepared a new conductive PU IPN material, which uses polytetrahydrofuran diol and MDI to form a network, and fullerene (C_{60}) and polyaniline to form another network. The electrical conductivity of the PU IPN can reach 2.0 S/cm at room temperature, and the original tensile strength and elongation at break of the PU matrix are retained. The thermal stability of the PU IPN is also greatly improved than that of the pure PU elastomer.

3. Nonlinear optical polymer

Nonlinear optical materials have been widely used in optical communication, optical signal processing and other important fields. However, almost all the nonlinear optical materials currently used are inorganic materials. Most of them cannot meet the requirements of high capacity, high speed, high frequency and wide bandwidth for the rapid development of these fields. Compared with inorganic nonlinear optical materials, polymer nonlinear optical materials have much larger nonresonance coefficient, low dielectric constant, good optical transparency and processability, especially their ultrafast reaction speed.

Xie Hongquan synthesized IPN PU/EP resin and IPN PA/EP resin containing nitrobenzene, azobenzene and alkylamine chromophores. The results of visible light absorption show that the polarization stability of IPN PU/EP is higher than that of IPN PA/EP at 120 °C. A kind of nonlinear optical material IPN was prepared by some researchers. The polarized solidified IPN sample was treated for more than 1,000 h at 110 °C, and its second-order nonlinear optical properties did not change significantly. This kind of nonlinear optical material has excellent long-term thermal stability and is due to its interpenetrating network structure.

4. Biomedical engineering materials

Biomedical engineering is an important application field of IPN technology. IPN materials containing polymethacrylate have long been used as dental materials; IPN materials prepared with polysiloxane and polyhydroxyethyl methacrylate have good biocompatibility, oxygen permeability, flexibility and transparency, and have been successfully used in the manufacture of contact lenses; IPN materials composed of rubber and hydrophilic polymers can also be used in the field of artificial blood vessels; IPN materials made from rubber and hydrophilic polymers can also be used in industrial production of thermoplastic IPN. Material Rimplast (polysiloxane/PU, polysiloxane/nylon, etc.) can be used in the manufacture of artificial organs, artificial joints, medical catheters and so on.

5. Drug-controlled release system

IPN technology is used to make hydrophobic polymers distribute in the outer layer of hydrophilic polymer pills with a certain concentration gradient, forming an IPN film, which can be used in a new drug-controlled release system.

Some researchers have designed a kind of controlled release system of percutaneous drugs by interpenetrating pressure-sensitive viscous and nonviscous polymers. The viscous polymer used is acrylate polymer, and the nonviscous polymer is block copolymer or cross-linker of PEO. The viscosity of IPN film is linear with the content of viscous polymer. Drugs used for testing pass through a certain area preferentially in the membrane. The uniqueness of this permeation is related to the experimental conditions, the interaction of component polymers and the drugs used. The permeation process and bonding strength can be controlled by using the diaphragm.

In order to obtain biodegradable hydrogels for embedding drug-controlled release systems, researchers have synthesized acrylate terminated polyethylene glycol (PEG) and polycaprolactone/PEG (PCL/PEG) half-IPN system. PEG monomer polymerized to obtain cross-linked gel, and linear PCL penetrates through PEG cross-linked network. Researchers have used IPN technology to make ionic polyelectrolyte gel an advanced controlled release system for percolating drugs. Ionizable drugs are placed in polyelectrolyte gel matrix by ionic bonding. The ionic functional groups in polyelectrolyte gel can maximize the drug loading of the system. Drug release rate is controlled in two ways. One is the breakup of ionic bonds between ionic and ionic compounds in polyelectrolyte gel matrix, and the two is the diffusion of drugs in the matrix.

6. Separation membrane

IPN has good selectivity for the transmission of various substances because IPN can make the two networks with different properties closely combined. For example, IPN composed of polyether with NCO terminal group and vinyl arsenopyrene or N-vinyl arsenopyranone on porous carrier has semipermeability and can be used for reverse osmosis, ultrafiltration, gas separation and so on. IPN formed by copolymer of polyether with NCO terminal group, N-vinylimidazole and N-vinylarsenolidone on porous polysulfone can be used as semipermeable membrane for sugar separation. The IPN film composed of polysiloxane/PU, the IPN film composed of PU and vinyl arsenidine copolymer and the PU/EP (mass ratio 50/50) IPN film can be used as oxygen-enriched membrane. The membrane made of IPN PU/PS can be used for alcohol–water separation. Biomedical membranes with high H^+ permeability can be obtained by plasma polymerization of acrylic acid on porous polypropylene.

7. Amphiphilic IPN

Amphoteric polymers are a new type of functional polymers, which can be used in waterproof and moisture permeable materials and biomedical materials. Generally, the compatibility between hydrophilic and lipophilic polymers is very poor, which brings some difficulties to the preparation. However, amphiphilic IPN materials can be prepared by IPN technology.

Some researchers have prepared IPN PU/PCPEG materials by synchronous IPN technology. PU is polyester PU and PCPEG is photo cross-linked PU containing PEG. The properties of the prepared materials are shown in Tab. 10.15.

Tab. 10.15: Performance comparison of amphiphilic IPN and corresponding block polymers.

Sample	Tensile strength (MPa)	Tear strength (kN/m^2)	Elongation at break (%)	Hygroscopicity ($\times 10^{10}$ (g/cm)·s·kPa)
PU/PCPEG IPN	12.1	39.8	320	36.6
Amphiphilic block PU	2.1	16.7	280	15.9

10.6.7 Adhesives and coatings

The application of IPN technology in adhesives and coatings has attracted much attention for its unique properties and effects. Since the 1970s, many IPN adhesives and coatings have appeared.

IPN adhesives containing PU components are the most widely used IPN adhesives at present. The varieties of IPN adhesives include PU/EPresin, PU/PA and PU/unsaturated polyester.

For example, researchers have used SIN PU/EP to bond aluminum sheets to HDPE, and measured the lap shear strength and peel strength. It is found that when the mass ratio of PU/EP in SINPU/EP is equal, the lap shear strength of the system has the maximum value, showing a significant synergistic effect.

When the molar ratio of NCO/OH and the mass fraction of PMMA in IPN castor oil PU/PMMA adhesives are 1.5% and 60%, the adhesives have the best properties, and the peeling strength of IPN castor oil PU/PMMA adhesives is double that of pure PU.

A PU-modified PA IPN pressure-sensitive adhesive was prepared by some researchers. PU mainly provides the heat resistance of pressure-sensitive adhesives, while PA provides the initial viscosity of pressure-sensitive adhesives. By adjusting the ratio of PU/PA, pressure-sensitive adhesives with excellent bonding and heat resistance can be prepared, and the maximum temperature can reach about 125 °C. When the mass ratio of PU/PA is 20/80, the peeling strength of PSA reaches 1,180 N/m. When the mass ratio of PU/PA is 20/80, the pressure-sensitive colloid system forms a microphase separation structure, and the PU component is partially separated into a rubber-like dispersed phase distributed in the continuous phase of the PA component. The dispersed phase of PU particles restricts the viscous flow of PA molecular chain at high temperature, and the elastic modulus of the whole system is difficult to decrease sharply, so the pressure-sensitive adhesive has good heat resistance.

A patented IPN high-temperature resistant adhesive is introduced, which can be stabilized for 1 h at 300 °C (insulated from oxygen for a longer time). The formulation of IPN adhesive is composed of alicyclic EP resin, cyanate ester resin, polyether and so on. It is prepared by step-by-step polymerization. The advantage of this kind of adhesive is that it can be made into "stage B" adhesive some components are precured at lower temperatures and have a certain initial adhesion. After curing at high temperatures, a strong curing adhesive can be obtained. The peeling strength of steel sheets bonded with this adhesive is 1,120 N/m, while that of PI adhesive is 440 N/m under the same conditions.

IPN adhesives can be widely used in the bonding of various materials, including glass, metal, polymer, cement and wood, and can also be used as sealants, waterproof materials. IPN as the matrix of composites can obviously improve the interfacial adhesion. In addition to good bonding properties, IPN damper binder has the characteristics of wide temperature damping. It is mainly used for bonding metal and plastic composite parts to make mechanical structural materials and components.

IPN is widely used in coating field. Many resin/rubber IPN systems can be used as coating binders with excellent properties, such as PU/PA, polysiloxane/PS,

PU/polymethyl methacrylate, PU/vinyl chloride copolymer, poly(butyl acrylate-acrylic acid)/poly(styrene-acrylic acid) LIPN and polyvinyl acetate/PA LIPN.

Some researchers have studied a temperature-dependent light transmission solar coating. It is a semi-IPN composed of polypropylene oxide and 2-hydroxyethyl methacrylate–St copolymer, in which PEO is a flexible component and copolymer is a stable component of mechanical properties. When the light transmits through the system and the temperature of the system rises to a certain value, the system appears phase separation. By adjusting the proportion of repetitive units or mixing composition of copolymers, the phase separation temperature can be changed at will (30–150 °C). As a protective coating for solar collectors, it has good application prospects.

Semi-IPN EP/PU coatings have excellent damping properties at room temperature. When the mass ratio of EP/PU is 70/30, the maximum value of tan δ is about 1. This material also has good corrosion resistance, so it is also an ideal base material for anticorrosion coatings.

Exercises

1. Please outline the definition of an IPN.

Answer
IPNs are systems in which two or more polymers form cross-linked networks through chemical bonds, while two cross-links penetrate each other and cannot be separated by simple physical methods.

2. What are the types and representations of IPN? (Classification by material morphology)

Answer
(1) Ideal IPN: Ideal IPN refers to IPN in which two polymers form interpenetrating networks penetrate evenly at the molecular level.
(2) Partial IPN: Partial IPN means that the two polymers forming interpenetrating networks are only partially compatible, so it is impossible for the two polymers to penetrate each other completely at the molecular level.
(3) Phase separation IPN: Phase separation IPN means that the two polymers forming interpenetrating networks are completely incompatible.

3. What are the preparation methods of IPN?

Answer
(1) Step-by-step methods
(2) Synchronized method
(3) Semi-interpenetrating method
(4) Emulsion method

4. IPN plays an important role in the theory and practice of polymer modification. Please outline the modification of IPN.

Answer
Good interfacial bonding in IPN materials is conducive to transfer and disperse stresses, reducing the possibility of interfacial defects and damage. And the interpenetrating entanglement of networks leads to the increase of the actual crosslinking density of the system, so the modulus and strength of IPN materials are often higher than the arithmetic average of the corresponding performance indicators of its component polymers, showing significant synergy. When used, sometimes even elongation at break also shows synergistic effect.

Index

https://doi.org/10.1515/9783110596335-011

www.ingramcontent.com/pod-product-compliance
Lightning Source LLC
Chambersburg PA
CBHW080658220326
41598CB00033B/5248